The Draining of the Fens

Johns Hopkins Studies in the History of Technology

Merritt Roe Smith, Series Editor

THE
DRAINING OF
THE FENS

Projectors, Popular Politics, and State Building
in Early Modern England

ERIC H. ASH

Johns Hopkins University Press
Baltimore

Johns Hopkins Paperback edition, 2022
9 8 7 6 5 4 3 2 1

Johns Hopkins University Press
2715 North Charles Street
Baltimore, Maryland 21218-4363
www.press.jhu.edu

The Library of Congress has cataloged the hardcover edition of this book as follows:

Names: Ash, Eric H., 1972– author.
Title: The draining of the Fens : projectors, popular politics, and state building in early modern
    England / Eric H. Ash.
Description: Baltimore : Johns Hopkins University Press, 2017. | Series: Johns Hopkins studies in
    the history of technology | Includes bibliographical references and index.
Identifiers: LCCN 2016028456 | ISBN 9781421422008 (hardcover : acid-free paper) |
    ISBN 142142200X (hardcover : acid-free paper) | ISBN 9781421422015 (electronic) |
    1421422018 (electronic)
Subjects: LCSH: Fens, The (England)—History—16th century. | Fens, The (England)—
    History—17th century. | Drainage—England—Fens, The—History. | Reclamation of land—
    England—Fens, The—History. | Fens, The (England)—Environmental conditions. | Fens,
    The (England)—Politics and government. | Political development—Social aspects—
    England—History. | Government, Resistance to—England—History. | Great Britain—Politics
    and government—1558–1603. | Great Britain—Politics and government—1603–1649 |
    BISAC: SCIENCE / History. | HISTORY / Europe / Great Britain. | TECHNOLOGY &
    ENGINEERING / History. | HISTORY / Renaissance.
Classification: LCC DA670.F33 A79 2017 | DDC 942.6—dc23
LC record available at https://lccn.loc.gov/2016028456

A catalog record for this book is available from the British Library.

ISBN 978-1-4214-4330-0

Special discounts are available for bulk purchases of this book.
For more information, please contact Special Sales at specialsales@jh.edu.

For Catherine, Claire, and Henry, with love

Contents

Acknowledgments

Many years ago, as I was finishing up work on my doctoral dissertation, my graduate adviser, Tony Grafton, told me that when I was ready to start thinking about a second book project, I might want to look into the draining of the Fens. He seemed to remember that there were abundant sources for such a project, he said, and thought it might be right up my alley. Thus it was that I acquired the first of many debts of gratitude in the course of researching and writing this book.

I must gratefully acknowledge the generous financial and institutional support I have received from various sources throughout the years. The archival research for this book was undertaken over many trips to the United Kingdom, supported by a grant from the National Science Foundation in 2002–4 (SES-0301851), a Wayne State University Career Development Chair in 2007–8, and a WSU Humanities Center faculty fellowship in 2009–10. I first began work on the project during a postdoctoral fellowship at the Dibner Institute for the History of Science at the Massachusetts Institute of Technology in 2001–2 (an institution, regrettably, no longer in existence). I was a WSU Humanities Center resident scholar in 2008, when I drafted the first few chapters, and I want to thank the center's director, Dr. Walter Edwards, for his long-standing support of this project and of all humanities research at WSU. The bulk of the writing was done at the Huntington Library, where I was fortunate enough to be a long-term Dibner Fellow in the History of Science in 2013–14—I cannot think of a more conducive, supportive, or pleasant place to work. I am particularly grateful to Steve Hindle, Juan Gomez, and Catherine Wehrey-Miller for all of their assistance during my year there. My colleagues in the WSU Department of History have supported my work in so many ways, and I especially want to thank Marc Kruman and Liz Faue, the former and current chairs, for their help, advice, and friendship.

Early modern fen drainage manuscripts are voluminous and impressively well distributed; it was impossible to see everything I might have wanted. My research was primarily undertaken at the British Library (Manuscripts, Rare Books, and

Maps collections), the National Archives at Kew (Chancery Papers, Exchequer Papers, State Papers, Domestic Series, and Maps and Plans), the Cambridge University Library (including the Ely Diocesan Records, now housed there), and the Cambridgeshire County Record Office (where the Bedford Level Corporation papers are held). I also examined various documents and collections at the Bodleian Library at Oxford University, the Parliamentary Archives, the Norfolk Record Office, the University of Nottingham (Manuscripts and Special Collections), and the Huntington Library. I am very grateful to the knowledgeable and courteous staff at all of these institutions for their assistance—they help to make archival research a joy. The WSU Library has been a real ally in helping me to locate and obtain hard-to-find items, and I wish to thank Cindy Krolikowski, the Library Liaison for the History Department, for all of her help. I have made frequent use of the University of Michigan and Michigan State Libraries, particularly the microform collections in each place, as well as the Michigan Electronic Library system, which is really a wonderful statewide resource. The photographic reproduction staff at the Huntington Library, Houghton Library at Harvard University, the British Library, Cambridge University Library, and the National Portrait Gallery all helped me to navigate the sometimes tricky process of acquiring and reproducing the illustrations in this book.

I have presented portions of this project at numerous symposia, colloquia, and conferences, including the History of Science Colloquium at Johns Hopkins University; the Cabinet of Natural History seminar series at Cambridge University; the History Seminar at Ludwig-Maximilians-Universität in Munich; the History of Science, Technology and Medicine Colloquium at UCLA; the Seminar on History and Philosophy of Science at Caltech; the History of Science Society annual meeting; the North American Conference on British Studies annual meeting; the Midwest Conference on British Studies annual meeting; and the American Society of Environmental History annual meeting. I have received valuable feedback from auditors at each forum, who have helped me to make this a better project. I also wish to thank the editorial staff at Johns Hopkins University Press, especially Elizabeth Demers, Meagan Szekely, and Bob Brugger for all of their support and guidance during the long process of revising and publishing the manuscript. Bob Cronan at Lucidity Information Design did a wonderful job on all of the custom-made maps in the book. It was a pleasure to work with Beth Gianfagna at Log House Editorial Services, whose careful attention in copyediting the book saved me from making numerous errors.

I have also benefited greatly from the advice, support, hospitality, and friendship of numerous individuals. Julie Bowring was kind enough to share a number of digital images with me from the Bedford Level Corporation papers, allowing me to complete several weeks' worth of research at home—I am very grateful to her. Tony Grafton, Pamela O. Long, and Peter Dear have all advocated for this project in my behalf and have opened many doors for me. Will Ashworth, Andre Wakefield, and Stanley Shapiro have pushed me out of my comfortable assumptions on numerous occasions, as good friends do. I have gained much from conversations with my "fellow fellows" and other scholars at the Huntington Library, especially Alison Games, Sarah Grossman, Paul Hammer, Steve Hindle, Fred Hoxie, Rupa Mishra, Lindsay O'Neill, and Keith Pluymers; this is certainly a much better book for all of their wisdom and advice. My colleagues in the Group for Early Modern Studies at WSU, including Simone Chess, Jaime Goodrich, Lisa Maruca, and several of the regular graduate student attendees, have read or heard drafts of several chapters and given me much encouragement and good advice. David Chan Smith gave me some valuable feedback on chapter 3, regarding the legal debates over fen drainage, and Moti Feingold inspired me to think more precisely about the motivations of the Hartlib Circle in chapter 8. All of these scholars have helped me to hone my argument and avoid mistakes both large and small; any remaining shortcomings in the book are, of course, mine alone. Erica Charters, Susanne Friedrich, Lauren Kassell, Frances Willmoth, and Jacob and Lauren Sager Weinstein have all made me feel welcome and at home during my trips overseas and have assisted me in more ways than I can count.

My wonderful family has without fail been loving, supportive, and, above all, patient with me as I struggled to bring this project to fruition. They make all good times better, all hard times easier, all adventures more rewarding. I would be lost and mired in the Fens without them.

Note on the Text

All dates in this book refer to the Old Style (Julian) calendar, used in England throughout the sixteenth and seventeenth centuries, save that the new year is understood to begin on 1 January, rather than the common early modern convention of starting the new year on Lady Day (25 March).

In all quotations, I have altered the spelling and capitalization to conform to modern American English usage and have expanded all contractions. In a few cases, where original punctuation was especially obscure or confusing, I have altered it slightly to reflect more modern usage. With respect to titles of works and documents, I have retained the original orthography.

The Draining of the Fens

The Unrecovered Country

Draining the Land, Building the State

> *Merecraft*: Then take one proposition more, and hear it
> As past exception.
> *Fitzdotterel*: What's that?
> *Merecraft*: To be
> Duke of those lands, you shall recover: take
> Your title, thence, sir, Duke of the Drowned-lands,
> Or Drowned-land.
> *Fitzdotterel*: Ha? That last has a good sound!
> I like it well . . .
> —Ben Jonson, *The Devil Is an Ass*, act 2, scene 4

The possibility of draining the English Fens had been considered and debated for at least a generation by 1616, when Ben Jonson satirized the whole idea by having Merecraft pitch it to the greedy and buffoonish Norfolk gentleman Fitzdotterel as a surefire, get-rich-quick scheme in *The Devil Is an Ass*.[1] Merecraft was Jonson's caricature of a *projector*, an entrepreneurial and opportunistic promoter of various projects, and a figure with a somewhat shady reputation in early modern England. Land drainage was only one of Merecraft's many schemes, but it was perhaps the one most closely identified with real contemporary events. Drainage projects had attracted the interest of the Crown and its advisers since Queen Elizabeth I's reign (1558–1603), and in fact Jonson got into some trouble at court for the passage quoted above. Although King James I did not object to the satire of fen drainage projects per se, he apparently considered the reference to the "Duke of Drowned-land" a bit too sensitive and told the author to remove it.[2] James's interest in drainage was due in part to the Crown's being the single largest owner of land in the Fens, so improving its landholdings there would help to boost royal revenues at a time when additional income was sorely needed. But beyond their potential to help refill the depleted royal coffers, drainage projects were part of the broader trends of state building in seventeenth-century England. They were an expression of changing early modern attitudes toward the proper management and exploitation of the natural environment and the state's role in facilitating such.

In the first decade of the seventeenth century (probably in 1606), while the first of many large-scale drainage proposals was under discussion among local officials in the Fens and in the House of Commons, two anonymous commentators weighed in on opposite sides of the debate. The first, claiming to write on behalf of the fenland inhabitants, offered a variety of objections to the drainage. There was no precedent, he wrote, for the improvement and enclosure of such a vast area of common waste lands without the commoners' consent.[3] The Fens' abundant common wastes had long enabled smallhold farmers and even landless cottagers to keep small herds of livestock, which were fed on the plentiful grass that the moist, rich soil provided. Even the poorest inhabitants, with no livestock to pasture, could still scratch out a living by harvesting the natural produce of the wetlands: catching waterfowl, fish, and eels to furnish the tables of wealthier urban consumers; collecting reeds and sedge for thatching roofs; and digging out peat, which was dried and burned for fuel. Even if the drainage should prove a success—an outcome he considered "very improbable"—it would only lessen the yield of grass, diminish the common wastes through enclosure, and deprive the nearby market towns of valuable wetland goods. Most husbandmen would be hurt by the scheme, and poor cottagers left without any employment, so that "no way else able to put bread into their mouths," they would be forced to "live on alms, beg, or starve."[4]

In concluding his objections, the author criticized the "justices and chief gentlemen of the fen countries," whom the king had commissioned to consider the project, and who had pronounced it "a work of great commodity to the common weal." While acknowledging that they might be "wise, discrete gentlemen" all, the author argued that since none of the commissioners depended on the fenlands for their very livelihoods, they could not "understand the commodities and discommodities of the fens as those who only live by them." They misguidedly sought to make the Fens more productive and to stamp out the supposed idleness of fenland inhabitants, yet in reality they threatened to make it impossible for the large majority of fenlanders to earn a living there. Rather than thinking of the "common weal" as they were supposed to, they thought only of the propertied elite who stood to gain by improving and enclosing the wastes on their own estates. Most were large landowners themselves, moreover, and would "not speak all they know [about the project] because the drainage may be more beneficial to some one of them, than to ten thousand of others." Because the commissioners were ignorant of how best to exploit the flooded Fens and may have had a private interest in seeing them drained, they could never provide a sound verdict on the putative merits of such a proposal, either for the great majority of fenlanders or for the commonwealth.

The pro-drainage respondent to these objections refuted them point for point, accusing the objector and his fellows of obstinately opposing "a work that is for the general good of themselves & the common weal." There was no precedent for improving and enclosing so much common waste all at once, he observed, because such a bold and ambitious drainage scheme had never been attempted before. Far from being "very improbable," the proposed project was held by knowledgeable men "to be both possible & profitable." Yet if it should prove otherwise, only the investors stood to lose anything by it, since they alone bore the costs and would gain no reward unless it was successful. Draining the Fens would increase both the quantity and quality of pastureland for everyone; although some commons would be enclosed to pay for the project, the remainder would be much richer and more productive than before. The notion that drier soil would yield less grass than waterlogged soil was dismissed as "a very ridiculous observation," so that "if the poorer sort make no benefit it is through their own laziness & idleness that they get no stock about them."

With respect to the diverse natural produce of the flooded fenlands, the pro-drainage author accused his opponent of being "merely ignorant" of the true situation. Once they had been drained and improved, the Fens would not only support more livestock than ever before, they would also yield marketable crops on newly arable land, fetching a much greater profit per acre than wetland goods ever could. Any assertion to the contrary was "a gross and palpable error," based on a ludicrous overvaluation of eels, reeds, and peat turves. As for the poor fenlanders who claimed to make a decent living from harvesting such things, he noted that "for the one half & more of the year, they live merely idle, and have no means to set them on work. . . . [T]he beggarly life that the poor idle wretches do lead, do manifest of what commodity it is to them." Nor would the drained fens be destitute of these same wetland resources; in fact, they should be easier to harvest when they were concentrated around drainage ditches and receptacles. The improved, navigable rivers and superior value of market crops, meanwhile, would allow the entire region to prosper through increased trade, exporting their various commodities to new urban markets and importing goods "from such places as they never dealt in before."

Like the complainant whom he was refuting, the respondent saved his ad hominem attacks for the end. The men commissioned by the king to consider the
project were all "gentlemen and others bordering upon the fens . . . such as have
been employed in matters of this kind." They were therefore well acquainted
with the flooded lands and the pitiful, hardscrabble living to be made from
them. They acted not from any private interest, except insofar as they sought to
improve the Fens as landowners with rights of common there themselves. As
knowledgeable and disinterested judges they were "all of one constant mind
that [the drainage] was both possible & profitable to the common weal." Nor were
the commissioners alone in their opinion; when they met at Wisbech in 1605, they
had invited any commoners who wished to be heard to come and voice their
concerns. More than five hundred turned up, and their chosen representatives
stated their objections fully; but once they "received such satisfaction upon the
answers as they gave no reply unto it," they "gave approbation" to the project.

It was neither the drainage projectors nor the commissioners, therefore, who
opposed the good of the commonwealth, but the few "malicious" men who continued to resist the undertaking. Their objections were "framed out of a most
seditious and dangerous head, to infuse discontentment into the heads of poor,
ignorant people, who are not capable of their own good, but . . . are rather carried
away by false bruits, than satisfied with good reason." The anonymous complainant did not represent the true views of fenland commoners at all, but "hath done
this out of a very malignant spirit, willing to set the people into a combustion."
The author concluded ominously that this was "a matter fit to be looked unto in
time, lest himself (for other respects) become a hatcher of a greater mischief."

The arguments presented in this early exchange were in no way unique;
indeed, there is some difficulty in dating the document because each of the various points made on either side of the question was articulated many times over
throughout the first half of the seventeenth century.[5] I have highlighted it
because it touches on so many of the complex and interconnected issues at stake
in the drainage of the Fens to be examined in this book: divergent understandings
of what the Fens actually were and alternative visions for what they ought to be;
the rational management and optimal exploitation of natural resources; the
tension between Crown authority and local customs in a burgeoning English
state; the changing economic relations between English landowners and commoners; the development of a wider international market for both agricultural
and manufactured goods, and England's growing participation in it; the ongoing
debate over the improvement and enclosure of common waste lands; the evolv-
ing discourse of the good of the "commonwealth," as opposed to selfish, private

interests; and the pivotal role of projectors in shaping the future and building the state.

~

One of the most characteristic driving impulses of state formation in early modern Europe was the need to quantify, rationalize, and exploit the natural environment. This was not a new problem, of course, as human societies have always and everywhere sought to make use of the water, soil, minerals, plants, and animals around them, for personal consumption and sometimes for profit. In the early modern period, however, Europe's rapidly growing population and the rise of an ever more integrated continental and even global market economy placed new pressures on natural resources and gave rise to new attitudes toward exploiting them. Burgeoning states strove to catalog and control whatever natural resources existed within their borders and to acquire more, both in neighboring territories and in distant colonies.[6] Ecological thought in early modern Europe thus cannot be separated from the political and economic circumstances of the time. Shifting perceptions of the natural world were linked to the rise of powerful, unitary states, the emergence of a global mercantilist economy, and the dawn of an imperial age in which the acquisition and exploitation of natural resources was of paramount importance. New approaches to the study of natural philosophy that emphasized the utility of natural knowledge both reflected and reinforced these trends.[7]

The impulse toward a more rational organization and exploitation of the environment was particularly evident with respect to schemes for better water management—providing clean water for growing cities, improving inland navigation, irrigating arid lands, and draining wetlands. Projects of varying scale were vigorously pursued throughout medieval and early modern Europe in an ongoing effort to control water more effectively for productive ends, most notably in Italy and the Low Countries but also in France, Germany, and Denmark.[8] By the end of the sixteenth century, both native and foreign drainage projectors had set their sights on draining England's many wetlands, particularly the eastern Fens. A key element in making fen drainage seem both feasible and desirable was the rapid shift in understanding of the Fens that took place after 1570 or so, from a challenging but functional landscape that supported a peculiar local economy, to a wasted natural resource that might be remade for the good of the commonwealth. One's perception of the Fens largely determined one's attitude toward the land and its inhabitants and how both ought to be managed; the power to define the Fens in the present was the power to determine their future.

The pre-drainage Fens were admittedly not an easy place to live. The land, by definition, was prone to annual flooding of unpredictable duration and severity.

The soil was mostly too wet to support cereal agriculture, and diseases such as malaria were endemic. Non-natives found the region especially unwholesome and forbidding, and tended to suffer inordinately from fevers and agues. It was difficult to keep clergy and schoolmasters in residence and in good health, and the paucity of wealthy lords and gentry living in the Fens appeared to confirm the poverty of the region. Among those English commentators who were not native to them, therefore, the Fens had a reputation as a poor, unhealthy, and uncivilized quagmire.[9]

Yet the lived reality of the Fens was not nearly so bleak for most; throughout the Middle Ages the fenlanders had developed a number of ways to live and even prosper within their watery world. Though the floodwaters could be unpredictable in any given year, they were reasonably dependable over time and could usually be relied on to follow a certain pattern, arriving in the autumn and receding in most areas by late spring. Nor, for all the trouble they sometimes caused, were they without benefit: while most of the land could not grow grain, the receding floods left riverine silt deposits that promoted the growth of abundant grass on meadows spanning hundreds of thousands of acres. Fenland agriculture was thus based primarily on livestock grazing, and in a good year the region even supported stock from neighboring uplands, whose owners could then put their drier, arable lands to more profitable use. Those fenland areas too wet for grazing yielded other valuable goods, including foodstuffs, building materials, and fuel. All of these were freely available on the large common wastes and could be harvested and sold to non-fenland consumers to supplement the livelihoods of the poorer inhabitants.[10]

The pre-drainage fenland economy, then, was complex, diversified, and well integrated into the broader regional economy: drier uplands and wetter lowlands were specialized in terms of the agriculture practiced on them, allowing both to be used more intensively, while the unique produce of the wettest areas found its way to urban markets throughout southeastern England. The annual inundations provided the water and silt deposits that produced so much rich grass in the summer and gave rise to all of the other wetland goods upon which the inhabitants depended. The traditional fenland economy did not exist in spite of the recurrent floods, but *because* of them. The key to managing and profiting from the Fens was to regulate and control the floods, rendering them as stable and predictable as possible, rather than eliminating them altogether. Specialized administrative bodies called "commissions of sewers" were created for this purpose during the Middle Ages.[11]

Early commissions of sewers were assembled intermittently, empowered to deal only with specific drainage problems that disrupted a given area's normal flooding patterns. Composed of local gentry and the more substantial yeomen, and advised by juries of more modest husbandmen, the commissions functioned as royal courts of record, appointed and authorized by the Crown to determine what was causing the problem and to take action to correct it. Their jurisdiction was bounded both temporally and geographically: a commission was granted for a certain delineated area (sometimes as small as a single fen, common, or meadow), and its decrees initially stood in force only as long as was necessary to solve the problem that had created the need for it in the first place. Commissions of sewers were therefore fundamentally local and conservative in nature. Their mandate was to preserve the extant drainage network of a particular place, allowing the inhabitants to take full advantage of the opportunities and resources offered by the flooded landscape, exploiting the Fens on their own terms. For this the commissioners relied on their own extensive knowledge as local landowners and the collective life experience of the jurors who advised them.

Although commissions of sewers were issued by the Crown and their powers were codified over time by parliamentary statute, they were the institutional embodiment of a unique legal and political culture that arose in the medieval Fens and other wetlands of England specifically to manage a challenging landscape. The constant need to maintain and repair their drainage networks and to oversee and mediate complicated use rights in the vast common wastes (often shared among several bordering communities), had given rise to a series of local customs and arrangements that enabled fenlanders to conserve and control their own natural resources. The first formal commissions of sewers, issued in the thirteenth century, thus reflected, institutionalized, and built on what were already longstanding practices in England's wetlands. Over the centuries, a working culture of environmental management had evolved around the commissions that was deeply rooted in custom, founded on local knowledge, and emphasized conservative solutions to local problems.

It is important to stress that the commissions were both legal and political institutions, rather like the county benches of justices of the peace (and many fenland justices served on the sewer commissions as well). They functioned formally as courts of law, but also wielded real political power in the Fens. The fact that the commissioners, their concerns, and the limits of their jurisdiction were all firmly rooted in their localities does not make their activities any less political.[12] They articulated and advocated for local interests when necessary; they engaged in frequent disputes with rival commissions in neighboring communities; and at times they interacted directly with the Crown. The long-standing legal and political culture of the commissions of sewers gave them the confidence and the institutional wherewithal to resist external incursions into their jurisdictions. When projectors with royal backing sought to drain the whole of the Fens after 1600 and threatened to alter profoundly both the landscape and the traditional economy it supported, the commissions struggled to assert and defend their customary rights, with some success.[13] Before the Fens could be drained once and for all, the commissions' power first had to be broken, co-opted, or supplanted so that the local interests of fenland commoners and smallholders could be subordinated to the more market-oriented interests of drainage projectors, the larger landowners who supported them, and the Crown.

During the early seventeenth century, as the English state worked to become more centralized and unitary, and its economy more integrated into the emerging globalized markets of Europe, a fundamental shift took place in elite perceptions of the Fens. The traditional customs, resources, and institutions of the region were no longer deemed adequate for meeting the needs of the English commonwealth, and both the landscape and its management came to be viewed as compelling problems of state, not least because of the Crown's large landholdings there. The Fens had previously been considered (when they were considered at all) as a stable and modestly productive region; so long as local authorities were empowered to maintain the extant drainage networks, the land was problematic to be sure, but capable nevertheless of supporting a modest alternative agriculture.

By 1600, however, political elites in London had begun to perceive the Fens as a region in need of drastic human intervention to bring it to a more acceptable level of productivity. Commentators used different metaphors to characterize the land as broken and in need of repair; diseased, and in need of a cure; fallen, and in need of redemption. The Fens had somehow become a waste of potentially valuable land, an embarrassment to a civilized state, and an unreasonable burden on the commonwealth.[14] Fenland pastures were unreliable, insufficient, and unhealthy, and they would be more profitable as arable land in any case. The traditional wetland produce of the region was little more than worthless trash, compared with grain or other marketable crops such as coleseed (rape). And the inhabitants were no better than the muddy land on which they slogged to make such a pitiful living—they were idle, backward, unruly, sickly, and too few in number.

In order to be made productive and prosperous, the Fens were not merely to be managed and maintained as wetlands, but transformed altogether into a landscape much more closely resembling the surrounding uplands. To achieve this, the geography, ecology, and economy of the region would all have to be permanently altered: the land must be recovered, its chronic defects corrected, and new crops and farming techniques introduced. Once this was achieved, the character of the inhabitants would also begin to improve; they would cease living as lazy, barbarous, semi-amphibious herders and hunter-gatherers and learn instead to be honest, hardworking farm laborers. Such a massive undertaking could never be accomplished piecemeal, one village or common fen at a time; instead, a broader, regional approach was called for, one that approached the vast levels of the Fens as a single entity, with a common problem and in need of a common solution. Such coordination would require an unprecedented degree of involvement on the part of the Crown, both as an interested party and as a referee, working to overcome the many competing local interests arrayed in opposition, while facilitating the construction of extensive (and expensive) new drainage works.

To enable such a sweeping and profound improvement, the traditional, conservative, and provincial culture of the commissions of sewers would also have to go. Their myopic goal of preserving each individual common waste so as to make the best use of it in its flooded state was now viewed as a problem in itself, a case of stubborn and backward fenlanders failing to recognize what was clearly in their best interest. A more dynamic and innovative outlook was called for, a regional vision of *what could be* rather than *what is and has been*. The commissioners' horizons were simply too narrow, too conservative, and above all too local. Because the commissions were each so deeply rooted in their own communities, altogether they represented a multitude of contradictory local interests that could only serve as obstacles to a broader vision of a drained Fens, an entire region of waste transformed into a productive asset. Private interests and provincial prejudices could no longer be suffered to stand in the way of incorporating the Fens more profitably into the wider economy, nor could hidebound and contentious local institutions be allowed to prevent the region from being woven more tightly into the English commonwealth.

Before the last half of the sixteenth century, the English Crown could not even consider undertaking such a large and expensive project. It was too small and weak to compel local cooperation and either unwilling or unable to spend the capital needed to hire so many laborers for such a protracted period, with little hope of immediate return on that investment. Long before 1600, however, the Tudor dynasty was already working to expand and consolidate royal powers, co-opting or undermining regional legal and political institutions that interfered with royal governance and erecting more tractable alternatives. Late-sixteenth- and seventeenth-century England experienced an intense and creative phase in the long-run processes of state formation. The Crown strove in a number of ways to govern the polity ever more centrally, from collecting taxes to training a militia, from relieving the poor to enforcing religious orthodoxy.[15]

In addition to these core missions of centripetal governance, the English Crown (like other early modern states) sought to initiate and oversee the execution of various technical projects—promoting the development of new mines, shipyards, foundries, harbors, and forts, as well as innovative surveying and navigation techniques—while also supporting new manufactures, trades, and agricultural improvements. When successfully coordinated and carried out, such projects bolstered the real power and wealth of the early modern state: they produced materiel for warfare, increased agricultural yields and customs revenues, and created employment opportunities for rapidly growing populations. Moreover, demonstrating mastery over such valuable specialized knowledge also reinforced the legitimacy of early modern states as they consolidated their power and exerted their expanding authority.[16]

In order to locate, acquire, oversee, and control the technical know-how it demanded, the Crown relied on certain agents who were capable of mediating between the royal advisers who commissioned such projects and the various skilled and unskilled laborers who carried them out. In previous work I have examined the important role of experts and expertise in building the early modern English state. Experts in this context, I have argued, were not merely craftsmen with extensive empirical experience; while such experience was clearly constitutive of expertise, during the early modern period an "expert" was also one who could claim to possess a broader, more theoretical understanding of the field in question. Expert knowledge might come from prior experience, to be sure, but it might also be founded on extensive book learning, travel abroad, or a strong knowledge of mathematics, for example. The meaning of expertise, in short, was starting to shift from having done something many times to knowing why it worked the way it did and being able to apply that knowledge in novel circumstances. Such knowledge was neither possessed nor claimed by most common practitioners, and thus it marked experts as being of a separate and higher caste. In their ability to traverse comfortably the gulf between the royal court and the active worksite, experts served as crucial brokers of both technical knowledge and Crown authority. They placed their skills and knowledge in the service of the early modern state, gaining legitimacy for themselves in the process.[17]

Other historians have concentrated on the character of the early modern *projector*, one who proposed various projects to those in power, usually in the hope of receiving some reward in the execution of them.[18] A product in part of the new system of royal patents developed in England under Elizabeth I and her principal adviser William Cecil, Lord Burghley, projectors had a decidedly mixed reputation in the seventeenth century (as did the patents themselves). While some saw them as entrepreneurial and visionary in their promises to bring forth great profits from resources yet untapped, others derided them as either benighted fools or parasitic frauds, as Ben Jonson's depiction of Merecraft would suggest. In his *Essay on Projects* (1697), Daniel Defoe sought to rehabilitate the term by drawing a distinction "between the honest and the dishonest" projects: "between improvement of manufactures or lands, which tend to the immediate benefit of the public, and employing of the poor" on the one hand, and "projects fram'd by subtle heads . . . to bring people to run needless and unusual hazards" on the other. Yet even Defoe was quick to point out that it could be very hard for a prospective investor to tell the difference between the two. While the first sort were obviously to be preferred, "yet success has so sanctifi'd some of those other sorts of projects, that 'twou'd be a kind of blasphemy against fortune to disallow 'em."[19] Projects, no matter what they were, always involved risk, and risky investments could destroy resources as well as multiply them. As Defoe suggested, projects were not so easily separated into good and bad, honest and dishonest, but rather involved a spectrum of risk—the promised "immediate benefit of the public" and the "needless and unusual hazards" were really two sides of the same coin.

Projectors may be seen as a personification of the drive of early modern states to rationalize and better exploit their natural resources, through any number of schemes for agricultural improvement or new manufactures. As states sought to promote economic development through novel means, projectors seemed to offer a bewildering array of possibilities. They were also associated, both positively and negatively, with the new approach to natural philosophy espoused by Francis Bacon and his followers, which stressed the utility of natural knowledge as the greatest indicator of its fundamental truth.[20] Adherents of the Baconian program for producing operative knowledge of the natural world trumpeted its potential for improving the human condition, yet they were often characterized by their critics as the worst sort of projectors—well-meaning, perhaps, but ultimately absurd, even dangerous. In his famous satire of the Royal Society of London, for example, Jonathan Swift lampooned the "Academy of Projectors in Lagado," who earnestly sought to "contrive new rules and methods of agriculture and building, and new instruments and tools for all trades and manufactures." Though this sounded promising enough, the academy actually squandered its resources in pointless ventures such as extracting sunbeams from cucumbers, mixing paint by smell, or turning excrement into food. Though the projectors' ludicrous schemes had left the country around Lagado in ruins and its people destitute, this only redoubled their enthusiasm, "driven equally on by hope and despair."[21]

The categories of "projector" and "expert" overlapped at times; many a projector was a would-be "expert," seeking recognition for his rare and putatively valuable skills and knowledge, and reward for deploying them in the service of the Crown and commonwealth. Both figures, I suggest, played an important role in the formation of the early modern English state, and in England's growing participation in Europe's market economy. With respect to the Fens, the Crown relied on projectors to carry out the drainage: they were supposed to provide the vital expertise needed to design and construct a massive new drainage network. Most of them claimed to have prior experience in land drainage and the ability to generalize and adapt their knowledge to novel conditions—this was especially true of Dutch drainage projectors, such as Cornelius Vermuyden. In many cases they were also responsible for raising the funds needed for the project, convincing investors to fund it up front in return for a share of the soon-to-be drained lands. Projectors played a key role in convincing the English state (both the Crown and, during the civil war and Interregnum periods (1642–60), the rulers of the Commonwealth and Protectorate regimes) that fen drainage would be both possible and profitable, stressing the financial and social benefits of improving and enclosing the land.

Projectors were part of a movement in early modern England that promoted myriad schemes for improvement of all kinds—economic, cultural, educational, and spiritual—and by the mid-seventeenth century a thriving culture and literature of improvement was taking root. Improvement was a vital topic of the day, especially during the tumultuous years of the civil war and Interregnum, and agricultural improvement projects were one of its earliest and most successful manifestations. With respect to agriculture, the English concept of "improvement" had evolved a great deal from the early sixteenth century through the middle of the seventeenth. What began as a limited (and rather unsavory) practice of landlords finding new ways to enhance their rents, gradually became a much broader mission to transform, rationalize, and optimize the organization and exploitation of all lands, both in England and in English overseas plantations in Ireland and the New World. The desire for profit always remained a principal motivation for pursuing agricultural improvement schemes, but promoters of such projects argued that they also benefited the commonwealth. They were a productive way to harness private interest to the public good, bringing order and reason to an unruly landscape, and were pleasing to God; even if a given improvement failed to yield the expected financial returns, well cultivated land remained an intrinsic good.[22] Through the rhetoric of improvement, projectors asserted that draining the Fens would make the land more rational, reliable, and productive; the inhabitants more prosperous, healthy, and civilized; the commonwealth larger and richer; and the realm more easily governable. Draining the Fens, in short, would literally and figuratively expand the purview of the English state, bringing more power to those who governed it and more wealth to those who lived in it.

Standing in the way of all these grand visions for improvement, of course, were the obstinate fenlanders, with their traditional customs and use rights, their local economy that relied on the periodic flooding of the land, and their well-developed institutions for keeping things as they were. Their efforts to preserve their watery world and the way of life they led within it, and the efforts of the state and various outside projectors to "improve" it in spite of them, are the subject of this book. It is a narrative that brings together many vital issues in the history of seventeenth-century England. The drainage of the Fens is a story not only of the physical transformation of a landscape, but also of the increasing reach of the early modern state into local governance, of a new understanding of nature premised on its capacity for improvement and exploitation, and of the role played by projectors and experts in bringing about these changes.

This book covers the period from the 1570s, when the Crown first began to entertain proposals to undertake a regionwide drainage of the Fens, through the 1650s, when a massive project to drain the Great Level of the Fens was completed. This was obviously a momentous period not just for the Fens, but for all of England. My principal goal is to use the drainage projects to connect the broader political, economic, social, and environmental developments of the era. What took place in the Fens—not just the drainage, but the popular resistance it incited and the state's reaction to it—makes a revealing study of the complex, turbulent history of seventeenth-century England. What follows is divided into two parts.

The first, "Popular Politics, Crown Authority, and the Rise of the Projector," examines the period of rising tensions (c. 1570–1625) between the fenland commoners, the commissions of sewers, and the Crown over how best to prevent and ameliorate fenland flooding, which was growing noticeably more severe. The first chapter discusses the ecology of the pre-drainage Fens, as well as the local economy and political culture that had evolved there to manage and exploit the region's natural resources as effectively as possible. Chapter 2 examines some of the earliest proposals to drain the Great Level, including the project promoted by Sir John Popham in 1605, and why that undertaking failed despite a promising start. Chapter 3 provides an analysis of a serious legal challenge to the traditional powers of the commissions of sewers in the early 1610s, which pitted those who saw the institution as a fundamentally conservative one against those who viewed it as a potential agent of transformation in the region. Chapter 4 looks at Crown efforts to impose a regionwide drainage project in the Fens in 1617–21, which succeeded only in uniting the fractious and squabbling commissioners of sewers against the undertaking.

The second part, "Drainage Projects, Violent Resistance, and State Building," examines two of the largest drainage projects carried out between 1625 and 1660, as well as the violent resistance they provoked. Chapter 5 discusses the drainage of the Hatfield Level in the decade after 1625, led by the Dutch drainage projector Cornelius Vermuyden; this project set the pattern for later projects in the Great Level and elsewhere, and also featured the earliest instances of anti-drainage rioting. Chapter 6 looks at the first partially successful attempt to drain the Great Level during the 1630s and the power struggle that ensued between the original investors in the project, including many local landowners led by the 4th Earl of Bedford, and the Crown for control of the project. Chapter 7 returns to the Hatfield Level to examine the remarkable series of highly politicized anti-drainage riots that broke out in the Isle of Axholme throughout the civil war and Interregnum, as well as the efforts of the Commonwealth and Protectorate regimes to suppress them. Chapter 8 looks at the massive undertaking of the 5th Earl of Bedford and his partners to drain the Great Level once and for all during the early 1650s and the vital role of the Commonwealth state in facilitating the project and quashing local opposition to it. The book concludes with an epilogue, examining the environmental, economic, and social consequences of the Great Level drainage after 1660.

PART I

Popular Politics, Crown Authority, and the Rise of the Projector

CHAPTER 1

Land and Life in the Pre-drainage Fens

[The Crowland Fens, in Lincolnshire] are an equilibrian plantation of earth & water . . . in an unhealthful, raw, & muddy land, whither no people of fashion have recourse, but to their ducking sport in molting time. . . . This sport had need give good content (as indeed it doth) for the beastly nasty town, stinking diet, the rugged condition & debauch'd manners of the people give but little, all alike neither sweet, clean, nor good. . . . I know not what to make of them, I think they be half fish, half flesh, for they drink like fishes, & sleep like hogs, & if the men be such creatures, judge what their women are . . . but how both men & women are able to subsist in winter exceeds my reach. Their climate is so infinitely cold, & watery; their habitations so poor, and mean; their means so small, & scant; their diet so course, and sluttish; & their bodies so lazy, and intemperate, that in spite of all these, they live to verify the old proverb for their namesake, No carrion will kill a crow.

—"A Relation of a Short Survey of the Westerne Counties . . . ," 1635

Negative contemporary depictions of the early modern fens were commonplace; the above passage, while more expressive than most, is representative of the genre. Such descriptions emphasized the region's waterlogged and putatively barren soil, the foul and unhealthy air, and the poor, rude, and sickly inhabitants. Another feature most share, however, is that they were written by outsiders—non-native observers who were traveling through the fenlands, with no thought of staying there. The above, for example, was written in September 1635 by an anonymous traveler from Norwich who was returning home through Crowland after a seven weeks' tour of England's western counties. By the time of his journey, great changes were already underway in the Fens, and the traveler expressed his hope that the "lusty, stout sweating pioneers" he saw at nearby Guyhirn, "hard at work, digging, delving, casting up, and quartering out, new streams, and rivers," would soon succeed in making "that large continent of vast, foggy, miry, rotten, and unfruitful soil," into something "useful, fruitful, & beneficial, & for the advantage of the commonwealth."[1]

Most native fenlanders, those who actually lived and made a living in the region, had quite a different perspective. Another anonymous commentator, identified only as "The Anti-Projector," complained in the early 1650s that drainage projectors "have always vilified the Fens, and have misinformed many Parliament men, that all the Fens is a mere quagmire . . . of little or no value." Such a pessimistic view, he believed, arose from the ignorance and inexperience of outsiders, for "those which live in the Fens, and are neighbors to it, know the contrary." He listed several valuable agricultural products to be found in the undrained Fens, including various livestock and tilled crops, as well as the traditional wetland goods he called "the rich ore of the common-wealth." He concluded by pleading for the livelihoods of the poor fenland cottagers who depended on the common wastes. The drainage projects then under way, he argued, would "destroy not only our pastures and corn ground, but also our poor, and utterly disable us to relieve them," so that they "must go a begging."[2]

The Fens were something of a paradox for non-native observers because the very thing that made them seem most outlandish and appalling—the endemic floodwaters—was also the source of their agricultural abundance and the basis for the inhabitants' livelihood. Occasionally a commentator struggled to capture both the positive and negative aspects of life in such a soggy landscape and the tension between the two perspectives. In his poetic description of Lincolnshire, printed in 1622, Michael Drayton divided the county into its three administrative districts—Holland, Kesteven, and Lindsey—and composed a verse for each of them. Holland—the wettest district, where Crowland is located—sang first and boasted for nearly four full pages about the multitudes of different waterfowl and fish that were to be found and harvested in its rich fenlands and the nearby sea. Drayton portrayed the Holland fens as a thriving and diverse cornucopia, full of beauty, pleasures, and profit for those who knew how to find them:

> The toiling fisher here is tewing of his net:
> The fowler is employed his limed twigs to set.
> One underneath his horse, to get a shoot doth stalk;
> Another over dikes upon his stilts doth walk:
> There other with their spades, the peats are squaring out,
> And others from their carrs, are busily about,
> To draw out sedge and reed, for thatch and stover fit,

That whosoever would a [landscape] rightly hit,
Beholding but my Fens, shall with more shapes be stor'd,
Than Germany, or France, or Tuscan can afford.[3]

When the upland region of Kesteven followed, however, it immediately disparaged Holland as little more than "a foul woosy marsh," its "soggy fens" plagued by "unwholesome air, and more unwholesome soil." Though waterfowl might be plentiful there, they were tainted by "that so rammish taste of her most fulsome mud," so that "[t]he cook doth cast them out, as most unsavory things." The so-called grass of Holland "so blady is, and harsh,/As cuts the cattle's mouths, constrain'd thereon to feed." And the fish to be had "within her muddy moor,/Are of so earthy taste, as that the ravenous crow/Will rather starve, thereon her stomach than bestow." Lindsey's concluding verse was no kinder, likewise asserting that the fowl to be had in Holland were unhealthy and unfit for consumption: "the fowl which she doth breed:/She in her foggy fens, so moorishly doth feed,/That physic oft forbids the patient them for food."[4]

William Camden presented a similar paradox of fenland desolation and abundance in his *Britain, or a Chorographicall Description* . . . , first printed in English in 1610. In his oft-quoted depiction of Cambridgeshire, Camden described the *"Fen-men,* or *Fen-dwellers"* in harsh terms: "A kind of people according to the nature of the place where they dwell, rude, uncivil, and envious to all others whom they call Upland-men. . . ." Yet in the next breath, he portrayed them hard at work in exploiting their watery environs: ". . . who stalking on high upon stilts, apply their minds, to grazing, fishing, and fowling." He described Huntingdon as suffering from "the offensive noisomeness of meres and the unwholesome air of the fens," but he also stressed that "for the unhealthiness of the place, whereunto only strangers, and not the natives there are subject, who live long and healthfully, there is amends made, as they account it, by the commodity of fishing, the plentiful feeding, and the abundance of turf gotten for fuel." And the Holland division of Lincolnshire he found "so thoroughly wet in most places with waters, that a man's foot is ready to sink into it . . . foul and slabby quavemires, yea and most troublesome fens, which the very inhabitants themselves for all their stilts cannot stalk through." But despite its sogginess, the land was far from barren: "The ground bringeth forth but small store of corn, but plenty of grass, and is replenished abundantly with fish and waterfowl. . . . In regard of this their taking of fish and fowl [in Crowland] they paid yearly in times past to the Abbot, as now they do to the King, three hundred pounds of our money." And as for the quality of the fish and fowl, Camden had no doubts: "All this *tract-over* at certain seasons, good God, what store of fowls (to say nothing of fishes) is here to be found! . . . the very delicate dainties, indeed, of service, meats for the demigods, and greatly sought for by these that love the tooth so well."[5]

The ambivalence of Drayton and Camden toward the pre-drainage Fens captures the diverse, adaptable alternative agriculture practiced in the region, as well as the revulsion of outsiders who neither understood nor appreciated what it had to offer. The flooded Fens did not conform to the agricultural conditions and practices of the surrounding uplands, and they could be a challenging place to live and prosper, but their reputation as a poor and barren wasteland was mostly a creation of non-native observers. That reputation endured, in part, because it supported the rhetoric of improvement espoused by projectors, large landowners, and state officials who sought to profit from the Fens by draining and enclosing them.[6] The lived reality of the Fens was probably much closer to Camden's balanced *Chorographicall Description* than the callous portrayal of the anonymous Norwich traveler. This chapter offers an introduction to the landscape, economy, and management of the medieval and early modern Fens, before the seventeenth-century drainage projectors profoundly altered all three.

FENLAND GEOGRAPHY: PEAT FENS, SILT MARSHES, AND MEANDERING RIVERS

The region in England known today as the Fens or the fenlands occupies a large, crescent-shaped area of land between East Anglia and the eastern Midlands, surrounding a large bay called the Wash, of which it was once a part.[7] The Fens stretch from Lincoln in the north to the outskirts of Cambridge in the south, a distance of nearly seventy-five miles; and from Peterborough in the west to Brandon in the east, roughly thirty-six miles (fig. 1.1). In all, they span some 1,100 square miles, encompassing more than 700,000 acres, or roughly 283,000 hectares. Wedged between elevated areas of clay, chalk, and limestone, the Fens themselves rest on a sunken bed of clay that was once part of the North Sea floor. As sea levels rose and fell over millennia with the expansion and contraction of glacial ice, the fenland basin was gradually filled in, partly with marine silt deposited by the tides and partly with peat formed along the wide floodplains of the freshwater rivers. Even after the land rose definitively above sea level, the region remained subject to frequent floods caused by strong marine tides, storm surges, and swollen upland rivers. The underlying clay surface is rather uneven, with occasional rises or "isles" of higher elevation dotting the otherwise flat landscape, which are spared all but the most severe floods. The largest of these is the Isle of Ely, located in the northern part of Cambridgeshire and home to the city

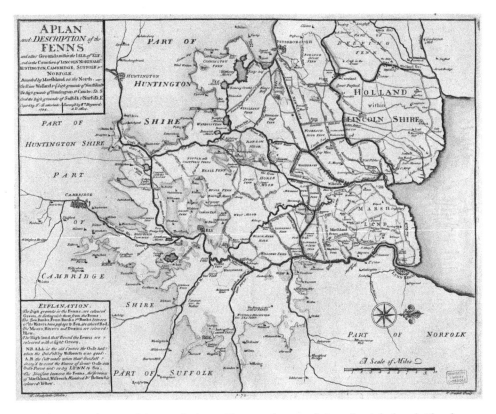

Figure 1.1. A map of the pre-drainage Fens, c. 1605; north is to the right-hand side of the image. This map was printed in Thomas Badeslade's *The History of the Ancient and Present State of the Navigation of the Port of King's-Lyn . . .* (London, 1725), but it is one of many based on an original manuscript map (now lost) produced by the surveyor William Hayward in connection with his 1604–5 survey of the Fens. Reproduced by kind permission of the Syndics of Cambridge University Library, Cam-a-725-1.

and episcopal see of the same name. Today such "isles" are surrounded by rich farmland, but before the drainage they would have been isolated outcroppings of (mostly) dry land, often surrounded by floodwaters.

In a pattern common to coastal wetlands across northern Europe, the Fens actually comprise two distinct regions and soil types, arranged concentrically: a wide band of silt marsh toward the coast, surrounded by an inland fen composed of peat, abutting the uplands (fig. 1.2). The silt marshes accumulated through the action of the North Sea tides in the Wash, as marine silt deposits gradually rose above the level of the mean high tide, which explains their seaward location.

Figure 1.2. A map of the pre-drainage Fens, including both the Great Level and the Hatfield Level.

Sometimes this process was natural, with silt collecting around an impediment of some kind, or else through a slight drop in sea levels. In other cases it was assisted through human effort, as tidal salt marshes were "inned," or cut off from the sea by an artificial bank and kept dry with a drainage dike and sluice to let water flow out at low tide.[8] The peat fens, in contrast, formed in the areas commonly flooded with fresh water from the upland rivers; the peat was composed of partially decayed plant matter, including mosses, reeds, sedge, shrubs, and sometimes trees. The floodwaters impeded the flow of oxygen so that dead plants did not fully decompose; instead, they formed a layer of peat that accumulated and deepened over time, becoming several feet thick in the centuries following the Iron Age. Any neat division of silt marsh and peat fen is complicated by the fact that periodic rise and fall of sea levels shifted the boundaries between them, which caused alternating layers of silt and peat to be deposited throughout the zone where the marsh and fen meet.

Technically, then, not all of the East Anglian wetlands collectively known as "the Fens" are truly fenland; the term *fen* more properly denotes an inland, freshwater wetland composed of peat, while a coastal wetland composed of marine silt is a *marsh*.[9] Though most people are not so precise in their terminology, the difference is of considerable importance not only in terms of topography and ecology, but also for the history and local economy of each area. Despite being closer to the sea, silt marshes are typically somewhat higher in elevation than peat fens and are composed of firmer soil, so they tend to be a bit less prone to regular flooding and more responsive to limited, small-scale drainage efforts. In periods of lower sea levels, silt marshes could support permanent settlements and were more amenable to agricultural exploitation, allowing them to be colonized long before the peat fens. Marsh inhabitants often constructed walls and banks on either side of their settlements to protect their lands both from the tidal surges of the sea and from the riverine floods of the peat fens. They also worked to augment the natural silting process, inning their seaward marshes so that their lands would continue to grow over time—though erosion and rising seas could and did sometimes take back the land that was so hard won.[10]

The peat fens sit rather lower in elevation and are bounded by the uplands on one side and the silt marshes on the other. They thus form a natural sink or basin, where the surging upland rivers regularly overflowed their banks in wetter seasons, passing onward to the sea only slowly and with difficulty. The recurrent floods allowed the peat to grow and deepen over many centuries, right up to the early modern period. Because they were always prone to flooding, the peat fens were ill suited for farming or permanent settlement. Such villages as dotted the region were located only along the periphery or on one of the raised isles within, even as the nearby silt marshes were becoming well settled. The histories of each landscape, while obviously linked in many ways, were thus divergent in others.

Especially given the vernacular imprecision of terms such as *fenland* and *the Fens*, it is important to specify the focus of inquiry for the present study. This book concentrates almost exclusively on the inland peat fens toward the western and southern parts of the region, looking specifically at two areas drained during the middle of the seventeenth century, between 1625 and 1655. The primary focus is on the Great Level, known today as the Bedford Level, after the 4th and 5th Earls of Bedford, who were the chief landowners and promoters of the drainage there. The Great Level contains more than 300,000 acres (c. 122,000 hectares), a little less than half of the total area of the Fens. It is located in the southern part of the region, between the uplands of Peterborough to the west, Huntingdon to the southwest, Cambridge to the south, and Thetford to the east—and with the silt marshes of Spalding, Wisbech, and King's Lynn to the north. It spans parts of what were once six counties: Lincolnshire, Northamptonshire, Huntingdonshire, Cambridgeshire, Suffolk, and Norfolk, though the bulk of it lies in the Isle of Ely district in Cambridgeshire.[11] The Great Level was by far the largest area of fenland to be drained as a single project; efforts to drain it spanned from Queen Elizabeth I's reign through the first half of the seventeenth century. It was finally accomplished during the 1650s by a group of investors led by the 5th Earl of Bedford, under the direction of the Dutch drainage projector Sir Cornelius Vermuyden. Currents of early modern England's political, social, economic, and environmental history all intersect within the Great Level, like the many rivers and drainage channels that now crisscross it.

In addition to the Great Level, this book focuses on a second region of peat fen known as the Hatfield Level. This area, comprising nearly seventy thousand acres (c. twenty-eight thousand hectares) lies at the opposite, far northwestern, corner of Lincolnshire and also includes portions of Nottinghamshire and the West Riding of Yorkshire, a few miles east of Doncaster and southwest of Hull. The Hatfield Level is not contiguous with the Great Level to the south, but it has a very similar topography: a broad swath of peat fens sandwiched between the uplands and the coastal marshes, dotted by occasional "isles." It was drained in the 1620s–30s by a group of mostly Dutch investors led by Vermuyden, whose approach to draining the Great Level during the 1650s owed much to his experience at Hatfield Level a quarter-century earlier. The two projects thus share many links, and their similarities and differences help to illuminate each of them more fully.

Along with the peat fens and silt marshes that composed the soil of the region, the Fens were defined most of all by water, salt and fresh, ebbing and flowing. Indeed, the land literally *could not be* without the water, because neither the silt nor the peat would ever have accumulated absent the water that made them possible. The main rivers flowing through the region—the Great Ouse, the Nene, and the Welland, together with their numerous tributaries—collectively drain more than six thousand square miles of the surrounding uplands, roughly an eighth of all England.[12] Other than the scattered "isles," the landscape of the Fens is unremittingly flat, and it was only slightly above sea level before the drainage. When faster-flowing upland rivers descended into the region, they lost velocity and meandered over many miles through both peat and silt before finally reaching the sea. They carried eroded soil from the uplands, which settled out as they slowed in the peat fens, so that the riverbeds tended to rise gradually above the surrounding landscape. The principal rivers were also tidal for a considerable distance from the coast, which impeded their flow during flood tides and increased riverine silting. During ebb tides their outflow increased, helping to scour away the newly deposited silt, though it still tended to accumulate over time. Also, just as some portions of the underlying bed of clay rose unevenly above their surroundings to form "isles," other parts sat somewhat lower and thus were permanently underwater, forming shallow, freshwater lakes known as meres.

When a high volume of water entered the region—through a tidal surge, heavy rainfall, or (worst of all) a combination of both—the rivers would top their banks. The utter flatness of the topography and the elevated position of the riverbeds ensured that the resulting floods always spread far and wide and were slow to dissipate. This was especially true in the peat fens, given their slightly lower elevation vis-à-vis the silt marshes that separated them from the sea. Such floods were commonplace before the early modern drainage, varying in duration and severity but occurring on a regular, usually annual basis. The artificial drainage channels and elevated riverbanks and levees constructed since the early modern period have lessened the frequency of fenland floods, but they still tend to be severe when they do happen (figs. 1.3, 1.4).

FENLAND SETTLEMENT AND HABITATION

Early settlement of the Fens generally coincided with changes in sea levels; a receding ocean left more of the region dry and open to habitation, while a subsequent rise resulted in at least a partial retreat. The region was inhabited in prehistoric times, when Britain was still connected by land with the Continent, though at that point it was probably not yet a wetland. By the time of the Iron

Figure 1.3. A photograph taken by the author in the Wicken Fen National Nature
Reserve in Cambridgeshire, March 2016. Wicken Fen is unusual insofar as the area
was left by the drainage projectors as a receptacle for flood surges, and so the land
was never drained. It is perhaps the best surviving example of what the pre-drainage
peat Fens may have looked like.

Age, when regular flooding was definitely a feature of the landscape, settlement
was limited to the higher silt marshes, the periphery of the peat fens, and the raised
isles within them. The Iron Age inhabitants are not known to have constructed any
drainage works, but they used dug-out canoes to navigate the waterways and built
extensive timber causeways linking their various settlements to one another.[13]

The Romans first arrived in the Fens during a period of relatively low sea levels;
they colonized the region and used their extensive engineering skills to improve the
drainage there. Among other works, the Romans are believed to have constructed a
massive artificial watercourse along the western edge of the Fens, running for nearly
ninety miles between Lincoln and Cambridge, called Car Dyke. This probably
served as a canal for the transportation of fenland goods to market, though it may
also have acted as a catchwater drain to prevent upland rivers from flooding the
peat fens. The Romans also built a road directly through the heart of the Fens,
from the Midlands to the Norfolk coast. As did their Iron Age predecessors, the

Figure 1.4. A photograph taken by the author in the Lakenheath Fen Nature Reserve on the Suffolk-Norfolk border, March 2016. Lakenheath Fen was drained in the mid-seventeenth century and served as heavily farmed arable land until 1995, when it was purchased by the Royal Society for the Protection of Birds and allowed to flood again, in order to create a wetland wildlife sanctuary. The reserve has rapidly recovered much of its former wetland ecology.

Romans settled and cultivated the silt marshes most intensively and probably only inhabited the outskirts of the peat fens. In addition to grazing livestock, the Romans also introduced cultivated crops on the drier portions of the land.

The Romans' fenland settlements probably declined before their ultimate departure from Britain, though the reasons for this are uncertain; rising seas may have made the area more difficult to live and prosper in, though sociopolitical factors likely also played a role, if only in making it harder to maintain the drainage works built there. The Fens were never totally abandoned even after the Romans left, but were inhabited in turn by Saxons and Danes. Early written records of the region date to the eighth century, by which time several monastic houses had been founded among the isles there, including some that would number among England's wealthiest and most prominent, such as Ely, Ramsey, Thorney, and Crowland. By the time of the Domesday survey in 1086, there were some fifty villages in the Fens, all located in an arc surrounding the Wash, along a narrow strip of the highest land in the silt marshes. The peat fens remained uninhabited, except for the edges and isles, where the monasteries dominated. The marsh villages were protected by massive banks built up along both the seaward and landward sides, each roughly fifty miles long and probably constructed after the Romans had left. The region was comparatively poor, with a lower population density and degree of wealth than the neighboring uplands. The land was farmed less intensively than in Roman times, and the amount of arable land declined.

Falling sea levels during the High Middle Ages led to another boom in fenland settlement. Population rose steadily as more land became suitable for pasture and even tillage, and by the fourteenth century the Fens had become one of the more prosperous agricultural regions in the realm, as indicated not only by the recorded tax rolls, but also by the many fine medieval stone churches built there. While the region remained rather sparsely populated when considered as a whole, this was because the peat fens were still liable to serious flooding and were thus uninhabitable; the silt marshes were in fact quite densely settled. The economy of the medieval Fens was little altered from previous ages, except in the scale of exploitation. Natural wetland resources were harvested with greater intensity and organization, especially with respect to digging peat, and eels were so abundant and valuable that they were used locally to pay rents and tithes. Tillage was maintained and expanded wherever land was dry enough to support it, so that virtually all of the isles in the peat fens were under cultivation. But by far the most important component of the fenland agricultural economy was its abundant grassland, and the huge herds of livestock that were maintained on it throughout the year. Fenland pasture was so rich that neighboring upland communities such as Cambridge could afford to commit all of their own land to tillage, paying a fee to graze their livestock in the Fens instead (a practice known as brovage). The landscape historian Oliver Rackham has suggested that during the fourteenth century the Fens were not only "fully used" but were being used in such a way that "we see men working with the distinctive landscape of the Fens, using it to complement the upland, instead of against it."[14]

The medieval fenlanders did not simply collect, use, and sell whatever the Fens had to offer; they worked to shape the landscape, maximizing its advantages and minimizing its inconveniences. They maintained the great banks that protected the silt marshes from both marine and riverine floodwaters, and constructed new ones to reclaim even more land from the sea. Along the periphery and in the isles of the peat fens, marginal lands were reclaimed by digging trenches to drain them and erecting banks to protect them. The best-documented drainage works of the time were undertaken by the monastic houses to improve their considerable landholdings, but smaller, secular landowners may well have attempted similar projects, without leaving the written records that the monasteries preserved of their own doings. Yet while medieval drainage works were fairly extensive when taken altogether, individual projects were almost invariably local and piecemeal, concentrating only on a single plot of flooded land at a time and accumulating over decades. There was little on the scale of the early modern drainage projects, which sought to drain tens or even hundreds of thousands of acres all at once in the span of just a few years.

The sole exception to this was the new river constructed in the latter half of the fifteenth century by order of John Morton, later archbishop of Canterbury and lord chancellor to Henry VII, but at the time serving as bishop of Ely. This drain, known as Morton's Leam, was intended to make the most of the land's limited gradient by redirecting the winding River Nene through an artificial channel, forty feet wide and more than ten miles long, running in a straight line from Peterborough to Guyhirn in the Isle of Ely. It was hoped that the straighter course would speed the river's flow and recover the Wisbech outfall by scouring out the sand and silt that had accumulated in it. Morton's goal was to improve both the drainage of the surrounding land and the river's navigability, and to provide much needed relief for the town of Wisbech, which was slowly losing its access to the sea as its river was choked with silt. Though the exact date and other details of its construction are unknown, the new river was probably completed near 1490, but it failed to accomplish the recovery of the Wisbech outfall, and by 1600 the whole project had fallen into decay. Nevertheless, Morton's Leam stands as the first large-scale attempt to alter the drainage network of the Fens since Roman times, an exception to the medieval pattern of small, piecemeal drainage works. It was also a harbinger of the early modern projects to come, which would likewise construct massive, razor-straight rivers across the landscape.[15]

Two points must be stressed concerning fenland geography and settlement. The first is that the Fens were never a static landscape or ecology. The alternating layers of freshwater peat and marine silt blanketing large portions of what became the peat fens are testament to the periodic advance and retreat of the North Sea, as ocean levels fluctuated. The waxing and waning of fenland settlement throughout human history depended in part on the frequency and severity of floods, which depended in turn on changes in mean sea levels. Such changes were neither recognized nor understood at the time but played a critical role in determining what agricultural activities one might hope to pursue in the region.

They also shaped human understanding of the Fens and of the prospect of reclaiming them. As flooding grew more problematic during the sixteenth century, monastic chronicles and cultural memories of fenland prosperity during the warmer and drier Middle Ages helped to convince some that a large-scale drainage was both possible and desirable.

The second point is that human inhabitants have always modified the landscape of the Fens in numerous ways to better suit their needs. Today, looking at their endless tilled fields and entirely artificial drainage network, it is very hard to imagine the Fens independent of human interference—what a "natural" fen might look like, or where a fenland river would flow "on its own." But human beings have molded, guided, and restricted the rivers and tides of the Fens from the earliest days of their settlement there, through deliberate interventions and the incidental effects of their use of the land. From Iron Age timber causeways to the Romans' Car Dyke to medieval trenches and banks, human beings worked to shape the fenland environment and its ecology for centuries before the early modern drainage projects. If "natural" is meant to imply the state of things before or without human involvement, then for all intents and purposes there has never been a "natural" fenland, and indeed such a term is virtually meaningless.

Both of these points must be borne in mind when examining the many controversies, lawsuits, and riots over draining the Fens that took place during the seventeenth century. Environmental historians (as well as ecologists) in recent years have moved away from a static model or understanding of nature, in which a stable "climax community" of particular plant and animal species would evolve and be sustained in a given environment "naturally," in the absence of any human "interference." Rather, it is increasingly recognized that no environment or ecology ever remains entirely stable. Fluctuations in climate and sea levels render the "natural" environment an ever-moving target, to which successive waves of plant and animal species must always adapt or be replaced by others. Nor has it been helpful for historians to posit an environment devoid of human presence and participation; human interactions with nature must be considered not as external interference, but as an integral part of the wider ecology.[16]

As the climate of northern Europe has changed through the centuries, Europe's coastal wetlands have changed with it, and human habitation in the East Anglian Fens became more or less viable with these changes. Yet the shifting environment was only one factor in determining how humans perceived and used the fenlands; political, economic, and demographic concerns were every bit as important. Nonenvironmental factors such as fluctuations in human population, the availability of labor, and the demand for various agricultural products had a direct impact on the socioeconomic desirability, the cost-benefit analysis, of struggling to cultivate such marginal land. These variables collectively shaped Britons' understanding of the landscape, what it was and what it might become. Ultimately they sought to transform the land altogether, but the environment also "pushed back" against their efforts. There were limits to what early modern technologies could achieve, and the unintended consequences of human intervention led to surprising changes in the region, creating new conditions with which the inhabitants were forced to struggle. The Fens, in short, have always been a dynamic landscape and ecology, and human beings have been active participants in that dynamism from the beginning.

FENLAND LIFE: DIVERSIFIED ALTERNATIVE AGRICULTURE

Through the processes of peat formation and marine siltation, and with some assistance from human inhabitants, a vast basin once covered by the North Sea was gradually filled in to become the Fens. By 1500, the area was home to a comparatively sparse population, densely clustered along the silt marshes near the Wash, with a long-established and flourishing pastoral economy. As we have already seen, the perception native fenlanders had of their country was so divergent from the view of outsiders that the two were almost unrecognizable to one another: the Fens were either an abundant, well settled, and prosperous pasture or a barren, empty, and impoverished wasteland; its inhabitants were either a thriving community of comfortable smallhold farmers or a ragged population of idle, sickly, and ill-mannered beggars. The negative characterizations of the region were a reaction to a landscape and economy that were foreign to the nearby Midlands, on the part of non-native observers unable to appreciate what they saw there on its own terms. Still, not all of the upland clichés regarding the sad state of the Fens were altogether untrue.[17]

Farming is never an easy way to make a living, and the Fens presented challenges not faced by farmers elsewhere in England. Most immediately, of course, there was the ever-present threat of flooding. In any given year, a strong storm surge or heavy rainfall could easily wipe out crops, drown livestock, damage property, and threaten lives. Even moderate floods could be devastating if the waters did not recede in time for planting crops or mowing hay. The Fens were also known to be an unhealthy region, in which malaria was endemic—this was particularly true in the coastal marshes, where dangerous populations of mosquitoes bred in stagnant pools of salt water. Agues and fevers were quite common, and while not usually virulent enough to be fatal in and of themselves, they did weaken the body and its immune system, providing an opening for other opportunistic infections that often proved deadly. Average life expectancy was noticeably lower in coastal wetland parishes than it was in higher and drier parts even of the same counties, with especially appalling rates of infant mortality.[18]

On the other hand, life in the flooded Fens may have been easier in some ways than it was in the drier parts of England because of the diversity, adaptability, and natural abundance of the fenland agricultural economy. The peat fens may have been too wet and flood-prone to allow for permanent settlement or cultivation, but they were nevertheless productive, valuable, and valued. Landowners and commoners did not live in the fenny wastes, but they parceled them out with care, exploited their resources intensively, and defended their use rights from anyone who sought to encroach on them.

The greatest treasure of the pre-drainage Fens, in any period, was grass. When floods covered the land each winter, they provided ample fresh water to nourish plant life, and new deposits of upland silt helped to replenish nutrients in the soil. At a time when more enterprising English farmers were just beginning to experiment with the expensive process of "floating" their meadows, flooding them deliberately to improve their yields of hay, most land in the Fens accomplished this naturally each year. The abundant grasses allowed large herds of livestock to be kept year round, including cattle, sheep, horses, and sometimes pigs. Cattle predominated in the sixteenth century—the region had a well-established dairy industry, producing milk, butter and cheese for local consumption and for the markets of urban centers such as Cambridge, Peterborough, Norwich, and London. Beef cattle were also fatted in fenland pasture, yielding hides and tallow as well as meat. Sheep came next in importance; in addition to their wool, which was of comparatively low quality, they yielded milk, meat, and sheepskins. These valuable products were exported to the rest of England and beyond via the region's many rivers, which constituted one of the most extensive inland waterways in northern Europe.

The ability to keep livestock throughout the year was unusual in sixteenth-century England and was possible in the Fens only because of the sheer abundance of grass there. Most fenland communities reserved their prime grasslands for meadow, leaving them ungrazed so that they could be mown for hay to support livestock through the winter months, when much of the land would be underwater. During the summertime they pastured their stock on the enormous stretches of nearby peat fen, which were treated as common wastes. The customary right to use common wastes was the cornerstone of the entire fenland agricultural economy. Smallhold farms predominated throughout the region, with most comprising just five acres or less of arable land; indeed, it was not uncommon for farmers in the region to possess no arable ground whatsoever. Virtually all farmers, however, had at least a modest herd of stock—the average husbandman in Lincolnshire boasted roughly ten cattle and perhaps twenty sheep, along with a few pigs and horses. Wealthier farmers owned much larger herds, of course, but even most cottagers kept a few cattle and sold the cheese and butter they produced to purchase their bread grain at market. The wastes were so extensive that smallholders in most fenland communities enjoyed the right to pasture stock there without stint or other limits, while many sixteenth-century landlords also exercised brovage rights, taking in additional stock from the neighboring uplands with the herds' owners paying a fee to graze them in the Fens.[19]

Tillage played a much smaller role in fenland agriculture, though virtually all settlements had at least some land that was dry enough to grow crops, and any land that could be cultivated in this way generally was. The chief crop grown in the Fens was barley, which served as the principal bread grain of the region, provided winter fodder for livestock, and was also used for brewing beer. Other crops varied widely from place to place but might include wheat (for local human consumption), peas and beans (for winter fodder), and hemp and flax (in support of a small local weaving industry). Fenlanders were exceptionally open to the idea of varied uses for the same land, depending on local conditions and needs; in dry years they would put more land under the plow and allow it to revert to pasture again in wetter years when flooding made tillage impossible. Though arable ground usually made up only a small percentage of a village's total land base, it tended to be extraordinarily fertile and productive in the Fens because of the ample manure it received from the large herds of livestock on the nearby pastures.

Pasture, meadow, and arable terrain were not the only valuable resources to be had in the Fens, especially for landless peasants and cottagers; the pre-drainage Fens presented a cornucopia of options for the poor to earn a modest living. Wetter areas of the peat fens abounded in fish, eels, and waterfowl, and fishing rights in the shallow meres were valuable enough to be jealously guarded and fought over by communities claiming a share in them. The natural vegetation of the more waterlogged soils—reeds and sedge—were cut annually, either for fuel, to be burned in the ovens of local bakers, maltsters, and brewers, or for roof thatching. The peat itself was also a valuable fuel source, especially in a region with so little woodland; peat was dug out in blocks called *turves*, dried, and then burned.

While much of this wetland bounty was consumed locally, a significant share was transported and sold in urban markets across southeastern England and beyond. All of these wetland goods were valuable; all (except the peat) were renewable; all were freely available on the common wastes to anyone who had use rights there (and in practice, to many who did not); all of them provided a vital source of income, especially to the poorer fenlanders who had no arable land and at most a few cattle to call their own; and all of them, it must be emphasized, existed in the Fens only because the land there flooded on a regular basis.

Anti-drainage advocates often invoked the plight of poor cottagers who depended on the natural wetland resources of the common waste for their survival. In 1597, for example, Lord Willoughby wrote to the Earl of Essex criticizing a bill then before Parliament to facilitate large drainage projects in the Fens. Willoughby believed that such projects would primarily benefit wealthy landowners and that the poorest fenlanders would be far worse off in trading their common waste for drained pasture. He claimed that even in the wettest areas, "a poor man may . . . make more commodity of a fen full of fish, fowl, and reed, rented for little or nothing, than of ground made pasture and improved to high rent, as the charges of draining will require." An enterprising man, he told Essex, "will easily get 16 [shillings] a week" cutting reeds, "and likewise three or four shillings a week in fish and fowl . . . which I speak not of hearsay but of mine own knowledge."[20]

The pre-drainage Fens thus supported perhaps the most underappreciated alternative agriculture in early modern England. Long before the seventeenth-century drainage projects, the Fens were already comparatively prosperous and egalitarian. They boasted a higher proportion of smallhold farmers, a less extreme distribution of wealth, and a less hierarchical social structure than was to be found in the more conventionally farmed Midlands just to the west of them. The regional economy was diversified and flexible, thoroughly adapted to take maximum advantage of the prevailing ecological conditions, and to react to changing local circumstances from year to year. It was also well integrated into the broader English economy, shipping goods to and from distant markets via its extensive network of inland waterways. Despite the widespread stories of abject poverty and ill health in the region, the Fens had one of the fastest growing populations in sixteenth-century England. Much of the increase was driven by landless laborers who migrated to the Fens in the hope of squatting in or near the common waste and taking illicit advantage of its availability and natural abundance. Taken altogether, the pre-drainage Fens might not be considered wealthy, but they were very far from poor; if resident gentry and grand estates were rare in the region, so were starving beggars.[21]

FENLAND MANAGEMENT: THE COMMISSIONS OF SEWERS

Fenlanders had a reputation among outsiders for being independent, isolated, stubborn, and even lawless; their alleged antisocial tendencies were ascribed to their living in such a deficient and inhospitable land. The Fens were often characterized as the haunt of bandits and rebels, a festering morass of hostility to authority, and a threat to the good order of the realm. There was perhaps some truth to this view, though it was largely superficial. The Fens did see more than their share of medieval rebellions, but in most instances the rebels had fled into the Fens to escape the king's justice, as royal troops could not easily be deployed in a wetland. The fenlanders themselves seldom joined in the rebellions whose participants sought refuge among them. Moreover, the Fens were in no way isolated, except from a military point of view; while it was exceedingly difficult to march an army through the region, the rivers had always provided an extensive transportation network for both fenland goods and passengers. Even cities located many miles from the sea, such as Lincoln and Cambridge, could engage directly in both coastal and maritime commerce.[22]

Nor could fenlanders accurately be portrayed as unusually independent or unwilling to cooperate with their neighbors; if this had been the case, they could never have maintained their traditional economy and way of life. The basic socioeconomic structure of fenland society was rooted not in the manor but in the larger framework of the village. In contrast with other parts of England, the boundaries of fenland manors did not generally coincide with parish or village boundaries, and many villages included some or all of several manors. Manorial control over local agriculture, and especially the all-important regulation of access to the common wastes, was therefore weak. Control was administered instead at the village level, requiring multiple manors to coordinate their activities together.[23] Fenland villages also had to learn to cooperate with one another; many stretches of common waste were so vast that they were bounded by several communities, each of which enjoyed some use rights within them. Villages sharing a common waste developed elaborate customs for "intercommoning" to determine who was entitled to put how many head of livestock on which parts of the fen at which times. Rather than being isolated, independent, uncooperative, and hostile to authority, fenland inhabitants were adept at negotiating, compromising, legislating, and disputing peacefully with their neighbors, because their common livelihoods depended on it.

Of even greater importance than coordinating the exploitation of the Fens was the need to cooperate in maintaining them. During the Middle Ages, both the silt marshes and the peat fens had been much modified by their human inhabitants to maximize the amount and quality of both arable land and pasture. The myriad drainage ditches, riverbanks, and seawalls crisscrossing the Fens had added considerably to the productivity of the land, but these works were expensive to build and maintain, and failing to keep them in good repair would lead inexorably to the loss of the hard-won benefits they provided. Even when an individual landowner was obviously responsible for maintaining a particular drain or bank on his property, such responsibilities had to be collectively enforced, since one man's private neglect could easily prove catastrophic for an entire community. To meet this need, commissions of sewers were created.[24]

By the mid-thirteenth century, in virtually all of England's wetlands, the inhabitants had developed institutions to survey local drainage works, make arrangements for their regular maintenance, and oversee their repair. Early evidence for their working methods is scanty, but during the 1250s, local customs of drainage maintenance began to be recorded. These customs were already well established by the time of their recording and were described as having operated time out of mind. The earliest known and most influential customs and institutions were those of Romney Marsh in Kent, in which twenty-four "jurats" were elected by the local landowners to survey the region's drainage works and assess responsibility for maintaining them. In the event that a landowner defaulted, the jurats were empowered to undertake the necessary repairs and fine the owner double the cost. The "laws of Romney Marsh," as they later became known, were first systematically recorded by Henry de Bathe, one of the king's chief justices, by order of King Henry III in 1258, and they became a touchstone for subsequent drainage and sewer law throughout England. Less well known, though similar in many respects, were the contemporary "dikereeves" of the Lincolnshire Fens. Like the Romney jurats, the dikereeves were responsible for surveying the area's drainage works and were empowered to levy and collect taxes to pay for the cleansing of the sewers there.

The Crown's dawning interest in drainage maintenance was reflected in a new royal policy of granting commissions to trusted advisers, charging them to work with local officials to ensure the timely and efficient repair of defective drains and banks. The first such commission was issued in 1258 to Henry de Bathe—the same judge who first recorded the laws of Romney Marsh, and in the same year. The commission ordered de Bathe to work with the sheriff of Lincolnshire (where he had previously served as an assize judge) to oversee the repair of dikes, banks, and bridges that had been damaged by recent flooding. The Crown issued similar commissions for other wetlands in time of need, and by 1280, the procedure was becoming frequent and well established—these may be seen as the first formal commissions of sewers.

Between 1280 and 1427, more than eight hundred royal commissions were issued for the purpose of coordinating drainage maintenance and repair throughout England; a large proportion of those addressed problems in the counties of the eastern Fens.[25] The Crown usually selected commissioners from among those who either owned land or had served in office in the county where the commission was directed, so they could be assumed to have some personal knowledge of local affairs. The commissions were temporary bodies; they were usually active for just a few months, though they might serve for many years, depending on the nature of the problem they were to address. Most were primarily concerned with the prevention of flooding and the repair of drainage works, though some were charged with ensuring the continued viability of river navigation.

Although commissions of sewers had become a regular tool of royal governance by the beginning of the fifteenth century, they remained ad hoc and irregular. In 1427, Parliament passed the first Act of Sewers (6 Hen. VI, c. 5) to better define their powers and make them more uniform and institutional. Commissions were to be issued by the Crown through the lord chancellor. The commissioners were empowered to make a survey of all drainage works within the bounds of their commission; to empanel a jury to give presentments under oath as to who was responsible for maintaining which sewers and banks; to hear testimony and make determinations in the event of a dispute; to employ laborers and purchase supplies for carrying out necessary repairs; to levy taxes to pay for needed works; to fine landowners for their negligence, distraining their goods if necessary; and to make new local sewer laws, as long as these were in accordance with the laws and customs of Romney Marsh. "In brief, the Commissioners were to survey, enquire, distrain, repair, employ and appoint, make statutes, hear and determine, and punish."[26]

The 1427 Act of Sewers was not a grand innovation in English sewer law, but a reflection and codification of what was probably more or less accepted practice by that time. It provided a sound and uniform basis, a fixed structure, and an additional legitimacy to what was already becoming a standard procedure. It was founded on nearly two centuries of royal precedent, which was in turn based on local customs that went back much farther. The act was originally meant to be temporary, lasting only ten years. It was renewed many times over the following century, though it was also allowed to lapse at several points. Remarkably, even during the lapses, Chancery continued to issue new royal commissions with no legal challenge. Commissions of sewers had become so routine, and their work so necessary, that they functioned equally well with or without specific statutory authority, just as they had prior to 1427—their powers were rooted deep in local customs, from a time long before either Parliament or the Crown took any interest in the details of drainage maintenance. The source of the commissions' authority and legitimacy was thus a complicated legal question—it derived partly from statute law, partly from Crown prerogative, and partly from centuries of customary local practices.

The most significant updating of the parliamentary sewer laws before the twentieth century came in 1532, under King Henry VIII (23 Hen. VIII, c. 5).[27] This act provided a considerable elaboration of the commissions' established structure and practices, delineating their composition, powers, and responsibilities much more explicitly. According to the new act, commissions of sewers would still be issued by the Crown through the Court of Chancery, with the lord chancellor, lord treasurer, and the two chief justices nominating the commissioners. These were to be men of standing, who either owned lands worth at least forty marks per annum (raised to £40 in an act of 1571, 13 Eliz. I, c. 9); were free of a city or borough and held a personal wealth of at least £100; or had been admitted to the bar in one of the Inns of Court. In practice, commissioners tended to be prominent local landowners and other members of the region's sociopolitical elite—gentry and wealthier yeomen, as well as prominent figures from the fenland towns. Many commissioners also served as justices of the peace and as members of Parliament. Those named to a commission were required to serve, on pain of a heavy fine, but only within their own county. All commissioners were required to take an oath prior to serving, the precise form of which was spelled out in the act itself. Commissions would ordinarily sit for no more than three years but could be renewed if necessary; this was later extended to five years, and then to ten years in 1571. The act was originally supposed to lapse after twenty years unless renewed, but it was eventually made permanent. Subsequent laws made only minor changes to the 1532 statute, which became the key reference point for all English sewer law thereafter, until the entire structure was abolished in 1930.

In their classic multivolume history of the institutions of English local government, Sidney and Beatrice Webb placed the commissions of sewers in a sort of miscellaneous category, alongside turnpike trusts and guardians of the poor, which they called "Statutory Authorities for Special Purposes."[28] The commissions represent something of an ambiguity in early modern English local governance. They were strictly limited by statute in terms of their jurisdiction and were empowered only to address specific concerns within a carefully delimited geographic area and only for a fixed number of years. Yet at the same time, within the bounds of their defined jurisdiction they wielded remarkable power, acting with judicial, executive, and legislative authority simultaneously, answerable only to Parliament and the Crown.

Commissions of sewers were supposed to hold sessions with at least six members in attendance, of which at least three had to be from the quorum.[29] While in session, the commission constituted a "court of sewers," and contemporary sources often refer to them as courts. Acting as justices of the court of sewers, the commissioners were authorized to conduct their own survey of the drainage works within the bounds of their commission and to determine what was in need of repair and who should be responsible for it. They could also instruct the sheriff to empanel a jury of local husbandmen, who made their own presentments under oath regarding the state of local drainage works and who was customarily responsible for keeping them in good repair. There was no consistent pattern for who carried out the crucial surveys; some commissioners did so themselves, others relied solely on the jury presentments. In any case, sewers courts were considered to be courts of record, meaning they had the power to levy fines, distrain property, and even imprison violators in cases where they deemed the negligence or fault to be deliberate and willful.

Having made their determinations, the commissioners also wielded executive powers to enforce their decrees. Besides punishing individual defaulters, they were authorized to take action themselves when the responsibility for repairs was deemed to be a collective one. This was usually the case when the work in question spanned several estates and villages, but also when a bank otherwise kept in good repair had been damaged by excessive storms and floodwaters, so that a given landowner was not to blame. In such cases the commission could levy sewer taxes on entire communities however they saw fit, usually in accordance with the value of the land each individual owned and how much he stood to gain from the required works. If anyone failed to pay his assessed share, the commission could distrain and sell that person's livestock or even a portion of his lands. The commission could also appoint officers, such as receivers to collect the sewer taxes and surveyors to oversee mandated repairs; hire laborers to perform the work required; and purchase the materials needed to carry it out.

Finally, a commission of sewers had the legislative power to make new sewer laws for the area delineated within its commission. The 1532 act stipulated that new laws were supposed to be framed "after the laws and customs of Romney Marsh," but it also granted the commissioners considerable latitude to depart from those customs in order to meet local needs and circumstances: "... or otherwise by any ways or means after your own wisdoms and discretions."[30] Most new sewer laws were intended to address particular situations, such as an especially negligent landowner who was to be held explicitly accountable for maintaining a given drainage work—these laws named specific people and places, and usually lapsed with the commission that had decreed them. Some laws were much broader, however, meant to define normative practice throughout the area and binding on everyone living within it. In such cases, the 1532 act also provided a means for making them permanent. If the commissioners certified a new law into the Court of Chancery under seal, provided it received royal assent, it was permanently binding and could only be overturned by a subsequent sewer law certified into Chancery, or by act of Parliament.

The 1532 act may be seen as part of early Tudor efforts to consolidate royal and local governance, though in some ways it runs counter to the broader trends of sixteenth-century England's increasingly centralized and unitary monarchy. Whereas other local practices and institutions were co-opted or subsumed in the process of Tudor state building, sewer commissions were deemed to require little more than a greater elaboration of their long-standing powers and methods.[31] They were still the institutional descendants of customary bodies that had evolved to meet the needs of wetland communities centuries earlier. The detailed definition and amplification of their powers via statute brought the commissions more explicitly under the purview of the Crown and confirmed them as power-ful royal instruments of local governance, but the irreplaceable local knowledge and experience that enabled the medieval commissions to function so effectively were left intact. Despite their sweeping constitutional powers, Tudor commis-sions of sewers remained conservative in their approach and local in their focus, even in a time of broader centralization under the Crown. Instead of abolishing or replacing them, the Crown simply folded the established local customs and institutions into its state building agenda.

The powers wielded by the commissions of sewers—to levy and collect taxes, pass new and permanently binding laws, fine negligent landowners, and even imprison those who resisted their decrees—were unmistakably political powers. Over the course of the medieval and early modern periods, the commissions came to embody a local political culture native to the Fens and to other English wetlands.[32] Fenland commissions of sewers frequently came into conflict with one another, especially during the latter half of the sixteenth century. As commis-sioners from different communities each asserted their own local interests against one another, the contentious process of negotiating acceptable solutions to mutual problems gave rise to a sort of regional fenland politics. At the same time, the Crown became steadily more involved in sewer commission business, with the Privy Council issuing instructions, giving advice, and helping to broker settlements to disagreements among rival commissions. The Elizabethan council was especially active, in part because of the growing number of intercommission disputes that threatened to disrupt drainage management throughout the region.[33] The councilors' inability to resolve these ongoing feuds frustrated them and ultimately led the Crown to consider large-scale fenland drainage projects as an alternative to the endless partisan bickering. This, in turn, forced the commissioners to articulate and defend their collective interests and traditional practices against the intrusion of outside drainage projectors with royal backing. They therefore engaged politically not only with one another, but also with the Crown.

Beyond their role as fenland political institutions, the commissions of sewers also embodied a particular understanding of the environment. Their approach to managing the landscape was a traditional one, founded on experience-based ecological knowledge and emphasizing conservative solutions to problems as they arose.[34] The medieval fenland economy relied on the predictable arrival and timely departure of floodwaters each year. A commission was usually charged to address a specific drainage problem within a given set of circumstances, for which the commissioners' local knowledge was ideally suited. Commissioners were above all *local* men pursuing local interests—they were rarely asked to serve outside their home county but were expected and required to serve within it because they possessed the local knowledge, experience, and stature needed for the job. They were supposed to maintain the integrity and operation of existing drainage works in order to preserve the status quo. Innovation was permitted, but was usually pursued only when ecological conditions had changed, making proper maintenance of the existing drainage network impracticable. The peculiar fenland ecology, economy, and political institutions were symbiotic, having evolved together over centuries, and they all served to reinforce one another.

This is certainly not to argue that the commissioners were early modern conservationists, who viewed the Fens as something like a national park to be preserved as a pristine, unspoiled landscape. On the contrary, the Fens were an unorthodox type of farmland to be exploited for both survival and profit, and the goal of the commissioners was to maximize the gains to be had from the land, taking it on its own terms. But before the seventeenth century, commissions of sewers were not intended, nor did they attempt, to permanently drain large areas of fenland or to alter the traditional arrangements that had worked for so long. Their mandate was not to drain the land, but to control the floods and render them as predictable and benign as possible—a conservative enterprise, not a transformative one. They sought no general, universal solutions to the "problem" of fenland flooding, because in normal years moderate flooding was not viewed as a problem. They may have had broad discretion to act as they saw fit, but that discretion was understood only to enable them to respond to new and unforeseen circumstances as they worked to restore the established drainage network. Most would have regarded the notion of a permanent, regional drainage as an impossibility, and there is no evidence that any such thing even occurred to anyone before the late sixteenth century.

The ideal commissioner of sewers possessed not only a deep personal knowledge of the lands and waters within the bounds of his commission, but also had years of experience in overseeing the drainage works there, with a long memory of how previous commissions had operated. He knew the land and its drains, knew how they worked and how they had always worked, knew their weaknesses and vulnerabilities, and knew which landowners were likely to be the most careless. What made a commissioner competent in his role was inseparable from the land, water, and people for which he was responsible. "Knowledge herein is gotten by hearing and seeing, and they carry it to the understanding, and hereof groweth knowledge," as one anonymous commissioner wrote in the early seventeenth century, staking his own claim to expertise on his unrivaled personal experience:

> Now for that a long time I have dealt in those business[es], both by mine own pains and view at diverse times and seasons, and also have procured many records and all concerning these causes which I could come by such as I am persuaded will give certain knowledge however things should be. I have been at every sessions of sewers these 20 years at the least, and can tell as much concerning sewers as any man now living, and know as well how, and am as willing to do good that way as anyone whosoever, and in my knowledge I affirm as thinking none can rightly gainsay it.[35]

Innovation was risky in such a world because it threatened to unravel a system that already functioned tolerably well, given regular maintenance and conscientious oversight. Indeed, for all their myriad powers and broad discretion, one of the most controversial legal questions during the early seventeenth century was whether sewer commissions could even order new drainage works to be built, or only maintain and repair existing ones. Proposals for large-scale drainage projects were regarded with skepticism by commissioners who doubted they could ever meet the specific needs and interests of each locality they affected—some areas might gain, but others would surely lose, and this would fly in the face of what the commissions were customarily supposed to achieve. Even if a regional drainage did prove feasible, permanent removal of the floodwaters would only undermine the traditional fenland economy, which relied on moderate annual flooding. When the Crown first began to promote large-scale drainage projects to the commissions of sewers at the end of the sixteenth century, many responded that such things were probably impossible, certainly unadvisable, and in any case well beyond their power to undertake or approve.

THINGS FALL APART: THE FENS UNDER STRAIN

Given the long-term efficacy of medieval commissions of sewers in managing their local drainage, the success of their conservative and customary approach in cultivating the land on its own terms, and the relative prosperity of the region, it is reasonable to ask, what happened? Why did proposals to redesign the drainage system and transform the ecology of the Fens begin to proliferate, and why did they do so toward the end of the sixteenth century? There are a number of answers to this question, touching on a series of ecological, social, and economic changes taking place both within the fenlands and across the whole of England and western Europe. Again, the environmental and ecological conditions of the Fens were not a constant, but a variable—one that shifted continually, with profound impacts for the human inhabitants. Some changes were very slow and might not even be perceptible to those living through them, but their cumulative effects were both real and profound. Other changes happened quite suddenly and required an immediate response.

A prime example of the ever-changing fenland geography is the silting up of the large outfall located north of the town of Wisbech. For centuries this had been the main outlet for the two largest fenland rivers—the River Nene and the greater branch of the River Great Ouse, both of which meandered toward it along their twisted courses—and its loss precipitated a series of cascading changes throughout the Fens. The precise timing and causes are uncertain, though a mix of natural processes and artificial interventions were probably involved. What is definitely known is that before the thirteenth century, the Wisbech outfall was the linchpin for the drainage of the entire southern fenland, and by virtue of the plentiful waters flowing through it to the North Sea, Wisbech had grown into a busy and prosperous fenland port town. During the middle of the thirteenth century, however, the outfall gradually became choked with silt deposits brought by the tides, and even the rivers' combined outflow was insufficient to counteract it. As their outfall decayed, the rivers became less stable in their courses and sought new outlets to the sea. Some fenland communities may have exacerbated the problem by trying to redirect the rivers for their own advantage—towns such as King's Lynn and Ely stood to gain from increased commerce if the waters could be made to flow their way.[36]

Whatever the cause or causes, by 1300 both the Great Ouse and a major branch of the Nene had shifted course entirely, bypassing Wisbech to flow instead past Ely to the once-minor outfall at King's Lynn. This may have boosted trade in the eastern part of the Great Level, but it had a long-term negative impact on fenland drainage. The loss of so much water at the Wisbech outfall caused it to silt up even faster, so the remaining waters seeking vent there had an ever-harder time in reaching the sea. The new river channels toward King's Lynn, meanwhile, were not always able to accommodate the much greater volume of water they now carried. The disruption caused by such major alterations necessitated a more deliberate human effort to manage the region's drainage. It is probably no coincidence that the period when these changes most likely took place, the mid-thirteenth century, was also precisely the time when the Crown took a much greater interest in wetland customs for maintaining local drainage works and began issuing commissions of sewers. The construction of Morton's Leam around 1490, by far the largest and most innovative fenland drainage work undertaken before the seventeenth century, was a failed effort to redirect the River Nene back through Wisbech to recover the outfall and revive the port there.

Climatic shifts probably also played an important role in shaping the geography of the Fens and the possibilities for human habitation and cultivation there. Recent ecological research has suggested that the period of peak prosperity in the Fens, the thirteenth and early fourteenth centuries, was also the end of a period of comparatively mild temperatures, sometimes known as the Medieval Warm Period. This may have helped to decrease the duration and severity of the annual floods and to open up a wider swath of land to permanent settlement, especially in the marshland. Medieval fenlanders, it seems, developed their prosperous pastoral economy during a period of extraordinary climatic mildness. By the latter half of the sixteenth century, the climate pendulum was swinging back with the so-called Little Ice Age: colder temperatures, rising sea levels, increased precipitation, and more severe storms all promoted more frequent and severe flooding.[37] As the climate became harsher, it made cultivation more difficult and exacerbated the problems caused by the silting up of the Wisbech outfall. By the end of the sixteenth century, the inhabitants of the Fens were struggling to preserve the hard-won gains their ancestors had achieved (fig. 1.5).[38]

Beyond the various climatic, ecological, and geographical changes that combined to make fenland drainage more precarious, a series of socioeconomic de-

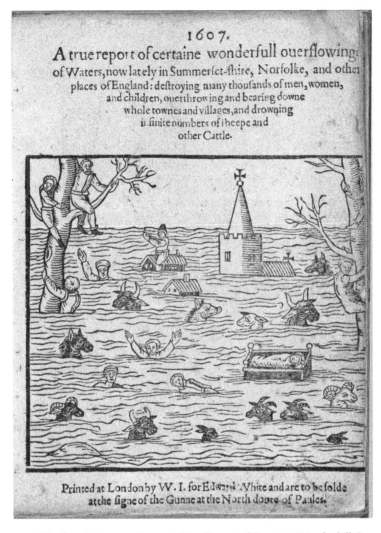

1607.

A true report of certaine wonderfull ouerflowing
of Waters, now lately in Summerſet-ſhire, Norfolke, and other
places of England: deſtroying many thouſands of men, women,
and children, ouerthrowing and bearing downe
whole townes and villages, and drowning
infinite numbers of ſheepe and
other Cattle.

Printed at London by W. I. for Edward White and are to be ſolde
at the ſigne of the Gunne at the North doore of Paules.

Figure 1.5. The frontispiece to Anon., *A True Report of Certaine Wonderfull Ouer-flowings of Waters* (London, 1607). It purports to show the trauma of an especially severe fenland flood in 1607; the striking image of the baby floating away in its cradle was allegedly based on true events.
RB 300920, The Huntington Library, San Marino, California.

velopments placed additional strain on the region. Regular oversight of the fenland drainage network may have been disrupted after 1540 or so by Henry VIII's dissolution of the many large monasteries that had previously dominated so much of the Fens as landowners. The monasteries' wealth and extensive landholdings were transferred to the Crown as the new Supreme Head of the Church of England, and the lands were then granted or sold to various royal favorites. The new owners then re-sold the estates they had acquired, leased them to tenants, or broke them up into smaller parcels and leased or sold them again, and so forth. All of this created a lot of turnover and disruption in fen landownership patterns, bringing a number of new landlords into the Fens from outside the region and badly confusing customary maintenance obligations.

H. C. Darby has suggested that the dissolution of the monasteries caused "uncertainty and disorder everywhere" throughout the Fens when it came to holding landowners responsible for sewer maintenance: "An enormous acreage of land was suddenly deprived of its traditional owners, and the great estates were divided and subdivided. . . . Liability for repairs changed hands frequently, or was subdivided and regrouped, thus obscuring and tending to obliterate defensive obligations."[39] The rapid turnover of so many landlords made determining responsibility for drainage maintenance a real challenge for commissioners and jurors of sewers, since their collective local knowledge and experience rapidly became obsolete. To illustrate the difficulty, Darby points to a series of presentments from juries of sewers in 1549, printed in William Dugdale's 1662 history of fen drainage, which in many cases could do no more than attribute responsibility to the (unknown) current owners of lands formerly held by the monastic houses.[40]

While a direct connection between the disruption caused by the dissolution of the monasteries and increased incidence of fenland flooding cannot be proven, at least some early modern fenlanders believed it was a factor. In an appeal to the Privy Council (c. 1600) for a series of sewer commissions to address deteriorating conditions throughout the Fens, an anonymous observer explicitly cited the dissolution of the monastic houses as the principal reason for the crisis. He lamented that "nothing hath been done effectually in sewers since the dissolution of the abbeys . . . and thereby our countries [are] greatly impoverished." The laws of sewers, he wrote, were "as necessary unto us as the observation of the peace, here or else where," and he hoped that new commissions of sewers would help to restore proper order in the region.[41] And in 1662, Dugdale asserted that "before the dissolution of the monasteries by King H.8 the passages for the water were kept with cleansing, and the banks with better repair, chiefly through the care and cost of those religious houses." Their loss, he wrote, had led to "the neglect of putting the laws of sewers in due execution in these latter times" and was a chief cause of "the total drowning of this great level (whereof we have in our times been eye witnesses)."[42]

The introduction of so many new landlords to the Fens all at once, most of whom were non-native and nonresident, also caused problems. Few would have known all of the customary responsibilities that came with ownership of any given estate, such as which banks and drainage ditches they were expected to maintain. An institutional, resident landlord such as a monastery would have preserved such knowledge over time, especially as the monks may well have been responsible for the original construction of the works in question. Although medieval complaints regarding the negligence of some monastic landlords show that they, like their secular neighbors, were often far from perfect in carrying out their duties, they would at least have known about them. And even if the new lords were aware of their customary responsibilities, they may not have cared. As non-natives, they would have had little understanding of the traditional fenland economy, the degree to which it relied on regular flooding, or the importance of carefully regulating those floods so far as possible. Nor would such men be expected to have much interest in looking after the poorer commoners whose livelihood depended on access to the vast common wastes. From the new lords' point of view, the common wastes were without obvious profit or advantage. After all, no one paid rent on those lands, and no one was ever going to lease them for as long as they remained in their flooded and common state. Such lands were a burden to the landlord, but one with the potential to become a generator of significant income, if only the land could be enclosed and improved.

Shifts in the early modern market for agricultural produce gave further impetus to the drive for enclosure and improvement. The sharp decline in Europe's population after the Black Death in the mid-fourteenth century had slackened demand for grain. Cereal agriculture became less profitable, so more marginal arable lands (such as those in the Fens) were allowed to revert to pasture, meadow, and waste, for which they were better suited in any case. Falling grain prices, in fact, had sparked a wave of enclosure and conversion of arable ground into sheep pasture during the fifteenth and early sixteenth centuries, a matter of great concern because of the resulting depopulation, as fewer peasants were needed to work the land. By the end of the sixteenth century, however, the market situation had reversed; rapid population growth throughout western Europe led to a steady increase in the demand for grain, and prices rose sharply. This, together with a series of poor harvests (probably related to the climatic cooling of the Little Ice Age), led to a renewed commitment to cereal agriculture among English landowners, including the enclosure and conversion of pasture into arable fields. In this market context, the flooded Fens came to be viewed by many as an unconscionable waste of potentially valuable, grain-producing land. This was especially true for the newer and larger landowners, who grew frustrated by their inability to profit from the whole of their estates. The steady deterioration of the Fens, which took place precisely in this period, only reinforced the impression that the existing drainage network was hopelessly inadequate and in dire need of a major improvement.

While the common wastes of the Fens may not have been profitable enough to suit their owners, the vast majority of the inhabitants there still depended on them for their survival. Unconcerned with the broader context of European grain markets, smallhold farmers and cottagers continued to graze their small herds and harvest the bounty of the wetlands, even as the annual floods became more troublesome. But here, too, new pressures were mounting during the sixteenth century. The very abundance of common wastes, and the ability of even a humble cottager to make a tolerable living from them, had attracted a significant in-migration of landless laborers to the Fens. The population of the region had thus been increasing even more rapidly than in the rest of rural England during the sixteenth century, with a large and growing proportion of cottagers, squatters, and "masterless men" who took advantage of the common wastes even though many did not have customary use rights there.

The additional strain caused by a swelling population was exacerbated by the ongoing practice of partible inheritance in many parts of the Fens. The relative prosperity of smallhold farmers diminished the incentive to consolidate and maintain large farms, so that primogeniture had not yet become the established norm. Since every smallholder enjoyed common use rights and grazed his own herd, the proliferation of smallhold farms led to more intensive exploitation of the waste. There is also considerable evidence that larger landowners were over-charging the commons with their stock, in order to take advantage of rising prices for foodstuffs and wool. Fenland gentry and wealthier yeomen grazed herds of several hundred of their own cattle and sheep on the commons and also exercised their brovage rights to bring in thousands more heads of livestock from the uplands to graze for a fee. The resources of the Fens were vast and renewable, but they were not infinite. The socioeconomic conditions of the sixteenth century had created a systemic shortage of pasture that was pushing the fenland economy to a point of crisis, especially in a time of worsening ecological conditions.[43]

By 1600, then, the traditional ecology and economy of the Fens were under mounting strain. A cooling climate and rising seas made the Fens ever more flood-prone; the dissolution of the monasteries had created upheaval and confusion, while also bringing new, non-native landowners to the region; a rising population and in-migration were putting unprecedented pressure on the available common wastes; and broader market forces were creating an insatiable demand for food, especially grains that could only be grown on well-drained arable land. It is little wonder that the established institutions and methods for managing the drainage and exploitation of the Fens began to buckle. The complicated customary systems of intercommoning between fenland communities were breaking down, as scarcity prompted neighboring villages to challenge one another's use rights within a shared fen. Larger landowners sought to enclose and improve portions of the common waste for their own profit but were vehemently resisted by commoners who feared that any reduction of the waste in an already precarious situation would ruin them. And as the floods grew more persistent and problematic, local commissions of sewers quarreled endlessly with one another over who was responsible for repairing drainage works that were proving ever more inadequate.

The first proposals to drain the Fens grew out of the many stresses that afflicted the region. The Fens appeared to be in decline, plagued by the very floods that had once nourished them but which could no longer be controlled. New landowners, to whom the traditional economy seemed alien and irrational, were alive to broader market trends that promised immense rewards for draining and plowing the land for grain. The Elizabethan Privy Council, frustrated by the apparent breakdown in the commissions of sewers, began to consider the possibility of a more aggressive regional approach to fenland drainage. Such a thing, they considered, might be undertaken by projectors, supported by the Crown, and paid for out of the newly drained lands themselves. Between 1570 and 1600, a period marked by severe flooding and economic hardship, the earliest proposals for draining the Fens were presented to the Crown, and they caught the attention of William Cecil, Lord Burghley, among others. As the financial difficulties of the Crown also continued to worsen after 1590, the queen's advisers considered the potential gains a large-scale drainage might yield for the royal treasury through the Crown's extensive landholdings in the Fens.

In sum, the medieval prosperity of the Fens grew increasingly difficult to sustain. The changing fenland environment forced its human inhabitants to respond, but demographic and market pressures constrained the options available to them. When the local customs and institutions that had served so well for so long proved incapable of addressing the new circumstances, it created an opening for forces outside the Fens—the centralizing monarchical state, the projectors who served it, and the investors who backed them—to intervene. These outsiders brought with them a radical new understanding of the fenland environment and an alternative model for its optimal management and exploitation.

CHAPTER 2

State Building in the Fens, 1570–1607

The improving of the fens, then is no miracle, yea rather unto such, as
with judgment do consider of them, it seemeth one of the wonders of
the world, that they have lain thus long neglected, in the midst as it
were of the realm, and among a nation of so politic a government.
—Humphrey Bradley to Lord Burghley, 1593

First it is generally said that the covetous bloody [John] Popham
offered to lend x. thousand pounds toward the draining of the fens,
thereby to take many poor men's commons from them for his own
profit. . . . [H]e is cursed of all the poor of that part of England and
swear they will kill him or such as shall be employed therein.
—Anonymous letter to James I, c. 1606

In July 1597, the Elizabethan Privy Council wrote to the commissioners of sewers
in "Lincoln, Norfolk and other counties adjoining," in support of a recently proposed drainage project slated to take place in the villages of Upwell and Outwell
in the Isle of Ely, and Denver in Norfolk. The projectors, they wrote, had already
reached agreements with the chief landlord and his tenants to drain their fens,
and had "laid forth great sums of money about that enterprise," but they now
wished to include Denver Common in their project. The council asked the commissioners to "call before you the lords and tenants of the said Denver Common
and to use your best endeavor to bring them to some good conformity . . . that
through their willfulness they be not the occasion of hindering any action which
will tend so greatly to the benefit of the common wealth." If any should object,
the commissioners were to refer them directly to the Privy Council, "that we
may otherwise deal with them." They stressed throughout the public good they
anticipated from the project and charged the commissioners to ensure that "so
good a begun work for the common wealth may not by the willfulness or obstinacy of any evil disposed person be overthrown or prejudiced." In the course of

one short letter, in fact, the council made reference to the good of the "common wealth" five times in all.[1]

Two months later, the Privy Council wrote once again to the commissioners of sewers in Norfolk regarding "the commendable and beneficial work of the common wealth" then under way at Denver and the protracted delays it had encountered. More than a thousand pounds had already been spent, and the drainage should have been completed long since "if the good endeavors had not been prejudiced by some of your fellow commissioners," who were obstructing the project for their own private interests. The intractable commissioners apparently feared that if the drainage were successful it would yield so much rich new pasture that the value of their own lands would be diminished. The council wrote to command them to move forward with the project immediately and to levy sewer taxes to pay for the remaining works. To "avoid partialities, which always arise where men are interested in property," the council ordered that "the deciding of such doubts or differences as may arise . . . be referred unto us at the Council Table or to the Court of Chancery, where the same shall be heard and ordered with all indifference and justice." They closed by charging the commissioners once again, "as you respect the good service of the common wealth," to permit no further delay.[2]

The many seventeenth-century projects to drain the Fens, as well as the obstacles and problems they encountered, have their roots in the reign of Elizabeth I (1558–1603). The late sixteenth century saw a gradual worsening of conditions in the Fens as flooding became more protracted and severe. Local commissions of sewers struggled to maintain their fragile network of drains and riverbanks, but they became deadlocked in a series of intraregional disputes to determine the best way to do that, as well as who should be financially responsible for needed repairs. The interests of one fenland community all too often conflicted with the interests of its neighbors, even though both might suffer from the same flooding. Some landowners' efforts to improve their estates created new and unforeseen crises farther up- or downriver; others tried to shift the customary financial burden for repairs onto their neighbors, denying their own responsibility or citing their inability to pay.

In the midst of these paralyzing disputes, and facing the pressing need to safeguard lives and property, the commissioners of sewers appealed to the Crown for assistance. The Privy Council's efforts to broker effective and lasting resolutions were rarely successful, however, and some councilors eventually became convinced that fenland flooding was a problem best tackled centrally, on a regional rather than a local level. They sought a new approach to fenland drainage that would overcome the squabbling among conflicting local interests to secure the entirety of the Fens from dangerous flooding, and they worked to bring the commissions of sewers into line with their regional vision. The council's early efforts to settle disputes and forge a consensus throughout the Great Level may be viewed as an aspect of sixteenth-century English state formation. Local elites were encouraged to work together and become active partners in promoting Crown priorities, enhancing both their own authority and that of the monarchical state.[3] This interpretation accords with the work of Mark Kennedy, who argues that cooperation between the Privy Council and the commissions of sewers under Elizabeth and James I reflected the existence of a "single, comprehensive polity," with the Crown serving as "an indispensible mechanism for alleviating the endemic social tension and political strife of provincial life." This harmonious relationship persisted, Kennedy suggests, until Charles I and his councilors abandoned it and "became partisans rather than mediators in local disputes."[4]

The policies of Charles I were undeniably more aggressive and unilateral than those of his predecessors, as will be seen. Yet long before his ascension to the throne in 1625, the Crown had already adopted a more partisan, interventionist stance. Unable to resolve local conflicts, build consensus, or persuade the various sewer commissions to coordinate their efforts under the Crown, the Privy Council sought instead to co-opt the commissions' authority and introduce a regional drainage scheme in the Fens around the turn of the seventeenth century. They hoped to rely on drainage projectors to accomplish their goals— entrepreneurs who proposed to drain the Fens using their own skills, plans, and funds, in return for a sizable share of the newly drained lands. The drainage projectors thereby became agents of the monarchical state, working with the Crown's full support to promote the interests of the commonwealth, while potentially reaping large rewards for themselves, their investors, and the Crown.[5] This was in keeping with other, more interventionist, facets of Tudor royal governance, where the Crown had grown frustrated by local officials' inability and/or unwillingness to handle their own affairs effectively.[6]

The Crown's attempt to manage fenland drainage regionally through projectors was not a voluntary approach, encouraging local elites to cooperate and "buy in" to Crown priorities, to the mutual benefit of both the center and the provinces—what may be termed *state formation*. It represented a much more deliberate, centralized, top-down, and at times coercive effort to bring local elites and institutions into line with the Crown's agenda—a process of *state building*. This change did not take place all at once but emerged as ecological conditions worsened and Crown frustrations mounted with the intransigence of the local sewer commissions. By the start of Lord Chief Justice John Popham's drainage project in 1605—the earliest attempt to drain the whole of the Great Level—the Crown was no longer a neutral arbiter in the Fens. It had become instead an active promoter of drainage projects, vesting the projectors with Crown authority and working to suppress local opposition for the (alleged) good of the commonwealth.

This move did not go unchallenged, however; it prompted a backlash in which fenlanders reasserted the primacy of their own local customs, knowledge, institutions, and interests, and questioned the authority of those attempting to undermine them. Their resistance to any regionwide drainage project, even one with Crown support, illustrates the political culture that had evolved in the Fens, one rooted in customary use rights, local knowledge and institutions, and conservative solutions to local problems. The fenlanders' political culture proved capable of articulating their interests, and engaging directly with national political institutions to advocate for them: Popham's unwelcome drainage project was halted in part through numerous fenland appeals to crown and Parliament.[7]

FENLAND DISPUTES AND CROWN INTERVENTION

Contemporary accounts indicate that the Fens experienced more severe and frequent flooding in the last decades of the sixteenth century. On the night of 5 October 1570, for example, "a terrible tempest of wind and rain both by sea and land" struck eastern England, resulting in one of the worst floods in memory in the Fens and elsewhere.[8] Dozens of people lost their lives; thousands of sheep, cattle, and horses were drowned in their pastures and stables; and a number of towns and estates were badly damaged. While most floods were not so destructive, moderate-to-severe episodes of fenland flooding were an ongoing problem, and commissions of sewers were exceptionally busy throughout the Fens from 1570 onward looking after local drainage concerns.[9]

The Crown also became more actively engaged in fenland drainage at this time. Numerous patents were requested and granted between 1570 and 1600 awarding exclusive rights to both Englishmen and foreigners to drain a specific area or to employ a novel drainage engine or technique—testament to a burgeoning interest in the potential for profit through land reclamation.[10] A parliamentary act was also passed in 1571 that enhanced the legal standing of commissions of sewers—their length of service was extended from five to ten years, and any laws they issued were to remain in effect indefinitely unless repealed by a successor commission, instead of lapsing with the end of the commission that had issued them, as formerly.[11]

Much more commonly, however, royal involvement in drainage affairs was reflected in the Privy Council's ongoing efforts to broker settlements between conflicting local interests about the best way to maintain the precarious drainage of fenland pastures, meadows, and common wastes. The main point of contact between the Crown and the commissions of sewers in this period was William Cecil, Lord Burghley (c. 1520–98). As Elizabeth's principal secretary and later lord treasurer, Burghley was one of the queen's closest advisers throughout most of her reign.[12] He had been involved in fenland drainage long before his rise to prominence in Elizabeth's government, especially in Lincolnshire, where the Cecil family's main landholdings were located. As a clerk and secretary at the court of King Edward VI, Cecil helped to nominate commissioners of sewers for his home county, and he served on that body numerous times himself throughout his political career.[13] Given his family connections in Lincolnshire, his personal experience as a sewer commissioner, and his stature within the royal government, Burghley was well placed to serve as an arbiter of drainage disputes between various fenland interests and communities.

The conflicts that Burghley was so often called on to help settle were usually either disputes over who should be required to pay for a proposed drainage repair or else situations in which one area's drainage "solution" appeared to cause unprecedented flooding elsewhere. One representative case that came to Burghley's attention in April 1571 involved a disagreement between the inhabitants of Deeping Fen and Holland Fen in Lincolnshire, concerning maintenance of a sewer called Star Fen Gote ("Starfengote") located in Holland Fen. The Deeping inhabitants believed the sewer was vital for their own drainage, but the Hollanders felt it did them more harm than good and neglected to keep it in working condition. The Lincolnshire commission of sewers could not break the impasse, having split into factions over the issue—each side had enough supporters to muster a working quorum and could therefore block or repeal any initiatives agreed on by their rivals. Burghley tried to broker a resolution, giving both sides a hearing, but there is no surviving record of the outcome.[14] A similar case arose in 1577, when Francis Russell, the 2nd Earl of Bedford, informed Burghley that some men of Wisbech had dammed up a common drain in the Isle of Ely. The new dam created a great nuisance for those living upriver from it, including Bedford and several tenants of the queen, "And it seemeth unreasonable," Bedford complained, "that the commodity of a few should be the hindrance of a whole country."[15]

Controversies such as these could be surprisingly bitter and protracted. Burghley and his fellow privy councilors were mired for years in an ugly dispute over responsibility for maintaining the River Nene near Wisbech. The area had been especially hard-hit by the 1570 floods, and various commissions of sewers called for an ambitious set of works to repair the decayed drainage network.[16] Yet little progress was made; most of the proposed works were never undertaken, and the commissioners could not compel local officials to carry out their decrees. The principal problem was determining who should pay for the works. According to the precedents of local sewer law, Wisbech ought to have been responsible for maintaining the depth and integrity of the River Nene because it flowed through the town's lands. Wisbech inhabitants, however, claimed that scouring the whole river was too expensive for them to handle alone, and they argued that anyone upriver from Wisbech who would also benefit from the work ought to contribute funds toward the project, making it more of a regional enterprise. Both sides also disagreed over which of the proposed new works were most likely to succeed in fixing the problem. The commissions of sewers in each of the five counties affected by the river's decay failed to reach a resolution, and matters continued to stagnate.

By 1576, several thousand acres near Wisbech were underwater, and the Privy Council was compelled to intervene. The council urged the sewer commissioners from the upriver counties to contribute to the scouring of the Nene, "whereby both they & their neighbors shall receive so general a benefit," but to no effect. In 1577, Burghley dispatched a special commission to Wisbech to broker a settlement, but having determined that the upriver counties should contribute to the required works, they lacked the authority to compel them to do so. The dispute persisted into the 1580s, by which time it had given rise to a protracted lawsuit in the Court of Exchequer, and the land in question remained subject to harmful flooding well into the seventeenth century.[17]

The River Nene dispute, and the many similar controversies of the same period, are important not only for what they show concerning the troubled condition of the sixteenth-century Fens, but also for the pattern they reveal of a crisis in local governance. Serious drainage problems, it seems, were hardly ever truly *local*; one area's actions (or inaction) could have a regional impact, for either good or ill. Finding an equitable settlement in such complicated cases was often beyond the capacity of the sewer commissioners, who were supposed to consider local interests and respond to local needs. The Privy Council intervened to arbitrate between rival commissions, and in doing so the councilors began to perceive that a truly *regional* drainage scheme, as opposed to a series of localized interventions, might be a more effective approach.

Burghley and the rest of the Privy Council soon sought to build a regional consensus in the Fens to address the ongoing problem of severe flooding. Subsequent disputes provide a good indication of how they went about it, such as the one that erupted during the 1580s in the Holland district of Lincolnshire. A new Crown lessee named George Carlton sought in 1580 to construct a drainage channel that he hoped would improve the condition of his leased lands. A number of his neighbors, local gentlemen who were also commissioners of sewers, vehemently opposed his plans because they believed Carlton's new drain would compromise Holland's existing drainage system. Carlton countered that the extant system did not keep his lands dry, which was precisely why a new drain was necessary. The dispute lasted throughout the decade, and at its lowest points involved vandalism and even physical violence.[18] The commissioners opposing Carlton's project were especially vexed because they viewed their rival as an outsider who was interfering in matters he did not understand, and with no regard for the greater experience of his neighbors. They complained to Burghley of having their existing drainage laws "overruled by strangers and foreigners contrary to truth and justice" and accused Carlton in court of acting "more for greediness of his own gain than to the general relief of the poor."[19]

As in previous Lincolnshire controversies, the county commission of sewers divided into factions that passed contradictory laws, creating a legal muddle and political stalemate that Burghley had to untangle. His approach in doing so illustrates his growing commitment to regionwide drainage management in the Fens. First, after declaring the recent flurry of new sewer laws and counterlaws to be null and void, Burghley dispatched a special committee of referees to examine the case and broker a resolution—all were fenlanders, but native to places outside the area in dispute, and were therefore presumed to be knowledgeable but disinterested arbiters.[20] This time the effort was successful; the committee recommended that a new drain be constructed but chose a different course for it than the one Carlton had proposed. Both sides agreed to accept the compromise proposal, and the work was finally begun.

Moreover, having watched the Lincolnshire commission of sewers become paralyzed "by occasion of much difference in opinions, and also for want of sufficient authority," the Privy Council decided to foster consensus by making that body not smaller and more exclusive, but larger. In 1589, the Crown issued a new "general commission" of sewers for all of Northamptonshire, Lincolnshire, Cambridgeshire, and Huntingdonshire, "being all counties adjoining one to other." With such a greatly expanded jurisdiction and membership, the new commission would have dozens of experienced and knowledgeable members, the large majority of whom would have no private interest in any given local dispute. The body could thus act "without partiality, indirect dealing or prejudice to any party, as near as may be, [so] that an equal course shall be taken for the relief of all the annoyances." It would also bring to bear a much broader, more regional perspective on drainage issues throughout the Fens and thus be able to function more effectively in each locality where problems arose.[21] Commissions of sewers spanning multiple counties were rare during the sixteenth century, but not unprecedented: in his younger days Burghley had been involved with two such commissions. After 1590, however, multicounty commissions became much more common, particularly in the vast expanse of the Great Level.[22]

THE GENERAL DRAINAGE ACT

Along with creating a new commission of sewers spanning most of the Great Level, the Privy Council also sought to gather more accurate and detailed information about the region so that a more comprehensive drainage solution might be found.[23] Having independent access to reliable information was critical for Burghley and the rest of the council if they hoped to be able to intervene successfully in mediating settlements to fenland drainage disputes. Otherwise they were forced to rely on the same interested parties who were at odds in them—the local commissioners of sewers. The council therefore ordered that a full survey and plat be made of "the several fens, marshes, and decayed grounds in the said four shires," including the descent of the rivers and drains and the condition of the soil throughout. They appointed three independent surveyors to the task: John Hexham of Huntingdon and Ralph Agas of Suffolk—each of whom already had experience in fenland surveying, having been involved with George Carlton's proposed drainage project—and also the Anglo-Dutch surveyor and engineer Humphrey Bradley of Bergen op Zoom in Brabant.[24]

Bradley was probably the son of an English merchant, John Bradley, and his Dutch wife. His involvement with the survey was prescient. Born and raised in the Low Countries, he was knowledgeable in matters of land drainage and hydraulic construction, though he most likely had no experience with projects on the scale of the Great Level before the 1600s. He came to England during the early 1580s and briefly served as a consultant for the rebuilding of Dover Harbor in 1584.[25] After leaving England again in 1594, he was appointed *maître des digues du royaume* in France in 1599 and went on to oversee extensive drainage works throughout that country until his death, sometime after 1625.[26] In 1589, the Privy Council commissioned him to perform a survey of "the several fens, marshes, and decayed grounds in the said four shires" of Northamptonshire, Lincolnshire,

Cambridgeshire, and Huntingdonshire, and he spent several months in the Fens carrying it out. When he finally made his report to Burghley in December 1589, he offered not only his survey, but a bold proposal for draining the entire Great Level.[27]

Bradley believed that the worsening conditions in the Fens were caused by the silting up of the region's principal rivers, a problem he felt could be easily corrected at low cost: "[T]he enterprise," he asserted, "does not need anything but some assistance given to nature. . . . To free [these marshes] from the water would not be a miracle." The reward for doing so, moreover, would be great indeed: ". . . a vague, deserted empire without population turned into a fertile region; and wild and useless products therefrom turned into an abundance of grain and pasturage; humble huts into a beautiful and opulent city, together with other benefits. With good regulation, the drained land will be a regal conquest, a new republic and complete state." The main obstacles to such a triumph, Bradley claimed, were not so much environmental as sociopolitical; the problem was not with the land, but with those who lived on it and managed it so poorly. "[O]wing to diverse opinions, disputes and disagreements," he wrote, "the lands are almost voluntarily lost; and really the greatest impediments to all good drainage projects lie in the minds and in the imaginations of men, and not in the facts [of nature] themselves." If the fenlanders could only be brought around to a more reasonable and cooperative point of view, they would see that the drainage was not only feasible but potentially of the greatest benefit for all, landowners and commoners alike.[28]

Despite Bradley's enthusiastic proposal for the drainage, nothing was done to pursue it. Not one to give up easily, Bradley wrote again to Burghley in April 1593 to promote his scheme with greater force and precision. He claimed that the whole of the Great Level, spanning some four hundred thousand acres, could be drained within a single summer, using no more than eight hundred laborers and at a total cost of less than £5,000, while yielding an annual profit to the Crown of at least £40,000. He based his wildly optimistic assessment on his recent survey, which concluded that the entire region lay above sea level even at high tide, "giving descent, to void the surplusage of waters, more than enough. . . . So that what furtherance, nature can afford, these fens have it." Given such favorable circumstances, Bradley marveled that such potentially profitable land had been suffered to lie waste for so long: "The improving of the fens, then is no miracle, yea rather unto such, as with judgment do consider of them, it seemeth one of the wonders of the world, that they have lain thus long neglected, in the midst as it were of the realm, and among a nation of so politic a government."[29]

The main obstacles Bradley perceived were once again not in the land, but in its inhabitants. There were simply too many stakeholders in the Fens whose interests were in conflict, so getting them all to support and contribute to such a massive undertaking would be all but impossible. This was especially true with respect to the bewildering diversity of landholding patterns and common use rights Bradley had observed throughout the region: "[M]any difficulties present themselves (by reason that the property, tenures, use and profit of the fens riseth diversely to diverse conditions of persons, some being lords . . . , some free-holders, some tenants for more and some for fewer years and some at will, some copyholders, some commoners and such like, whereby it is almost impossible to raise an equal and willing contribution)."[30] The only remedy, he argued, would be to coordinate the drainage centrally, taking it out of the hands of local sewer commissioners and giving it to an outside authority to undertake it instead—in a word, draining the Fens would require a projector.

Granting full responsibility for the drainage to an agent from outside the Fens, Bradley believed, could best be accomplished through an act of Parliament. Such an act would give the projector in question the unassailable legal standing needed to cut through the tangled thicket of complicated landholding claims and customary use rights, overcome the many and conflicting local interests involved, negotiate viable contracts with the major stakeholders, and allow the work to proceed at last. Moreover, a precedent for such an act already existed: a parliamentary act in 1563 had granted the Italian projector Jacob Acontius the right to undertake the drainage of some two thousand acres in Kent. The act specifically guaranteed Acontius's right to one-half of all the drained lands once the project was complete, to be taken proportionately from all of the landlords and tenants whose lands were improved.[31]

The Great Level drainage project Bradley envisioned was of a vastly different scale than that of Acontius, affecting an area some two hundred times larger and spanning several counties. Nevertheless, the underlying problem that Acontius's act was meant to address was the same: getting around the confusing variety of landholding claims and customary use rights that existed across the region and overcoming the many conflicting interests that would necessarily be involved in any large drainage scheme. Together these made it all but impossible for a would-be drainage projector to negotiate the requisite contracts with everyone involved and guarantee a just recompense for himself. A parliamentary act could obviate the need to reach independent agreements with each and every land-owner, tenant, and commoner by declaring the drainage to be a matter of state, to be undertaken for the good of the commonwealth. Once the work was completed, all those affected by it would realize such a profit from it that they could not but consider it a bargain, even after they had parted with a suitable portion of their lands. Bradley did not anticipate any complication in writing such an act, which he felt "may be drawn in few lines between this and tomorrow . . . seeing how beneficial and honorable the action is." He suggested that the queen might be named as the principal undertaker, in order to lend her royal imprimatur to the scheme.[32]

The Great Level drainage, Bradley argued, would obviously be a universal good for both the Crown and the commonwealth; it would "reduce this wilderness to a fruitful soil, better able to nourish in good state a hundred thousand families, than now one thousand in want." No other project could be "more profitable, honorable and worthy of Her Majesty's name, than this action."[33] Of course, Bradley was also alive to the possibilities for private gain, and he used these to strengthen his argument even further. In a letter separate from his formal proposal, he informed Burghley that in addition to the projected £40,000 annual profit for the queen's coffers, Burghley himself would stand to gain "two or three thousand pounds by year" as a Lincolnshire landowner. And if the queen were still unwilling to undertake the project for the good of the commonwealth, Bradley offered to "bring your lordship the names of certain gentlemen of worship and wealth, willing upon any reasonable conditions to perform the action."[34] While any distinction between "public" and "private" ventures can be notoriously difficult to make in the context of early modern projects supported by the Crown, Bradley's separate letter to Burghley underscores the fact that both the queen and her chief adviser were interested stakeholders in draining the Fens.[35]

No action was taken on Bradley's proposal in 1593, perhaps because the Fens were then enjoying a brief respite from severe flooding, pushing all thoughts of disruptive and expensive new drainage works into the background. Burghley's famously cautious temperament may have also played a role.[36] By the late 1590s, however, the "outrageous inundations and overflowing of waters" had returned to the Great Level, bringing with them yet another round of pitiable complaints from the fenlanders and a renewed interest among the Privy Council in finding a more permanent solution.[37] Mindful, perhaps, of Bradley's suggested tactic, Burghley first introduced a bill to facilitate the drainage of the Great Level during the parliamentary session of 1597. The bill's preamble asserted the fenland inhabitants were greatly impoverished by the continual flooding of their lands, leaving them "not able to endure the charge for the recovering of those grounds and bringing them to be firm and profitable." A regionwide drainage could yet be achieved "with reasonable charge and industry," to everyone's potential benefit, but ". . . by the reason of the multitude of the commoners and diversity of titles and interests of lords and others in the said grounds no certain conclusion can well be had or any good done by any that should undertake the same *if a few of the multitude should obstinately refuse to give their consents thereto.*"

The bill thus echoed Bradley's concern that any effort to drain a sizable portion of the Fens could be stymied by even a small number of landowners or commoners who might oppose it, for whatever reason. To overcome this obstacle, the bill would have required drainage projectors to negotiate agreements with all of the affected landowners, but only "the most part of the commoners." The minority who had not agreed to a contract would nevertheless be legally bound by the terms their neighbors had committed to. The projectors could then proceed to drain the whole region and maintain it at their own expense, in return for a suitable portion of the drained lands (or a cash settlement in lieu of that). The contracts would have the backing of parliamentary statute, giving all parties some assurance that their agreements could not easily be broken, gainsaid or overthrown.[38]

Though presented as a measure for the benefit of the commonwealth, and especially the poor fenland inhabitants who suffered from the recurring floods, the bill soon ran into opposition on the grounds that it was not the commonwealth but private interests who would benefit. Peregrine Bertie, Baron Willoughby de Eresby, a prominent lord and landowner from Lincolnshire, wrote to the Earl of Essex condemning the bill. While such a law might "pretendeth to enable and relieve a multitude of poor men for Her Majesty's service and the commonwealth's good," Willoughby suspected that "instead of helping the general poor, it would undo them and make those that are already rich far more rich." He was especially concerned that just a bare majority of commoners would be empowered to contract for the whole, disposing of the common lands and use rights of all without the consent of what might be a large minority of commoners. The dominating majority, moreover, would likely include "bailiffs, buyers of cattle, servants or fearful tenants" who would vote as their masters commanded, rather than in the commoners' true interest. Willoughby included with his letter a plan of the fens lying north of Peterborough and noted that Burghley and several of his clients were major landowners in the region, implying that Burghley's support for the bill was neither disinterested nor intended for the good of the commonwealth. In the end, the bill was dashed when the queen sent word to the House of Commons that it should not be read or discussed further.[39]

Damaging floods continued to plague the Fens through the late 1590s, underscoring the need for some sort of action on the part of the Crown to relieve them, the failure of the 1597 bill notwithstanding. The Isle of Ely complained to the Privy Council that flooding had reduced the whole region to destitution; many who had once given generously to relieve the poor "are now enforced to live of alms & many to beg & some to be starved & so die for hunger."[40] Commissioners of sewers throughout the Great Level informed the council that their yeomen and husbandmen were "utterly decayed, undone, and live in sharp penury," while the poorer sort "in great numbers . . . go on begging, and very many have this last year, for want of food, died." They implored the queen to allow "some course for draining, & inning of our country."[41]

In 1601, therefore, another bill for the general drainage was presented to Parliament. The new bill had much the same intent as the previous one: to empower landowners and a majority of commoners in the Fens to make legally binding agreements with anyone who offered to drain their lands in return for a negotiated recompense. Titled "An Act for the Recovery of Many Hundred Thousand Acres of Marshes and Other Grounds, Subject Commonly to Surrounding [i.e., flooding], within the Isle of Ely and the Counties of Cambridge, Huntingdon, Northampton, Lincoln, Norfolk, Suffolk, Sussex, Essex, Kent, and the County Palatine of Durham," the act enjoyed quick and easy passage through both houses of Parliament and was usually referred to thereafter as the General Drainage Act. Its preamble explained that whereas the flooded areas of the realm ought to be "recovered by skillful and able undertakers," to the great benefit of both the queen and her subjects, that recovery had been "chiefly hindered" by the fact that so much of the land was held in common. Insofar as "the commoners, in respect of their poverty, are unable to pay the great charges to such as should undertake the recovery of the same," they were likewise unable to use a portion of the lands to pay for the work because their customary common use rights could not "be extinguished or granted to bind others which should inhabit there afterwards."[42]

The new law decreed that the lords, owners of severals (privately held property, as opposed to commons), and "the most of the commoners for their particular commons" should be authorized to "contract or bargain" using a portion of their lands, awarding the same to "such person and persons which will undertake the draining and keeping dry perpetually the severals, wastes or commons of that quality." All such contracts would be "good and available in law to all constructions and purposes," and would bind "*all the commoners* and such as shall or might have common or interest there afterwards," not just "the most of" them. By barring a minority of commoners from obstructing the will of the majority, it was hoped that enterprising fenlanders would be able to attract drainage projectors into the Fens and use their extensive common wastes to bargain for their valuable services—a crucial condition for any large-scale drainage project to go forward in such a cash-poor region. Interestingly, the new act accorded no formal role to the traditional drainage authorities in the Fens, the commissions of sewers. All agreements were to be concluded directly between the drainage projectors on the one hand, and the lords and commoners on the other.[43]

The surprisingly concise law was meant to overcome all of the obstacles to a general drainage identified by Humphrey Bradley by providing the remedy he had recommended. If it worked as intended, willing and able projectors would drain the land and maintain it in perpetuity, realizing a handsome profit from their share of it; the Crown would benefit as a large landowner in the Fens (and perhaps also through the projectors' "gifts" in gratitude for royal backing); other landowners would profit by enclosing some of the improved commons on their estates, as was their right; commoners would prosper on the much-improved lands still remaining to them; the numerous local drainage disputes that had paralyzed the region would be rendered moot; and the Privy Council would be relieved from the frustrating burden of having to try to broker resolutions for them. The General Drainage Act was a prime example of late-Tudor methods of state formation, in which local elites were encouraged to align their own interests with the priorities of the monarchy and to engage with Crown-sponsored agents (in this case, drainage projectors) to achieve mutually desirable ends.

Not everyone was so optimistic. Projectors were already notorious and deeply controversial figures in the political economy of late-Elizabethan England. Fostered by Burghley throughout his political career, the policy of granting royal patents to would-be projectors was originally intended as a means to promote technological and entrepreneurial innovation by encouraging new inventions and attracting valuable expertise to the realm. By the end of the sixteenth century, however, projectors had already begun to acquire a reputation as deceiving charlatans, greedy opportunists, and royal favorites who exploited their connections at court to obtain monopolies on lucrative trades in return for a kickback to the Crown.[44] The benevolence of fenland drainage projectors and the good they might bring about for the commonwealth were far from being truths universally acknowledged— especially among the large numbers of commoners who resisted bargaining away their customary use rights for a drainage project they did not want.

STRANGERS AND FOREIGNERS: JOHN POPHAM'S PROJECT

Despite the hopes and expectations of those who had sponsored it, the General Drainage Act was a resounding failure because it did not reflect the priorities and desires of the fenland inhabitants it was ostensibly meant to help.[45] The act's promoters had assumed that once legal impediments were removed, willing parties would soon come forward to negotiate contracts for new drainage schemes. But the anticipated groundswell of demand never materialized in the Fens. Of willing projectors and flooded lands there was no shortage, but despite pleas from throughout the region in times of acute crisis, few fenlanders were really eager to barter away a sizable portion of their severals or commons. Some still wondered whether large-scale drainage projects were either necessary or wise, and many more doubted they were even possible.

The underwhelming response of the fenlanders did not discourage the entrepreneurial spirit or cupidity of the projectors, however, nor did it dispel the desire of the Privy Council to see a full, regional drainage of the Fens undertaken, preferably at little or no cost to the Crown. When the landowners and commoners of the Fens collectively failed to seize the "opportunity" presented by the act, alternative means were soon found to enable drainage projects to go forward with or without the fenlanders' consent. The Crown soon shifted from a more passive role of encouraging fenland inhabitants to agree to new drainage projects (that is, a process of state formation), and adopted a more proactive role of imposing such projects upon them, whether they liked it or not, for the good of the commonwealth (that is, a process of state building). In doing so, however, the Crown undermined its own credibility as a neutral arbiter in the Fens and provoked resentment. Each side understood the situation very differently: while the Crown aimed to promote the good of the entire realm over the selfish, myopic interests of a few stubborn fenlanders, the fenlanders were defending their lands, customary use rights, and traditional institutions from the depredations of greedy projectors.

The crown's new, more aggressive policy took advantage of the statutory powers of the commissions of sewers to force landowners and commoners to part with some of their lands to support vital drainage maintenance. Under the terms of the 1532 act that was the basis for virtually all English sewer law, commissions of sewers had the power to levy taxes on both commons and severals to pay for any drainage works they deemed necessary, and they could distrain and sell a portion of land or goods for nonpayment of any such taxes. These provisions were meant to ensure that commissions would be able to enforce their decrees and carry out essential repairs whenever individual landowners or entire communities were derelict in their duties. If broadly and aggressively applied, however, they could also be used to compel the sort of land-for-drainage swap that the 1601 act was supposed to facilitate, but without the need for voluntary cooperation of even a majority of the inhabitants.

A would-be drainage projector (with Crown backing) would only have to convince the commissioners of sewers to levy a general tax on all the lands he hoped to drain. The more exorbitant the tax and the more unreasonable the deadline for paying it, the better—collecting the money was not the point. Once the deadline for paying the tax had passed, with most of the country in default, the commission could then distrain a sufficient share of land to cover the delinquent assessment. In practice, this meant the commissioners could seize lands indiscriminately and use them to conclude their own bargain with a drainage projector, in spite of any and all local resistance to it—the owners' and commoners' objections became legally irrelevant once they had failed to pay their sewer taxes. The wealthier and more powerful landlords who continued to protest might be bought off by allowing them to invest in the project—in effect, buying back the lands they had lost. Unhappy commoners, however, would simply have to be content with their diminished, but improved, commons. Projectors still had to secure the cooperation of the commission of sewers, but since commissioners were appointed by the Crown, the institution was vulnerable to packing and manipulation in behalf of projectors who enjoyed royal support.[46]

This shrewd tactic had already been employed in a few small-scale drainage projects during the 1590s; it was originally devised, in fact, because there had been no other legal means for fenland commoners to extinguish their use rights and transfer a portion of common lands to a drainage projector prior to the 1601 act. In the 1596–97 project undertaken at Upwell, Outwell, and Denver, mentioned at the start of this chapter, the projectors had agreed with the local commission of sewers to drain the area in exchange for half of all the lands they improved. In order to make that possible, the commissioners levied an exorbitant blanket tax of ten shillings per acre and distrained the lands they needed when the tax went unpaid.[47] After the failure of the General Drainage Act, it became standard practice for sewer commissions to tax and distrain lands in order to negotiate drainage contracts. In 1604, a group of well-connected projectors began an undertaking that would employ the tactic on a much grander scale than ever before: they proposed to drain hundreds of thousands of acres in the Great Level, and they managed to obtain the enthusiastic support of the Crown. The project is usually associated with Sir John Popham, the lord chief justice and member of the Privy Council who eventually became the principal projector, though his direct involvement can only be documented roughly a year after the project's inception.

In July 1604, King James I wrote to the commission of sewers for the combined counties of the Great Level in behalf of John Hunt and Henry Totnall, two would-be drainage projectors. Though little is known of Henry Totnall, John

Hunt was a significant landowner in the Fens—he purchased George Carlton's lands after the latter had died. He was an active commissioner of sewers in both Cambridgeshire and Lincolnshire, and he was involved in several small-scale drainage projects during the 1590s and early 1600s. Most important, he was one of the commissioners who had assessed the tax of ten shillings per acre to drain the area around Upwell, Outwell, and Denver.[48]

Hunt and Totnall had informed the king that lands in the region totaling more than two hundred thousand acres "yield small gain" to the owners and commoners because of frequent flooding but might yet "be made of very great value" if they were drained. Proceeding according to the terms of the General Drainage Act, Hunt claimed to have made voluntary agreements with several landowners and commoners, who "have offered part of those commons and waste grounds to such as will undertake the recovering of the whole." The king, being himself a large landowner in the Great Level, was "desirous to advance and further so good an action," tending not only to his own benefit, "but also to the whole common wealth of this kingdom," and he declared his own willingness to surrender a portion of Crown lands in support of the project. He charged the commissioners of sewers to assist Hunt and Totnall in surveying the area and securing local agreements, and he expressed his expectation that "none of our loving subjects for any private respect will oppose themselves against so good an action."[49]

Before such a massive undertaking could be attempted, a much fuller sense of the land to be drained first had to be acquired, including its precise extent, condition, and value. This was no trivial problem. While most commissioners of sewers had a sound knowledge of their own localities, none had a firm grasp of the entire region—nor, in spite of Humphrey Bradley's survey, did the Crown. It was vital, therefore, to collect and compile detailed information on a regional scale.

In January 1605, Richard Atkyns of the fenland town of Outwell was commissioned to make a full survey of the Great Level; he was to account for all of the lands to be drained, the degree to which each was troubled by floods, and the depths and conditions of all rivers and drains running through the area. Atkyns was an experienced and respected commissioner of sewers as well as a skilled surveyor; the survey he returned in the spring of 1605 was careful, detailed, extensive, and thorough.[50] Running to nearly fifty manuscript pages, it included descriptions of every stretch of fen in the region, organized first according to which river drained it and then by town, manor, and/or lordship. It described the boundaries, lords of the soil, rights of common, and general condition for each plot, as well as the depth and condition of local rivers and drains.[51] In addition to

Atkyns's work, a second surveyor named William Hayward was commissioned to produce yet another survey and new map of the region. Like Atkyns, Hayward was probably from the village of Outwell, and by 1605 he already had extensive experience in fenland surveying.[52] He prepared his report in 1604–5 and determined that the area to be drained contained a total of 307,242 acres. He also created a large wall map of the region that was one of the earliest English maps to concentrate on a particular kind of topography (wetland) as the key organizational focus, rather than the more common sociopolitical division of the shire. It served as the template for maps of the Fens for decades thereafter.[53] Together, the Atkyns and Hayward surveys provide one of the most comprehensive surviving depictions of the pre-drainage Great Level. They also illustrate the Crown's new conception of fenland flooding as a truly regional problem, rather than a series of local ones.

While the surveyors collected their data, Hunt and Totnall continued to pursue contracts with landowners and commoners, but in spite of Hunt's optimistic report to the king, it was an uphill battle. One jury of sewers near Peterborough grumbled that the country had already been "mightily taxed and charged" to build and maintain local drainage works, which functioned quite well enough as they were. They therefore recommended to the sewer commissioners "that neither Mr. Hunt nor Mr. Totnall nor any other stranger or undertaker shall have to do to drain our fens." Some inhabitants of Ely informed the bishop there that they had "little cause to be displeased with the fruitfulness & commodity of our commons (but for some part of the year only surrounded)." They were willing to conform to the king's wishes, but expressed reservations about the projectors, "with whose project we are hitherto altogether unacquainted."[54]

The fenlanders' evident reluctance was countered by the king's growing enthusiasm. In April 1605, James wrote once again to the commission of sewers after having traveled through the Great Level himself. The king had witnessed with "our own eyes . . . the infinite loss" caused by the flooding, and he reaffirmed his desire to see the land drained, "whereby so many thousands of our loving subjects which now lack the same may be planted in convenient habitations." Draining the Fens, he believed, would improve both the condition of the land and the character of those who lived on it. He wanted to see "the same fens drained whereby the soil thereof may according to his own proper nature be made rich, fertile and profitable for the use of man . . . and not to be suffered to lie as they do . . . the means and occasion of all sloth and idleness." He ordered the commissioners to redouble their efforts to assist Hunt and Totnall and pledged for his part to yield fully one-half of all royal landholdings in the region.

He also advised them to carry on without regard to the "turbulent dispositions of any" who might oppose the work, "either of willfulness or through ignorance." If the commissioners could not bring such men around, they were to be referred to the king for further dealing.[55]

Four days after the king's letter, the Privy Council also wrote to the commission of sewers. They declared that if anyone "without respect to the general good shall oppose themselves or use any persuasions or means unto others to impugn an action so generally conceived to be to the good of the common weal," they would answer to the council. Moreover, any small-scale drainage works then in hand were to be suspended immediately, lest they become "a great prejudice and hindrance to this intended work of general draining." The integrity and success of the larger, regional project were paramount.[56]

Finally, the Crown issued a new commission of sewers for the Great Level that included several new members. The membership was further modified in the following weeks, as the commissioners heard and considered the projectors' proposals.[57] The commission's active cooperation was required in order to levy an exorbitant blanket sewer tax, opening the door for the project to proceed even without the consent of a majority of fenland commoners. Mark Kennedy has argued that the Privy Council's direct and ongoing involvement in "fine-tuning" the list of commissioners was an effort to ensure that the proposal would receive a sympathetic hearing from the body.[58]

The newly composed commission of sewers met at Huntingdon on 29 May, and "after long debate, and all objections heard," they declared with one voice that the drainage was both feasible and desirable. Any putative "inconveniences & dangers that are pretended or feared" on the part of certain landowners or commoners could be either reasonably accommodated or dismissed. The commissioners also took the opportunity to praise the undertaking to the skies; in a letter to the Privy Council, they called it "the most noble work for your [lordships] to use your honorable furtherance in & most beneficial for the countries interested to have good by, that ever was taken in hand of that kind in these our days." The letter was hand delivered to the council in London by Hunt and Totnall, who had attended the session.[59]

According to the 1532 Act of Sewers, before a commission could issue a new sewer law, the commissioners were supposed to make a personal survey of the area in question. This they did, 21–27 June 1605; fourteen commissioners surveyed all of the major rivers in the Great Level to observe their condition and determine the best means of improving them. As Dugdale relates in his history, John Hunt personally accompanied the survey party to advise them. His opinions were decisive, with the commission acting as little more than a rubber stamp: "Upon this view Mr. *Hunt* (who was the artist for the draining) represented to the said commissioners what cuts, banks, sluices, clows &c. would be in his judgment farther necessary in order to the perfecting this work; all which they signified under their hands to the lords of the council, together with their opinions how much it would tend to the honor and enriching of the kingdom."[60]

The commissioners declined to provide any estimate of the total cost of the work or the terms on which it might be undertaken, pleading that such vital issues were beyond their ability to consider: "we crave pardon, being ourselves few in number, the time prescribed for so large a survey very short, the matter of so great consequence, and not knowing how far this new project will be approved by the rest of the commissioners."[61] The "time prescribed" was indeed very short: the survey party had just one week to traverse the region's three major river systems, which together drained more than three hundred thousand acres of land. The 1532 statute required such a survey, however, because commissions of sewers were understood in that law to be *local* bodies. They were meant to address drainage problems affecting much smaller areas, for which a personal inspection would have been much less daunting—another indication that the Crown's adoption of a regional approach to fen drainage represented an innovation.

On 24 June, with the commissioners' personal survey still in progress, the king wrote once again urging "a more speedy prosecution of this business." To move things along, he dispatched Sir John Popham (c. 1531–1607), lord chief justice and a member of the Privy Council, to attend their next session in person and help guide it toward a successful outcome.[62] This marks Popham's first official involvement with the project, though given the zeal with which he took the matter in hand, he may have been more actively engaged behind the scenes—he was definitely one of the privy councilors with whom the king consulted when Hunt and Totnall first proposed to undertake the drainage (fig. 2.1).[63] Popham was renowned as one of the greatest lawyers of his generation. As attorney general and later chief justice, he was involved in several important state trials. He was also very active in Crown governance as a member of Parliament and a privy councilor. He had served as Speaker of the House of Commons and was probably one of the main architects of the landmark 1598 Poor Law.

In addition to his brilliant legal and political career, Popham was one of the great speculators, projectors, and fortune-seekers of the era. His personal fortune was immense, he was broadly interested in maritime affairs and overseas trade, and his more notable investments included shares in the Munster Plantation in Ireland and the New World colonization schemes of the Plymouth and

Figure 2.1. Sir John Popham, copied by George Perfect Harding; original artist unknown.
Portrait 2405, © National Portrait Gallery, London.

Virginia Companies. He was also no stranger to the management and drainage of wetlands—he had been born and raised in Huntsworth, near the Somerset marshes in western England. He purchased the fenland estate of Upwell in the Isle of Ely in 1562, and he was politically active in the Fens as a justice of the peace and a member of Parliament for various fenland districts.[64] He would certainly have been involved in the 1597 project to drain the area around Upwell, Outwell, and Denver, along with John Hunt, and he was already well aware of the details of the present project, having advised the king on it from the start. From June 1605, with the king's full support, Popham supplanted Hunt and Totnall as the principal undertaker of the Great Level drainage project and was its most powerful proponent at court.

The next general session of sewers convened at Cambridge on 28 June 1605, with forty-three commissioners in attendance as well as Popham, and the body formally approved the creation of a massive new drainage network for the entirety of the Great Level. The plan called for an extensive series of works to be built, including a new, straighter channel for the River Great Ouse from Earith Bridge to Salters Lode, near Denver; a new channel for the River Nene from Peterborough to Wisbech; a new channel for the Welland River; and a new drain from Salters Lode to the March River, in addition to several smaller drains and banks. The commissioners estimated the project would cost £120,000, and levied a general tax on the entire region to pay for it. The tax was completely unrealistic from a practical standpoint. It was assessed by county, apparently based on the proportion of lands in each that stood to benefit from the project. But there was no provision for making individual assessments, the details of which would have been exceedingly complicated to determine, involving hundreds of landowners and thousands of commoners claiming customary use rights. The entire sum was to be collected in just two weeks—an impossibly short time to raise so much capital, in a region known to be lacking in it—and yet no officers were appointed to collect it.[65] But of course, the tax was never supposed to be collected; rather, the unpaid assessments would serve as the legal basis for the commission to distrain tens of thousands of acres, both severals and commons, and conclude an agreement with Popham and his partners to drain the whole region.

Fenland commoners were hardly quiescent in the process; the commission of sewers received numerous complaints and objections to the proposed project throughout May and June. Petitioners from Cambridgeshire, for example, protested the incursion of "any foreign undertakers" into their local drainage affairs.[66] The inhabitants of Ely petitioned Popham directly, alleging their lands were already "for the most part continually good & profitable," and not in need of any drainage. Yet if the king insisted on seeing the Fens drained, they asked to be allowed to undertake the work themselves, rather than having to cede a portion of their lands to the projectors. "[B]eing bred & brought up in the country all our lifetime," they felt they had "more knowledge of the estate thereof & the means to drain the same" than any outsider could possibly have, and a far greater interest in ensuring a successful outcome.[67] The fenlanders' petitions illustrate their traditional understanding of land drainage as a conservative matter, rooted in local customs and responsibilities, a view at odds with the Crown's determination to pursue a more radical regional solution.

Fenland opposition to the project culminated at the next session of sewers, held at Wisbech on 13 July, just one day after the deadline had passed for paying the uncollectable tax. One source claims that at least five hundred commoners came to voice their objections, another says it was more than a thousand.[68] Yet after hearing and responding to the commoners' complaints for the entire afternoon, the commission nonetheless awarded the project to Popham and his associates.[69] The projectors were to drain the Great Level within seven years, at their own expense, and then maintain it in perpetuity; in return, they were to receive 130,000 acres (or just over 42 percent of the total acreage, as set down by William Hayward's survey) to be allotted from the wettest of the drained lands. The allotments were to be set out before work was begun, but the projectors could not take possession until it was completed, and the law included a number of pointed stipulations to ensure that they did all that they promised before they received their reward. If any of the inhabitants' remaining lands were to flood subsequently, they would be entitled to repossess a share of the projectors' allotted lands until the drainage was restored.[70]

The commission's new law seemed a clear victory for Popham and his partners, and indeed Popham considered it sufficient warrant for work to begin. On 5 August, John Hunt and Richard Atkyns laid out the proposed course for one of the smaller drains, later known as Popham's Eau, the construction of which was completed by the following December.[71] But the law also indicated a surprising degree of uncertainty on the part of the commissioners, perhaps in reaction to the hundreds of commoners who had attended the session to protest, or else the unprecedented scope of the transfer of lands they were being asked to approve. In any case, the commissioners apparently suffered a crisis of confidence, and they declined to let such a momentous law rest on their authority alone. They urged that "learned counsel" should review it "in respect of the weightiness of the said work to be performed," and then present it to Parliament for confirmation. Shortly after the session was concluded, a bill confirming the terms of the law was drafted for consideration in the next session of Parliament.[72]

This clause was a remarkable admission of doubt on the part of the commissioners; it undermined the legitimacy of their new law, rendering it more advisory than binding. Commissions of sewers had clear statutory authority to make new sewer laws and to submit them to Chancery for enrollment, and legal precedents existed for everything they had done (albeit not on such a huge scale). Yet the commissioners of sewers for the Great Level—handpicked by the Privy Council for the purpose—refused to take full responsibility for approving Popham's project. They supported it in principle but appealed to the higher authority of Parliament to review and confirm their decree. They were, in other words, not entirely comfortable in stretching fenland customs and institutions to accommodate such an unprecedented undertaking, nor in doing so in the face of strenuous local opposition. This may be seen as a sign of deference or weakness on the part of the commission, a diminishment of their local authority and autonomy. But it also represents an effort by fenland officials to engage politically with the state. Presenting their new law to Parliament for confirmation ensured that it would be more widely and thoroughly debated. It also made it much harder for the Crown and projectors to manipulate the process—getting the new law approved would now require much more than royal browbeating and packing a local commission with supporters. The commissioners raised the political stakes, and whether the law was ultimately approved or defeated they could not be held solely responsible for it.[73] Such proactive engagement was neither weak nor deferential.

"COVETOUS, BLOODY POPHAM"

As the question moved from the Fens to Westminster, the commoners' opposition to the project remained vigorous enough that the king moved to suppress it. James wrote to the commissioners of sewers on 23 July, admonishing them to "take special care to prevent the spreading of false rumors, to the intent to distaste the people with this proceeding and with the parties employed by us in this service." He further ordered them to investigate "all such discontents (if any be) and to punish the offenders as appertaineth: and to advertise us or our council of such as shall give way to any mutinous speech concerning this action."[74] But the fenlanders continued to express their "discontents" through petitions, rumors, and what might well be termed "mutinous speech," delaying and hampering the project.

Robert Cecil, Earl of Salisbury, was informed in October 1605 by a landowner in the Isle of Ely that the projectors had made little progress on the one drainage channel they had begun so far, for which he blamed "[t]he unwillingness of the people to the bettering of their estates, and their willfulness not to conceive of their own good."[75] The archbishop of Canterbury also informed Salisbury that fenland commoners were so "desperately bent" against the project that some were reportedly traveling "into the most parts of England exclaiming amongst the baser sort against the work, as tending to the utter undoing of them, their wives and children, and affirming that this course of taking away their commons without their consent was but a beginning to deprive all the poor of the realm of theirs." They also allegedly threatened that "if [the Chief Justice Popham] come into those parts any more about that business, he shall be met with."[76] Similar threats followed. In March 1606, an anonymous letter to the king complained of "covetous, bloody Popham" and his desire "to have many poorer men's commons from them for his own profit." The author went on to say that Popham "is cursed of all the poor of that part of England and swear they will kill him or such as shall be employed therein."[77] Passions were running high in the Fens, even if the muttered threats did not give way to actual violence.

Opposition to the drainage was multifaceted in both its justification and its approach. Some objected on the basis of the diversity in local circumstances and conditions: they were not opposed to draining the Fens per se, but believed their own lands were already in good shape and ought not to be included in, or forced to contribute to, the projectors' scheme. Several fen towns in Cambridgeshire and Suffolk, for example, explained in a petition to the king that their commons were located on higher ground than most of the region and were not subject to harmful flooding. They contrasted their own "firm grounds . . . being of good value" with "the low & bottom grounds . . . which be little or nothing worth" and complained that the proposed project would "little avail or profit the said upland fen grounds, but rather impair them and make them less profitable."[78] Another petition from the same area claimed that their lands were "nothing at all moorish, affording fine and good feeding grass," and were already worth between ten and thirty shillings per acre annually. Their estates had been unfairly lumped together with much less valuable lands "by reason of the common name of fens (as we conceive) which is generally used in regard of the level rather than for any other cause." Since their lands were "already in so good estate, they need not the help of any foreign undertakers," and they asked to be "exempted out of their design."[79]

The reference to "foreign undertakers," and the suggestion that outsiders failed to comprehend that not all fens were worthless and unproductive, are telling. The petitioners believed that those from outside the Fens—both the projectors and the Crown advisers who supported them—lacked sufficient knowledge of fenland conditions and were thus unable to assess and solve the region's problems. They also suggested that since the projectors only stood to gain by including as much land as possible in their scheme, especially land already of high quality and value, they could not possibly claim to have a disinterested perspective on what areas might truly be improved by their works. Drainage affairs therefore ought to be left in the hands of local officials, who alone had the knowledge, the experience, and the incentive to keep the land in the best condition possible.

This view was at odds with the Crown's—that destructive flooding in the Fens was largely the result of petty squabbling among local commissions of sewers, who had neglected for years to maintain their drainage works, and that matters would only be remedied by imposing a centrally planned, regionwide drainage scheme. Such a thing could only be accomplished by outsiders with no stake in the local disputes, under the auspices of the Crown as a disinterested broker. The Privy Council therefore refuted the petitioners by turning their accusations of ignorance and private interest back around on them. The councilors refused to believe that land anywhere in the Fens could be of such high value, when they knew the entire region to be a flooded waste, and they accused the petitioners of "not well understanding the state of the cause as we conceive." They ordered the commission of sewers to investigate "the true causes and circumstances" of the petitions and to report back to them concerning the real condition and value of the lands in question.[80] The commissioners responded as instructed, declaring that the project would create a general benefit for the entire region, but they dodged the central argument of the petitioners who had never denied that much of the Fens might be improved.[81] Their contention was that not all of "the Fens" were equally poor, unproductive, or in need of improvement, and that local circumstances had to be carefully considered when determining which lands would eventually be subject to partition and enclosure.

Alongside those who expressed particular objections to having their own lands included in any drainage scheme based on varying local conditions, others opposed any large-scale drainage project on general principles, believing it to be unnecessary, undesirable, and unfeasible. For example, one anonymous list of objections pointed to the obvious fertility of the undrained Fens as pasture and meadow, the value of which was manifest when upland farmers paid to graze their own livestock in the region. This fertility relied on annual flooding, and those who had already tried to drain their lands had soon regretted it when the drier soil yielded less fodder for their stock. Moreover, draining the Fens would deprive the poorer commoners there of the ability to make a living by harvesting the natural resources of the wetlands: fish, eels, waterfowl, reeds, and peat. Those who advocated a large-scale drainage neither understood the land and its bounty nor truly wanted to benefit its inhabitants—they sought only to profit themselves by seizing a large portion of the commons for their own private gain.[82]

The response to these more general objections, possibly written by one of the projectors, took much the same approach of turning the complainants' arguments around on them: it was not the projectors who were ignorant and self-interested, but those fenlanders who opposed them. The notion that the Fens were more valuable in their flooded state than they would be after being drained was "a very ridiculous observation," and anyone who believed otherwise was "merely ignorant," unable to comprehend what was manifestly in his own best interests.

The author, however, did not believe that his opponent was truly ignorant, but rather that he harbored a misguided desire to profit himself and to foment popular unrest in the region. The objections, he alleged, were "framed out of a most seditious and dangerous head, to infuse discontentment into the heads of the poor ignorant people." Anyone who disingenuously continued to oppose "a work that is for the general good of themselves and the common weal" could only be acting "out of a very malignant spirit, willing to set the people into a combustion," and ought to be investigated lest they "become a hatcher of greater mischief."[83]

But ultimately the most serious objections to Popham's project, and others to follow, were based on the law rather than the land; they attacked not the feasibility or desirability of the drainage, but the authority of the commission of sewers to authorize it. Such arguments had been brewing for some time by 1605, and Popham's project served as a lightning rod for them. If they held up in court, they had the potential to undermine the ability of all commissions of sewers to carry out what most saw as their traditional responsibilities.[84]

The surviving evidence in this case comes not from the complainants, but from Popham's defensive response to them. Thomas Lambert, a commissioner of sewers who had supported the drainage project, wrote to Popham in September 1605 to inform him of the legal challenges being floated by fenland commoners. Lambert's original letter no longer survives, but Popham's reply does, in which he responded to Lambert's concerns point for point. The most serious objections addressed three main issues. The first was whether the all-important 1532 statute gave commissions of sewers the power to order and erect new drainage works or limited them to repairing and maintaining extant works only. This was especially troubling because the wording of the statute was ambiguous: the law empowered commissions to "amend and repair" drainage works, as well as to "reform and correct" them, "and the same to make new." The second issue was whether or not a commission of sewers could levy taxes on the population of an entire village, fen, or county generally, or if they could only assess specific individuals by name according to each man's unique landholdings, common rights, and liabilities. The third issue was whether the actions and decrees of commissions of sewers could be appealed or challenged by lawsuit in common law courts.[85]

It is important to emphasize that the three issues raised in the legal critique of the sewer commissions' powers—the authority to order new drainage works, to levy taxes generally, and to act with broad discretion without being subject to review or appeal under common law—were precisely the sewer commission powers exploited by Popham and his partners in pursuing their drainage project. The commissions' traditional and legitimate power to levy targeted taxes for the maintenance of local drainage works had allegedly been co-opted to deprive thousands of landowners and commoners of a large portion of their lands to pay for an enormous regional drainage project that few wanted or needed. This was being done not for the benefit of the fenland inhabitants, but for the profit of John Popham and his partners. The fenlanders' legal critique was as serious as it was shocking; it constituted a direct assault on the ability of the commissions of sewers to carry out their duties. Each of the commission's powers brought into question had numerous precedents, dating back to the medieval origins of English sewer law.[86] Yet the unusual circumstances and unprecedented scale of Popham's project had prompted some to call for a broad reconsideration of the entire legal authority of the commissions and the limits to their powers.

In Popham's reply to Lambert, the Lord Chief Justice made it clear that he was not troubled by any of the putative legal challenges to his project: "I am glad that you are so desirous to be satisfied where objections are made," he wrote, "but I think no man professed in the law will make the doubts that you speak of." Concerning the authority to order the construction of new drainage works, Popham explained that in order for a commission to be able to address the unforeseen crises caused by severe flooding, the language of the statute had to be interpreted broadly and flexibly. Since the statute was intended "for the safety and preservation, relieving, and draining of all surrounded grounds," commissions of sewers were "not to be restrained to the letter" of the law. With respect to assessing sewer taxes generally, on entire communities rather than individual landholders, Popham assured Lambert that the 1532 statute was clear: "you may assess or charge after the rate by such ways and means & in such manner and form as to you shall seem meet." Nor were commissions' decrees open to challenge in common law courts; so long as they acted appropriately and according to due process, their jurisdiction in drainage matters was all but absolute. "Laws of sewers are like acts of Parliament," Popham wrote, "and therefore not examinable by any other authority elsewhere. . . . [N]o court in England did or would take the cause out of the hands of the commissioners of sewers or once examine or meddle with the law by them made."[87]

Despite Popham's confidence in his interpretation of sewer law, the challenges reported to him by Lambert proved to be remarkably persistent. As will be seen in the following chapter, lawsuits grounded on very similar arguments soon implicated various commissions of sewers in the common law courts, snarling up the business of land drainage in the Fens for the better part of a decade and creating a serious legal crisis. With respect to Popham's project, however, the more immediate challenge was getting the new sewer law confirmed in Parliament, since an act authorizing the undertaking would render all of the legal objections moot. A bill for the drainage was submitted to the House of Commons on 18 February 1606. In order to garner support and mollify any opposition, Popham and his partners made several concessions: among other things, they agreed to slash their demand for recompense from 130,000 acres to 112,000; they set limits on the number of livestock that landlords could graze on the remaining common lands; they made some provision for poor cottagers who had no legal claim to common use rights; they offered to organize a permanent corporation to maintain the drainage in perpetuity, at their own expense; and they agreed that if flooding did recur, they would pay for all damages.[88]

The concessions were not enough to do the job, however. The House received many petitions against the bill from throughout the Fens, and an amendment was introduced that would have exempted all lands worth at least 6s.-8d. per acre annually from being included in the project. The proposed amendment reflected the localist position that not all fens were created equal, and lands already of significant value ought not to be charged for a drainage project they did not need. The projectors and their supporters responded that no large-scale drainage project could ever be carried out unless it encompassed the entire region: "[E]xcept the whole [Great] Level be dealt withal, the work cannot proceed."[89] The projectors won the battle in defeating the amendment, but they lost the war: Popham's bill was rejected in the House of Commons, 93 to 116.[90]

The defeat sent Popham back to square one. He still had the sewer law passed by the commissioners the previous July, but that law stated explicitly that confirmation should be granted by an act of Parliament, which he still lacked. The bill was reintroduced at a new session of Parliament on 24 February 1607, but it made slow progress. It was hotly contested, with "[m]any speeches, *pro et contra*," and during the debate one opponent of the bill claimed that "he hath been threatened to be complained on to the king" by the projectors, suggesting that the bill's sponsors were prepared to use more than sound reason and eloquence to secure its passage. The bill was ultimately sent back to committee but was never reported back to the full House and never voted on.[91] On 10 June 1607, John Popham died; without its most illustrious investor to push the matter forward, the project dissolved. All that remained of it in physical terms was the single drainage channel near March called Popham's Eau, which by that time was already in poor condition and in need of repair.[92]

More symbolically, however, the project's legacy extended throughout the seventeenth century and down to the present day. Popham's was the first serious proposal for a project to drain the entire Great Level using a single, newly

designed drainage scheme, financed by projectors and investors (most from outside the Fens), supported by the Crown, and centrally managed by a corporation
of governors. Fifty years later, both the new drainage scheme originally proposed
by John Hunt and the management structure envisioned for it had largely come
to pass through the efforts of the 5th Earl of Bedford and his partners, who became the Bedford Level Corporation. Moreover, the legal device of using the
sewer commissions' powers of taxation as a means to confiscate private and common lands to recompense the projectors became the standard tactic for facilitating fenland drainage projects. On the other hand, all of the arguments heard in
opposition to the scheme also continued to flourish throughout the century and
beyond, making the debate over Popham's project a paradigm for nearly all
future drainage disputes.

The late sixteenth century saw not only a rise in the frequency and severity of
flooding across the Fens, but also a crisis in the local management of drainage
affairs. As ecological conditions grew wetter, the need for positive action to
preserve life and property became ever more apparent, yet for more than three
decades after 1570, nearly every effort to improve the region's drainage was stymied by infighting among the various commissions of sewers. The worsening
floods and the local stalemate over how to combat them suggested that the Fens
might be better managed not as a series of distinct localities but as a single region
suffering from a common problem and in need of a common solution. In spite of
some of the local place names, no fenland community was truly an "isle," either
geographically or politically. One landowner's relief could create real trouble
for his neighbors, and a small village could not be expected to afford huge and
expensive maintenance works all by itself just because the worksite happened
to be on its lands, when the work in question would benefit dozens of other communities. As a result of the intractable disputes, few repairs or improvements were
made, and the destructive flooding persisted.

To break the impasse, the Crown tried to intervene as a neutral arbiter, working to broker effective resolutions between opposing factions with an eye toward
the greater good of both the Fens and the commonwealth. After years of failing
to reach a consensus, however, some privy councilors began to perceive the need
for a more centralized, unitary approach, and in support of that end they sought
to introduce drainage projectors to the Fens. Projectors seemed to offer a solution for every problem. They promised to design and undertake an expensive
large-scale drainage and to fund the work themselves, in return for a share of the
drained lands. They would cut through the petty local disputes and make real

progress at last—in part because they were outsiders with no prior stake in insular squabbles. Through projectors, the Crown saw a means to implement a more rational approach to fenland water management, leading toward a more effectively drained, more prosperous, and more governable fenland. The General Drainage Act of 1601 was supposed to open the door for projectors to take the Fens in hand, to everyone's apparent benefit, while relieving the Privy Council of a burdensome administrative headache.

But when the act failed to give rise to the expected series of voluntary drainage agreements, the Crown soon adopted a more proactive policy, especially after King James's ascension to the throne in 1603. Acting for the good of both the Fens and the commonwealth, as they saw it, the king and his Privy Council manipulated and browbeat the commissions of sewers; co-opted their formidable statutory powers, placing them at the disposal of the drainage projectors; and worked to suppress any fenland opposition, viewing it as the product of ignorance, intransigence, and private interest. In adopting such an aggressive posture, promoting radical new drainage schemes in spite of persistent and widespread protests in the Fens, the Crown began to shift from a cooperative process of state formation to a more coercive process of state building.

In 1605–6, however, the attempt at state building was unsuccessful. Instead of securing the drainage of the Fens, it provoked a wave of popular resistance manifested through the region's local political culture. Hundreds of commoners protested to the commission of sewers, petitioned both Crown and Parliament, spread troubling rumors, and hampered the construction works. Their complaints— particular, general, and legal—were all part of an effort to preserve their customary use rights, as well as the traditional methods and institutions that had long protected them. Anti-drainage petitioners emphasized the superiority of their own local knowledge and experience, insisting that they were far better equipped than any "foreign undertakers" to manage their own drainage affairs. Even the commissioners of sewers, handpicked by the Privy Council to promote the project, refused to accept responsibility for it: they granted it their approval, but deferred to Parliament to confirm their decrees. Popham was unable to secure that confirmation, owing in part to the stubborn and vocal opposition of the political culture of the Fens, and his proposed drainage project died with him. As will be seen in the next chapter, however, both the idea for a large-scale drainage of the Great Level and the fenlanders' tactics for resisting it endured. In particular, the burgeoning legal challenge to the commissions' broad statutory powers proved so contentious that it badly undermined their institutional authority and forced the Crown to intervene aggressively in their behalf.

CHAPTER 3

The Crisis of Local Governance, 1609–1616

We hold them not be indifferent commissioners for the executing of
the commission of sewers within the Isle of Ely, who have great estates
there, and unto whom a great part of the ancient approved sewers &
drains within the said Isle (now in great decay) do belong, & ought to
be maintained. . . . A lamentable case, that many thousands of people
should be constrained to bestow their substance for trial of new
experiments only grounded upon the conceits of some few persons
young in years & of small experience in such works.
        —Petition from the commoners of Swaffham Prior, 1609

The commission of sewers is as necessary as may be, but [no]
commission is more abused than this is; for they do pretend the good
of the common wealth, but do intend their own proper good.
        —Sir Edward Coke, *Hetley v. Boyer,* 1614

The commissioners of sewers in the Isle of Ely faced a thorny administrative
challenge in the autumn of 1609, one that threatened to undermine their ability
to manage the region's drainage affairs. They had determined that the "ancient
drains" of the Isle had been "of long time neglected," and were thus "lost and
grown [i.e., silted] up, to the great loss & hurt of the said countries."[1] The old
drains were deemed to be irreparable, so in a series of sewer laws the commissioners ordered new ones to be constructed "at the costs and charges of all and every
person and persons, bodies politic or corporate, who shall or may receive benefit."[2] They also levied sewer taxes on all of the surrounding communities to pay
for the necessary works and issued warrants for their collection.[3]

These taxes were extraordinarily controversial, however, and the mood of the
fenland commoners was mutinous. A petition from several inhabitants of
Cambridgeshire and the Isle of Ely alleged that the proposed new work was "not
warranted by the commission & statute & also is needless, the ancient sewers

being sufficient."[4] Others stated that "they see no benefit like to come to them" from the work, and vowed that "they will not pay one penny . . . unless extremity of law will enforce them, & do choose rather to try it in a peaceful and legal course . . . than to pay those taxations."[5] This was no idle threat, and the commission had great difficulty in collecting the taxes they had levied. The town of Sutton, for example, paid only £4 of the £10 it had been assessed, and the commissioners ordered the bailiff of Ely to collect the remainder by distraint of goods if necessary, threatening him with a £10 fine if he should fail to do so.[6] Likewise, when the town of Soham did not pay any of the £60 it had been assessed, the commissioners ordered the sheriff of Cambridge to collect double that amount from the residents and threatened him with a £40 fine if he did not.[7]

If the need for punitive measures aimed at the towns illustrates the degree of popular opposition to the levy, still more revealing were the threats of heavy fines for local officers if they neglected their duties. Like all the various arms of royal government in early modern England, when a commission of sewers issued a fine, levied a tax, or called for a jury to be empaneled, the commissioners relied on an array of local officers such as sheriffs, bailiffs, and parish constables to carry out their orders; without them, the commission could not fulfill its executive functions. Such posts were all held by prominent local men, from the gentry who served as sheriffs to the relatively modest husbandmen who might take a turn as constables, all of whom served on a volunteer basis. While such offices both conferred and denoted local status and honor, they could also be burdensome and unpopular—taxation was an especially onerous and disliked duty, since it required individuals to assess their friends and neighbors and to distrain their goods for nonpayment, if necessary. If the taxes in question were deemed to be unreasonable, unwarranted, or unjust, the officers responsible for collecting them often found ways to mitigate or avoid the task.[8]

The commissioners of sewers at Ely could not afford to have their authority ignored. On 17 September 1609, after several Cambridgeshire constables had failed to collect the taxes levied on their respective communities, the commission wrote to the undersheriff of the county, William Bridges, and commanded him to collect them instead.[9] The sheriff, John Cage, later claimed that Bridges had carried out their orders "so far forth as the laws of the land would warrant, permit and suffer," but he had been "advised by his learned counsel" that the tax "could not lawfully be done and justified," and that collecting it would therefore be illegal.[10] At issue was the sewer commission's putative power to levy taxes generally to pay for new drainage works, as opposed to the more conservative and traditional practice of requiring only individual landowners or commoners to repair and maintain the existing drains on their lands. Given the questionable legality of the tax, and having "a will to pleasure and save the towns from the charges taxed upon them, and to defend their right," the sheriff also declined to collect it.[11]

In the face of such widespread local resistance, the commissioners appealed to the Privy Council for a gesture of Crown support. The council responded by ordering the sheriff to execute the commission's warrants immediately. To reassure him concerning the tax's legality, and the potential liability of himself and his deputies for collecting it, the councilors told Cage that "by the opinion of His Majesty's attorney general for the point of law . . . a tax may be set by the said commissioners as well for the making & maintaining of those new works according to the meaning of the statute as it might have been for the old."[12] But still Cage hesitated; at a session of sewers on 7 October, he returned a *tarde venit* to seven of the commission's warrants, claiming that he had not received them in sufficient time to execute them by the date specified. Given that the documents had been in his possession for more than three weeks, the commissioners rejected his claim and fined him £40 for each false return, a total of £280.[13]

The threat of such a heavy fine compelled Cage to back down, but his concerns were not unfounded. When he moved to collect the taxes at last, he was "threatened with multiplicity of suits" by those who still saw the tax as illegal; if the courts should ultimately decide in their favor, Cage would be personally liable. Fearing this, Cage told the commission of sewers that he would not execute their warrants unless they gave him "security for his indemnity in case such sums were recovered of him at the common law." The commissioners, having already hired men to build the new drainage works, with no money on hand to pay them, and fearing to leave the work uncompleted when the winter floods arrived, had no choice but to give Cage their security against any and all judgments. They tried to downplay the opposition they faced, complaining that the whole affair had been instigated by only "3 or 4 turbulent men of the inferior sort," but having already guaranteed Cage's indemnity, they began to worry about their own liability should they be "left to the nice points of the common law." They appealed once again to the Privy Council, begging that since they were "proceeding for a public good . . . to uphold so worthy a work begun," the Crown might indemnify them in turn from any threatened lawsuits, with all disputes to be settled through "an honorable hearing before your lordships" instead.[14]

⁓

The 1609 controversy over sewer taxes in the Isle of Ely was about the legal powers of the commissions of sewers and the legitimacy of their efforts to create a new drainage network at public expense. The first two decades of the seventeenth century saw a number of legal challenges to the commissions, which threatened their ability to carry out their sworn duties. Commissioners who pronounced the need for new drainage works were accused of acting beyond the limits of their statutory authority and of abusing their powers in pursuit of private gain. These challenges were remarkable because they flew in the face of long-established practice and precedent and made it much more difficult for sewer commissions throughout the Fens to improve the region's drainage in a time when serious floods were more frequent and new works may well have been needed to address the problem.

Both sides in these increasingly bitter disputes saw themselves as the true proponents of fenland drainage, a paradox brought about by different understandings of what had caused conditions to deteriorate in the first place. For the commissioners of sewers who advocated building extensive new works, the old drains were obviously decayed beyond repair, and the surrounding land could only be recovered through a massive effort to design and build a whole new drainage system—an aggressive, interventionist, and more regionally oriented approach. Yet according to those who opposed the new works as unnecessary and illegitimate, flooding in the Fens had only reached a crisis point because of the willful neglect of certain landowners who would not fulfill their customary duties in maintaining the extant drains on their lands. The list of offenders in this regard included the very commissioners of sewers who called for expensive new works to be built at a public expense, to relieve them as landowners of the private burden. The worsening floods, in other words, were not a purely natural occurrence but a result of greedy self-interest. New works were not only uncalled for but were potentially dangerous insofar as they would substitute an untested scheme for the proven reliability of the old drains. The Fens would be recovered only when the customary sewer laws were enforced and negligent landowners (including sewer commissioners) were punished for dodging the repairs for which they ought to be responsible. The opposition's approach to keeping the Fens drained was thus conservative, traditional, and locally oriented.

The dispute between these two contending ideas of how to drain the Fens and keep them dry was aggravated by the tendency of each side to cast its opponents as ignorant and self-interested parties. The commissioners calling for the construction of new drainage works claimed always to be mindful of, and acting for, the common good. They argued that promoting a more effective drainage throughout the Fens would make everyone's lands more productive and profitable, so the cost of the project ought justly to be borne by all who would benefit. Those refusing to pay their assessed sewer taxes lacked a sufficient perspective as to what was best for the entire region and were selfishly trying to evade their own financial obligations, legally imposed by the commissioners of sewers according to their statutory powers. Their opponents, meanwhile, pointed to what they saw as an obvious conflict of interest: the commissioners were shirking their customary obligations as landowners and then using the resulting floods as an excuse to foist the burden of paying for new works onto the public. Not surprisingly, many of the offending sewer commissioners—such as Sir Miles Sandys, one of the most prominent interventionists—were recent arrivals in the Fens, with little interest in maintaining the area's established customs and traditional political economy. Far from acting for the good of the commonwealth, these "improving" landlords sought only to profit themselves at their neighbors' expense.

Each side sought support from the Crown, Privy Council, and Parliament. The interventionist commissioners appealed to the Crown's long-standing desire to see a more permanent regional drainage in the Fens, as manifested in the king's support for Popham's project, and they cast their opponents as stubborn and ignorant commoners with little conception of the region's real plight. The more conservative fenlanders, meanwhile, rejected the accusation that they did not understand how to manage the lands they lived and depended on. They argued that they were *protecting* those lands from the depredations of dishonest and incompetent commissioners, who were abusing the legal powers they held in the king's name.[15]

The conservative fenland commoners' appeals for relief from rapacious landlords, corrupt commissioners, and unjust taxes were not restricted to petitions. Some sought relief through the common law courts by challenging the legal authority of the sewer commissions to build new drainage works at a general expense. Though their claims ran counter to legal precedent and threatened the ability of the commissioners to carry out their duties, their argument fell on fertile soil. As various sewer law cases came before Sir Edward Coke, lord chief justice of the Court of Common Pleas and later of King's Bench, he began to perceive a pattern of abuses of power on the part of the sewer commissions. He took up the matter in his printed *Reports* partly to stem the abuses but also as part of his broader effort to defend the jurisdiction of the common law courts from incursions by royal prerogative courts, such as Chancery—one of the most contentious legal issues of the day.[16] In the process, however, he severely hampered the commissions' ability to manage and prevent floods. The Privy Council was thus forced to intervene aggressively in defense of the commissions—both because they were vital in maintaining fenland drainage and because the Crown, under James I, could not abide Coke's challenge to the prerogative courts and the royal power they represented. What began as a dispute over the proper course of fenland water management thus became part of a much larger legal and political battle over the abuse of royal prerogative power and the jurisdiction and integrity of English common law.

COMMISSIONERS OF SEWERS, CONFLICT OF INTEREST

Allegations that certain commissioners of sewers were abusing their powers first arose in the spring of 1604, just a few months before the start of John Popham's project. In May of that year, a bill was presented to Parliament "for the more speedy recovery of many hundred thousand acres of marsh & other grounds," located throughout the Fens. The bill painted a grim picture: severe floods plagued vast areas of fenland, such that the landowners and commoners yielded no profit, to the detriment of the Fens and the commonwealth in general. New laws had been created in order to improve matters, including the General Drainage Act in 1601, but these had "wrought no good effect . . . neither are like to do in any speedy time hereafter, for that the charge is so great, that no one or reasonable number of persons can undertake the same." The only way to alleviate the flooding was to build an extensive series of new drains with high banks all along them to protect the surrounding country from sudden inundation. This would be labor intensive, and thus expensive, but the landowners of the region had neither the cash nor the incentive to undertake it because the lands in question were mostly common wastes that promised them no financial return.

The proposed bill would overcome these problems by dividing up all of the common wastes of the Fens and allotting them to be held in severalty instead. Special commissions, issued from the Court of Chancery and composed of local gentry and yeomen, would oversee the process. They would determine who received how much of each former common "after a just & equal rate & proportion . . . answerable to the quantity of the said grounds & to the benefit of common or other profit or commodity" that each stakeholder had enjoyed before. Lands would be allotted proportionally, with a large share going to the improving landlord, a more modest portion for each of his tenants, and a small parcel for every inhabitant who had previously enjoyed common use rights. Once the commissions had done their work, improving landlords would have both the legal mandate and the financial incentive to build the required drainage works, and the flooded land would "be speedily & perpetually recovered," to everyone's profit.[17]

Fenland opposition to the bill was vehement and comprehensive. Commoners from the Isle of Ely attacked virtually every aspect of it, beginning with what they saw as its true motivation: "The ground of the bill is private gain . . . covered with shows & shadows pointing at one thing & inferring the contrary." They condemned the attempt "to alter so long continued estates of so many 1000 people without their liking or privity," and "to make severals of common which is freely held without rents, services, customs or fines & to make it subject unto all these & many other inconveniences." They also alleged that the proposed allotment scheme would heavily favor landowners over their tenants and commoners, since the lords would almost certainly get to nominate the commissioners who would determine the allocation of land, and these men would either be their fellow lords, or tenants who were dependent on them. Finally, the scheme was unlikely to improve the land in any case but might well make things worse by imposing a novel, unproven drainage scheme instead of simply repairing the old drains. The proposed bill was "full of injury, innovations, oppressions, & infinite inconveniences. . . . It would be a great grief unto us if . . . so great calamity should befall the country."

The Ely commoners argued that there were already drains enough to keep the land dry, but these had fallen into disrepair because the established laws for maintaining them had for years gone unenforced by the commission of sewers. This was not a case of simple negligence, they alleged, but the result of a conflict of interest: some commissioners had deliberately failed to maintain the drains on their own lands so as to avoid their customary financial responsibilities, and they then tacitly agreed with their colleagues not to bring one another to account. The proposed bill would now allow corrupt commissioners to enclose the common wastes of the Fens and foist the cost of their own obligations onto all of their neighbors. They therefore petitioned that a different act be passed instead, "for the speedy execution of the laws of sewers, & scouring of the ancient drains whereby those countries may be reduced to their former good estate & good of the inhabitants & common wealth."[18]

The bill's proponents defended it, arguing that most fenland commoners would be much better off with a small plot of well-drained land to call their own than with their customary right to share a flooded common waste with all their neighbors. The poorer sort could still catch fish and fowl on their allotted lands if they wished, although "there were not many lived either honestly or richly by that trade, most of them walking in the night, idle in the day, & beggary all the year." They particularly objected to what they called a "most unjust & causeless suspicion of partial & corrupt dealing," insisting that they only ever sought to improve the Fens for the good of all, with the consent of their neighbors and at their own expense. In the end, however, the bill did not pass the House of

Commons. It was rewritten in committee to try to satisfy its many critics, but the House ordered that it be tabled until the next session, by which time attention had shifted to Sir John Popham's bill.[19]

While the sources provide no indication as to who was behind the 1604 bill, certain landlords in the Fens were already notorious for ruthlessly squeezing every penny they could from their lands, with little regard for their tenants, commoners, and neighbors. Many were newcomers to the region, who had acquired their estates after the dissolution of the monasteries.[20] They wanted to collect the highest possible return from their lands, and they saw particular potential in improvement projects, such as draining and enclosing their common wastes. The bill proposed in 1604 would have facilitated such schemes had it passed, but its failure, together with Popham's radical new drainage project in the following year, may have given some of the improving landlords an inspiration: by 1609, they had realized that they did not even need another statute. In their capacity as commissioners of sewers, they already had the power to build new drainage works and to finance them through general sewer taxes, if only they could convince enough of their fellow commissioners to go along.

One such improving landowner was Sir Miles Sandys (1563–1645), who soon became a leader among the more aggressive faction of fenland sewer commissioners. Sir Miles was the third son of Edwin Sandys, the Elizabethan archbishop of York, and a younger brother of the prominent parliamentarian, also called Edwin. During the last part of the sixteenth century and first decade of the seventeenth, Sir Miles acquired a number of fenland manors once held by the See of Ely, located to the south and southeast of the cathedral city. He was a proud and quarrelsome man, and a thorn in the side of many of his tenants and neighbors in the Fens, who complained bitterly that he had "for our punishment planted himself in this country." He was involved in a number of lawsuits with other area landowners, particularly over his efforts to alter both the extant drainage network and the customary use rights to common wastes on and around his estates.[21] Despite his unpopularity, he was appointed a commissioner of sewers for the Isle of Ely in 1608 and remained one of the most active, aggressive, and outspoken commissioners there for decades. He was very interested in drainage projects and was personally involved in many as an investor and promoter, ranging in size from a few hundred acres on his own lands to hundreds of thousands of acres stretching throughout the Fens. He eventually became one of the principal backers of the Earl of Bedford's project to drain the Great Level during the 1630s, an investment that ultimately cost him his fortune.[22]

Sandys's smaller drainage projects, those undertaken on his own manors, were deeply contentious and won him many enemies in the Fens. For example, in the early 1600s he unilaterally constructed a new dam known as Bathing Bank on his Stretham estate, preventing a troublesome and flood-prone stream from flowing through his lands, but causing unprecedented flooding upstream in the process, which generated a flurry of complaints. He also tried to cut a new sewer across Stretham in the late-1610s that would have improved his own drainage but would also have interfered with navigation on the River Great Ouse, thereby antagonizing the town and university of Cambridge, who relied on that river for trade. Sandys was known throughout the Isle of Ely for his ruthlessness in seeking to maximize his revenue as landlord, and he was perfectly willing to exploit his power and influence as a leading commissioner of sewers toward that end. If he was not himself a prime mover of the failed 1604 bill, he was certainly in total agreement with its goals.

In 1609, as a new commissioner of sewers, Sandys joined with several others in championing a novel and highly controversial drainage scheme for the Isle of Ely.[23] The commissioners declared that the "low and fenny grounds" of the Isle were flooded "for want of maintaining and repairing their ancient drains." Since the old drains were "of long time neglected" and had become completely irrecoverable, the construction of new drains was the only viable remedy. They therefore issued two laws: one called for the construction of three new sewers and a sluice to improve the flow of the River Great Ouse, while the other called for the restoration and completion of Popham's Eau, which had already become inoperable. In the latter law they made a point of praising Popham, who "to his great costs and charges began (in our judgment) an excellent piece of work in making a new lode or river."[24] Eight days later, the commission met again to issue warrants for the collection of sewer taxes from all the towns they presumed would benefit from the works, including communities both within and outside the Isle of Ely.[25]

The laws were controversial for two principal reasons: first, they called for the construction and financing of entirely new drainage works rather than repairing the existing drains; and second, they mandated that any and all communities that might benefit from the new drains should be required to contribute to the cost of building them, even if they were located outside the Isle of Ely (and thus were not represented by the commissioners who enacted the laws). The backlash of protest was intense, both within and outside the Isle, as inhabitants refused to pay the assessed taxes on the grounds that the new drains were "needless," "politically devised," and "not warranted by the commission & statute."[26] At least seventeen communities in Cambridgeshire and the Isle of Ely petitioned the Privy Council in 1609 for relief from what they saw as the illegal decrees of a corrupt commission of sewers. They admitted that the Fens were not as well drained as they ought to be but blamed the laxity of certain unethical sewer commissioners, who ought to enforce (and obey) the existing sewer laws rather than building a costly and potentially dangerous new drainage network at public expense.

The petition from the town of Swaffham Prior, located in Cambridgeshire outside the Isle of Ely, is representative. The Swaffham petitioners argued that neither they nor their ancestors had "at any time heretofore within memory of man" been charged for drainage works in the Isle, and they were "unwilling to charge ourselves to the present making, or our posterities and inheritances to the future maintaining of such new works." Their lands, they wrote, were untroubled by flooding and needed no improvement; nor had any sewer commissioner or juror from outside the Isle consented to the new works or the taxes levied to pay for them. They went on to assert, "We hold them not be indifferent commissioners . . . who have great estates there, and unto whom a great part of the ancient approved sewers & drains within the said Isle (now in great decay) do belong, & ought to be maintained." These corrupt commissioners were attempting "by these new works to disburden themselves of their due charge for the old, and would lay the charge upon the country for the new." They therefore refused to contribute anything toward the new works, "except by law we shall be forced thereunto," and concluded by condemning the entire scheme as being both unwise and unnecessary. "A lamentable case," they wrote, "that many thousands of people should be constrained to bestow their substance for trial of new experiments only grounded upon the conceits of some few persons young in years & of small experience in such works."[27]

Beyond the conservative appeals to customary sewer law and the ad hominem accusations regarding corrupt sewer commissioners, a more radical critique also began to emerge, similar to that provoked by Popham's project in 1605. This more fundamental criticism alleged that the powers being abused by the Isle of Ely commissioners were really illegal to begin with and ought not be wielded by any commissions of sewers for any reason. The argument is articulated in a list of "Doubts or Questions Moved at the Commission of Sewers, 1609," a lengthy diatribe against the new proposed drains and those who sponsored them—Sir Miles Sandys in particular. The anonymous author made the same points found in other petitions; he argued that the new drains were little more than "a policy to unburden Sir Miles & other great men, which are bound to maintain old drains, & to lay the charge & burden upon the country," and he condemned

Sandys and his allies for abusing their neighbors "with much cruel dealing, threatenings, imprisonings, & sale of distresses."

Yet the author of the "Doubts" went further in stressing that the offending commissioners were not just using their powers in a corrupt and unethical manner; they were *assuming* powers to which they had no legal right in the first place. Ordering the construction of new drainage works instead of repairing the old ones, or taxing whole communities generally rather than assessing each individual landowner for his own liability, were not powers granted to this, or any, commission of sewers, either by statute or by customary sewer law. The Isle of Ely commissioners, he wrote, ". . . politically and only (as we all in our consciences believe) to discharge themselves of their due charge in maintaining the old sewers . . . , have devised new rivers *against the law* to be made at the general & common charge of the country (*contrary to all equity*), pretending the old sewers to be needless & insufficient. . . ." They were attempting to do so, moreover, "by color of a commission of sewers, *wherein they err, as learned counsel advise, in form, in matter, in manner of proceeding.*" In an ominous concluding note, the author wondered, "What danger it is to do *an unlawful act* touching the body of a common wealth under color of a commission *when in truth they have no such authority?*"[28]

It was in the midst of this brewing legal uproar that John Cage, sheriff of Cambridgeshire, refused to collect the sewer taxes levied by the Isle of Ely commissioners on several communities within his county, sparking a crisis of local governance. The author of the "Doubts" may not have intended to undermine the authority of sewer commissions in general when he attacked the Isle of Ely commissioners and their novel scheme. Yet by advancing such a "radically conservative" interpretation of existing sewer law, his critique threatened the very foundation of the commissions. Throughout the sixteenth century, commissions of sewers had ordered the construction of new drainage works when necessary, taxed communities generally to pay for them, punished individuals who refused to obey their decrees, and otherwise exercised a broad discretion. Numerous precedents supported each of these practices, as did the legal opinions of several eminent jurists—including that of the recently deceased Lord Chief Justice, John Popham.[29] Without such powers, sewer commissions would be hard pressed to respond to a dangerous and worsening ecological situation, as severe fenland floods became more common.

The acute threat posed to fenland commoners by novel, and arguably predatory, drainage projects such as those proposed by Popham and Sandys had provoked a broad reconsideration of the sewer commissions' legal powers and

customary practices. The protests against any given project may have been conservative in their intent, seeking only to preserve the status quo within the Fens.
But by calling into question the ways in which sewer commissions had operated
over decades or centuries, they posed a radical legal challenge to the entire institution. When the protests eventually found their way into the central common
law courts, therefore, they had implications far beyond a single drainage scheme
in the Isle of Ely.

SIR EDWARD COKE AND *THE CASE OF THE ISLE OF ELY*

As fenland complaints against the proposed new drainage project piled up in
1609, the Privy Council referred the matter to Sir Edward Coke (1552–1634), at
that time lord chief justice of the Court of Common Pleas (fig. 3.1). They passed
along the many petitions they had already received and asked Coke and two of
his fellow judges to hear both sides of the dispute, learn the true state of the case,
and recommend an equitable settlement for the common good.[30] Coke was already familiar with the case. In addition to his post at Common Pleas, he was also
the assize judge for Cambridgeshire, and at the summer assizes of 1609, some of
the anti-drainage petitioners had appealed to him for relief. Coke's initial impulse at that time was to gather more information with an eye toward mediating
a settlement, and he ordered a survey of the existing drains to determine their
viability. The survey was to be carried out jointly by commissioners of sewers both
from within the Isle of Ely (who had sponsored the project) and from the rest of
Cambridgeshire (where it had provoked the fiercest opposition).[31]

The survey was conducted in July–August 1609 but failed to yield any hint of
an agreement or compromise as tempers flared on each side. The Cambridgeshire
men reported to Coke that "we [are] not satisfied in our consciences that their
rates were lawfully set for these new works," to which the Isle of Ely men had
responded that "they had made a law, & it was a perfect & sufficient law without
us & needed not our confirmation, & that they would proceed thereby to the end
of their works." The Cambridgeshire men also interviewed several inhabitants of
their district, and found that opposition to the new works ran broad and deep.
Most insisted that the proposed works were "but new devices laid upon the
country's charge only to ease the private charge of some of the commissioners &
others for making & keeping their old drains" and vowed that they would contribute not a penny toward them.[32]

When the Privy Council referred the matter to the Court of Common Pleas,
therefore, Coke already had doubts about the project that soon became apparent.
At the hearing held in December 1609, he began by asking "whether the new

Vera Effigies Viri
Equitis aurati nuper
ad Placita ·coram
clariſs.EDOARDI COKE
Capitalis Iuſticiarij
Rege tenenda aſsignati.

Figure 3.1. Sir Edward Coke, by David Loggan, 1666.
Portrait D26082, © National Portrait Gallery, London.

kind of draining decreed by the commissioners of sewers were for the general good of the common weal, & in that regard to be favored, *though otherwise not altogether lawful.*" When the Ely commissioners were forced to admit that their proposed new sewers would drain only forty thousand acres, whereas the old sewers (when operational) had drained an additional one hundred thousand, Coke declared that such a reduction "cannot be good for the common wealth." When some opponents of the new drainage works alleged that they would endanger the surrounding lands, Coke agreed that it would be far more sensible to repair the old drains than to rely on unproven new ones, "for it is not fit to lose a certain good in hope of an uncertain." Finally, in considering whether a commission of sewers had the power to levy taxes generally rather than on each individual landholder, and especially on communities located outside the bounds of their commission, Coke asserted that it was "in his opinion against law."[33]

Coke's assessment of the Ely sewer commissioners' case, based on his strict reading of the 1532 statute, was damning indeed. He denied their authority to build new drainage works unless these were manifestly superior to the old, and thus "for the general good of the common weal"—a high bar, which the proposed new drains had failed to clear.[34] He also denied their authority to levy sewer taxes except on an individual basis, looking at each landowner's particular liabilities and expected benefits, and only within the bounds of their commission. Having determined that the proposed project was "not warrantable by law or statute," Coke moved to terminate it. The December 1609 hearing was not a formal trial; the judges were charged with gathering information and recommending an appropriate course of action, so Coke's judgment in that forum was nonbinding. He was still the assize judge for Cambridgeshire, however, and in the 1610 assizes, at least two of the constables who had collected the commission's assessed taxes were indicted and prosecuted for doing so—the very fate John Cage had feared when he refused to collect the same taxes himself. Having little choice thereafter, since anyone enforcing their decrees would be vulnerable to prosecution, the commissioners repealed their former laws and suppressed the fines they had assessed.[35]

By the middle of 1610, then, the new drainage works proposed by Sir Miles Sandys and his fellow Ely commissioners appeared to be a dead letter, owing in large part to Coke's surprisingly conservative reaction against them and his active intervention as assize judge.[36] Yet his opinion in *The Case of the Isle of Ely* (as he later called it in his published *Reports*) was a measured one, focused only on the matter at hand. In the years that followed, Coke presided over more sewer law cases in which the outcome depended on the commissioners' statutory powers and how much discretion they were entitled to use in wielding (or stretching) them. He ultimately became convinced of the need for a more aggressive stance against what he perceived to be the rampant abuse of power among the sewer commissions.[37]

Coke heard two additional sewer law cases in 1609. The first, *The Case of Chester Mill upon the River of Dee*, was referred by the Privy Council to the chief justices of the Courts of King's Bench and Common Pleas and the chief baron of the Court of Exchequer to determine whether a commission of sewers could order a breach to be made in an existing mill causeway. The justices determined that a commission could not legally do so and indeed could do no more than order a conservative restoration of the status quo in any case where a causeway or similar work had been recently built up or modified in some detrimental way.[38] The second, *Keighley's Case,* was referred to the justices of Common Pleas by the Privy Council. The case turned on the question of who ought to pay for vital repairs to a damaged sea wall: Should all of the inhabitants be charged whose lands were protected by it, or only the individual landowner or tenant who was customarily responsible for it? The justices ruled that if the wall had been breached by violent storms or some other act of God, and through no fault or negligence on the part of the owner or tenant, then the cost of repair should be borne generally; but if the owner or tenant had negligently permitted the wall to fall into disrepair, he should bear the charge alone.[39] As with the *Isle of Ely* case, neither *Keighley's* nor the *Chester Mill* case provoked an overtly hostile reaction from Coke. Though he hewed to a conservative view regarding the powers of commissions of sewers to alter existing drainage works or to levy taxes generally, he did not censure them as an institution.

Other cases soon followed, however. Even as Coke sought to limit the sewer commissioners' discretionary powers, the ecological situation in the Fens continued to worsen. A particularly severe flood took place across Lincolnshire, the Isle of Ely, and Norfolk in 1613. Wisbech was especially hard hit, while in Norfolk "there were five towns together wholly underwater, and many people drowned," with survivors taking refuge in church steeples as they awaited rescue.[40] The ongoing crisis of fenland water management naturally led to a flurry of drainage-related activities, some of which were controversial and were challenged in court. In 1614, after his promotion to lord chief justice of the Court of King's Bench, Coke heard a case known as *Hetley v. Boyer* that finally galvanized him into addressing what he now saw not as unrelated episodes, but as a pattern of abuses by sewer commissions throughout England.

In 1611, a Northamptonshire commission of sewers fined a village within their jurisdiction the sum of £5, and for reasons unknown, they further decreed that the entire amount should be paid by a single inhabitant, William Hetley (sometimes "Heathley" or "Headley" in the sources). They then appointed two agents to distrain several of Hetley's cattle and sell them to pay the fine. Hetley sued the two men for trespass in the Court of King's Bench and obtained a judgment against them, but the commissioners then summoned him to answer for his impudence. They ordered him to release the judgment he had against their agents, and when he refused, they committed him to Peterborough jail, there to remain at their pleasure and without bail. The commissioners justified their actions by invoking their statutory power to imprison anyone who would not pay his taxes or fines, or who willfully disobeyed their decrees. Unwilling to back down, Hetley sued the commissioners in King's Bench for a writ of *habeas corpus*, moving that the court should release him from jail and grant an attachment against the commissioners for wrongful imprisonment.

It was the *habeas corpus* suit that came before Coke and the other justices of King's Bench in 1614; having heard the details of the case, Coke was furious with the commissioners. He ruled it "unjust and illegal" for them to commit Hetley to prison simply for refusing to release a judgment he had already won in King's Bench and thundered that "we will grant an attachment against all the commissioners, and set good fines on all their heads." The commissioners' high-handedness in the case was "not to be suffered to go without exemplary punishment . . . for that this is not sufferable, for them, thus to imprison one of the king's free subjects in such an unjust way, and manner." In the following term, when the matter was moved again with some of the offending commissioners present in his court, Coke's anger had not abated, and he declared their imprisoning of Hetley to be "a very great offense, and of a high nature." His fellow justices concurred: they found the commission's attempt to levy the town's entire fine on a single inhabitant to be "malice apparent" and agreed that "he might very well have his remedy for this at the common law." They therefore fined one of the commissioners present, John Boyer, the considerable sum of £200 "for his ignorance." They also made clear that they were being merciful in attributing Boyer's offence to ignorance: "otherwise for this so great an offense, he deserved to have been fined £500 and to be imprisoned till he pay his fine."[41]

Coke was not finished, however. Having made an example of Boyer, he took the opportunity to expound more generally on the commissions of sewers and the limits of their statutory powers. He was offended by the commissioners' apparent belief that they were authorized to do whatever they pleased, acting only according to their own discretions and immune from any common law challenge or oversight. This was a mistaken and potentially dangerous view, in urgent need of correction: "[T]hey cannot do what they will, but they are to follow and pursue their directions by the statute; they have not an absolute power, but they are bounded, and are to proceed, according to the rules of law (this being contrary to the common opinion, that they may do what they will)." The commissioners had strayed far beyond their legal and customary mandate, acting arbitrarily and abusing their statutory powers for their own benefit. "The commission of sewers is as necessary as may be," Coke declared, "but [no] commission is more abused than this is; for they do pretend the good of the common wealth, but do intend their own proper good."[42]

At another subsequent session of King's Bench, with two more of the offending Northamptonshire commissioners present, Coke raised the stakes yet again. He told the defendants, "[Y]ou by your commission, cannot restrain any man from having the privilege of the common law," warned them to beware of the statute of *praemunire*, "and to observe the danger, which happeneth to those, which do sue in any other court, to defeat, or impeach the judgments given in the king's court."[43] The reference to *praemunire* was not an idle threat. When Sir Anthony Mildmay, "the chief in the commission of sewers," failed to attend the trial in person, pleading old age and ill health, the court indicted him for *praemunire* for his role in committing Hetley to jail, and only a royal pardon freed him from Coke's wrath.[44]

Coke's invocation of *praemunire* in this case was highly provocative and must be understood in the context of the much larger legal and political battle he was waging in the mid-1610s, struggling to prevent the abuse of royal prerogative powers and to limit the king's interference in common law judgments.[45] Coke had long defended the superordinate jurisdiction of the central common law courts from incursions by royal prerogative courts such as Chancery, which was at that time "the supreme 'prerogative court'" and under the leadership of Coke's great rival, Thomas Egerton, Baron Ellesmere, the lord chancellor of England.[46] Among the most contentious issues was the Court of Chancery's authority to enjoin a litigant to halt a suit at common law and to have the matter determined in Chancery instead, under equity law.[47] Coke was especially incensed at Chancery's attempts to enjoin litigants even after they had already received judgment in their favor through a common law court such as King's Bench, so as to prevent execution of that judgment. This dispute did not originate with Coke—the jurisdictional rivalry between King's Bench and Chancery had been simmering for at least a century by 1614—but it assumed particular importance as he resisted James's repeated attempts to involve the Crown in the trial of common law cases.[48] The legal weapon Coke used to fight this battle was *praemunire*.

The first statute of *praemunire* was passed in 1353, during the reign of King Edward III. The law made it illegal for an English subject to appeal any case to an authority or institution outside the king's jurisdiction, and it was originally intended to prevent English subjects from appealing to Vatican ecclesiastical courts.[49] Coke, however, interpreted the statute very broadly as a means to prevent any other English court from nullifying a judgment in the Court of King's Bench. He viewed King's Bench as the premier institution of royal justice in

England and believed that its judgments could not be overturned or mitigated by anyone, with the sole exception of Parliament. He invoked *praemunire* against the Court of Chancery several times during the mid-1610s in order to establish the superior jurisdiction of King's Bench.

Although the commissions of sewers had their roots in local customary practices, from at least 1532 onward, they were formally agents of Crown governance. Commissioners were appointed through the Court of Chancery to wield extraordinary powers in the king's name to protect lives and property from the threat of severe floods, but like all Crown agents, they were vulnerable to corruption and self-interest. When Coke ruled in Hetley's favor, protecting him from unjust persecution by the Northamptonshire commission of sewers for refusing to release the judgment he had received against them in King's Bench, he made the case part of his broader struggle to defend the common law against incursions from royal prerogative courts. The main point, for Coke, was to demonstrate that Crown agents such as commissioners of sewers were not beyond the oversight of the common law; they were susceptible to legal challenge—as they had to be—in order to check their potentially corrupt, self-interested, or arbitrary behavior. And any common law judgments issued against them could not be gainsaid by royal prerogative courts, or even by the Crown itself, or else confidence in the king's justice would erode.

It was almost certainly *Hetley v. Boyer* that prompted Coke at last to revisit his earlier cases touching on the discretionary powers of commissions of sewers and to view them in a new light. What he saw alarmed him. Commissioners were charged in the king's name to protect the realm by preventing and mitigating dangerous floods, and both custom and statute law gave them considerable discretion to act in doing so. Yet while their mission was unquestionably a vital one, they were too often overreaching their statutory powers and apparently believed themselves immune from legal challenge—they had even unjustly committed men to prison without bail, simply for seeking redress at common law. For Coke, what the sewer commissioners were (and were not) empowered to do *in the king's name* was precisely the point, and it was an issue of enormous contemporary importance.

After he had ruled in Hetley's favor, fined the Northamptonshire sewer commissioners and rebuked them in court, and had Sir Anthony Mildmay indicted for *praemunire*, Coke once again took up his decisions in the three 1609 sewer law cases he had been involved with and used them to critique the overweening power of the sewer commissions in volume 10 of his *Reports*.[50] The thirteen volumes of Coke's *Reports* contain his collected notes and opinions for several of most the important law cases that he had either observed or participated in; they were intended as a pedagogical resource for English common lawyers.[51] Coke was venerated as one of the greatest lawyers of the age even in his own lifetime, and the influence of his *Reports* throughout the seventeenth century and beyond is difficult to overstate. When he finally addressed the subject of sewer law in 1614, therefore, he did so with a voice of commanding legal authority, in the midst of a running battle over the judicial independence and superiority of the common law at a time when such issues were extraordinarily sensitive.

Though based on a very conservative reading of the 1532 statute, Coke's reinterpretation of English sewer law was radical and sweeping. It ignored at least a century of precedent and placed strict new limits on the powers of commissions of sewers, while portraying it as a return to proper practices. The cornerstone of his analysis was *The Case of the Isle of Ely*. In his report on that case, Coke denied that commissions of sewers had any statutory power to order the construction of new drainage works. He parsed the precise language of the statute and found that while fifteenth-century commissions may have had such power, a key phrase in the 1532 statute had been altered in such a way as to revoke it.[52] But his opinion was not solely based on a slight (and perhaps accidental) change in wording; he also expressed concern that "to give power to commissioners to try new inventions at the charge of the country" might lead to "great inconvenience thereupon . . . for public damages," such as flooding lands not previously troubled or silting up a harbor. Some commissioners might even be tempted to abuse their power to innovate, and pursue "private lucre" at the expense of the commonwealth, for "sometime when the public good is pretended, a private benefit is intended."[53]

Coke did not close the door on new drainage works altogether. He suggested that if any proposed new works were "apparently profitable," then local landowners and commoners would presumably agree to authorize and fund them voluntarily, but they could not be *compelled* to do so by a commission of sewers. And if any truly beneficial new work were proposed but "yet no consent can be obtained for the making of it," then, Coke argued, "there is no remedy but to complain in Parliament, and there to provide relief." As a precedent he cited the 1605 drainage project of Sir John Popham, who (Coke claimed) had known perfectly well that "without an act of Parliament, none could be compelled by force of the commission of sewers, to contribute to such new attempt." He also did not fail to note that Popham's bill had been "utterly rejected," thereby ending his project.[54]

The second major determination in Coke's *Isle of Ely* report was that sewer commissions could not tax anyone even for the repair of existing drainage works "but those who had prejudice, damage, or disadvantage" from the defect and could expect to receive "benefit and profit" from the proposed repair. Moreover, sewer taxes had to be assessed proportionally, "according to the quantity of their lands, tenements, and rents, and by the numbers of acres and perches." The need for proportionality in each assessment meant that sewer taxes could only be levied with precision individually and that "the said tax generally of a several sum in gross upon a town is not warranted by their commission, but it ought to have been particular."[55]

The other two 1609 sewer law cases that Coke included in the tenth volume of his *Reports* were intended to limit the powers of commissions of sewers even further by addressing certain issues that did not arise in the *Isle of Ely* case. In *Keighley's Case*, Coke had ruled that sewer commissioners could not tax an entire community to pay for the repair of a given drainage work if the defect had been caused by the negligence of the landowner or tenant responsible for maintaining it. The whole community could be held collectively liable for repairs only if the defect had been caused through no fault of the individual legally responsible— for example, if a raging storm had breached a seawall that was otherwise in good repair. This report closed a loophole in the *Isle of Ely* case, whereby a commission might tax anyone they wished as long as they would benefit from the repair in question and the tax was levied proportionally. In practical terms, it barred commissioners from deliberately neglecting to maintain the drainage works on their own lands so that the cost of rebuilding the ruined structures could be passed along to all their neighbors (as Sir Miles Sandys and others had long been accused of doing).[56]

Finally, Coke used both *Keighley's Case* and *The Case of Chester Mill upon the River of Dee* to place strict limits on commissioners' discretionary powers. Although the 1532 statute had authorized commissioners to act "according to your wisdoms and discretions," Coke countered that such phrases must always be "intended and interpreted according to law and justice," and that "the discretion of the commissioners was limited . . . to proceed according to the statutes and ordinances before made." For Coke, legal "discretion" permitted judges, commissioners, and other officers of the state a bit of latitude to exercise their superior judgment when handling novel or unusual circumstances, but it did not constitute a license to do whatever they wished. Especially in cases involving penal statutes, which allowed for penalties such as fines and/or imprisonment (as the 1532 statute of sewers certainly did), he believed terms like "discretion" had to be read quite narrowly in order to prevent arbitrariness, injustice, and abuse.[57]

But how were Coke's strictures to be enforced? His own legal authority ran only as far as the Court of King's Bench, and it was by no means clear that the actions of commissioners of sewers should be subject to oversight in that court. After all, Lord Chief Justice John Popham had argued only ten years earlier that "the laws of sewers are like acts of Parliament and therefore not examinable by any other authority elsewhere. . . . [N]o court in England did or would take the cause out of the hands of the commissioners of sewers or once examine or meddle with the law by them made."[58] Coke's answer to this was the statute of *praemunire*; he had threatened the Northamptonshire commissioners with a *praemunire* indictment for committing Hetley to prison after he refused to release a judgment he had received against their agents in the Court of King's Bench. Coke saw this as an affront to the king's justice, a denial of the authority of the common law courts to rebuke even Crown officials who had abused their powers. By invoking *praemunire*, he declared that sewer commissioners were indeed subject to judicial oversight and susceptible to challenge under the common law.

Coke's *Reports* on the three 1609 sewer law cases made a strong argument that the statutory powers of commissions of sewers were both strictly delimited and fundamentally conservative. Commissions could not order new drainage works unless they were universally approved or sanctioned by an act of Parliament. They could levy sewer taxes only on an individual basis, proportionally for each owner, tenant, or commoner; only where the taxpayer could expect to receive some benefit; and only when the defect to be repaired was not caused by the negligence of whoever was customarily responsible. And their discretionary powers were to be read in the narrowest terms—commissioners had some flexibility in responding to new and unforeseen developments in order to restore the status quo, but they were not permitted to draw up new plans, rules, and policies whenever they pleased. If any commissioners exceeded these powers, their actions were open to legal challenge in the common law courts.

Coke had done his best in the tenth volume of his *Reports* to circumscribe the powers of greedy and overzealous commissioners of sewers throughout England, as part of his struggle to defend the independence and superiority of the common law. The pattern of abuses of power on the part of corrupt commissioners, he believed, undermined confidence in the rule of law in England, and he asserted the jurisdiction of the common law courts in curbing their misconduct. Looking at the alleged corruption of Sir Miles Sandys in the Isle of Ely, or the Northamptonshire commissioners' arbitrary imprisonment of William Hetley, it seems clear that abuses of power did exist and that some form of legal recourse against them was necessary. Coke's judicial assault on the overweening commissions tilted the balance toward the more conservative, customary approach to water management favored by a majority of fenland commoners, and his *Reports* had made it much harder for commissioners to introduce radical innovations at public expense. However, they also made it extraordinarily difficult for the commissioners to carry out even their basic duties without fear of lawsuits and thus dealt a potentially crippling blow to the institution in a time of ecological crisis.

RESTORING AUTHORITY IN THE FENS

By the fall of 1616, the full impact of Coke's reactionary reinterpretation of sewer law was being felt in the Fens. William Hetley, for one, immediately took advantage of Coke's rulings and sued Sir Anthony Mildmay for false imprisonment.[59] He was not alone; the Northamptonshire sewer commissioners had apparently offended a number of local landowners with their high-handedness, and they found themselves mired in an untenable situation, unable to enforce any of their decrees and facing several lawsuits. The sewer commissions' authority soon broke down in other fenland counties as well.[60] When the beleaguered commissioners complained to the Crown of "some hard measure they have received," which "discouraged [them] further to employ their travails, and endeavors therein," the Privy Council invited them to send representatives to court to report their "grievances" more fully. In the meantime, however, it was vital that they continue to carry out their duties actively, "according to the ancient customs, and such ordinances, as by the laws of this land have been prescribed, notwithstanding any inhibitions, or other obstacles which have been interposed to the hindrance of this service, which is held to bring so great benefit to the common wealth."[61]

When the commissioners' chosen representative, Sir Francis Fane, attended the Privy Council in October 1616 to deliver their collected grievances, the council opened the hearing in an ominous tone. They made reference to "diverse persons . . . refusing to obey such orders and decrees as the said commissioners had thought meet to set down for the good of the country" and who had "commenced diverse suits at the common law without privity or leave of the said commissioners, against some of the commissioners themselves." This state of affairs was "contrary to the ancient and usual course of proceeding by that commission, and to the great interruption and disturbance of the same, and unless some speedy remedy were taken, would tend much to the general inconvenience and prejudice of the whole country thereabouts." The council instructed Fane to consult with the attorney general, Sir Francis Bacon, regarding the "interruption and disturbance of the said commission by occasion of these late questions, and suits, so contrary to former usage and custom." Bacon would then advise the council as to "what course may be best taken for the settling of the authority of the said commission, and the avoiding of inconveniences that may grow to the country by their late questions." In the meantime, the councilors also decreed that those who had sued the commissioners or their appointed agents were to be committed to prison immediately.[62]

Less than two years after Coke ruled that a commission of sewers could not arbitrarily imprison William Hetley for seeking redress at common law against their decrees, the Privy Council now sent several men to prison (including Hetley) for doing exactly that. The council's handling of the matter was heavy-handed, to be sure. This was not a conciliar attempt at impartial mediation, nor an effort to broker a compromise; it was meant as a knockout blow against anyone looking to sue commissioners of sewers for taking aggressive action to preserve fenland drainage. The councilors were ruthless because, as they saw it, the urgency of the situation left them little choice. If the commissions could not carry out their sworn duties for fear of vexatious lawsuits, then they would cease to function, and the Fens would surely continue to deteriorate, endangering lives and property throughout the region. Furthermore, if royally appointed officials acting in accordance with parliamentary statute could be sued at common law merely for doing their jobs as they had always done them, it would have a chilling effect on the willingness of local magnates to assume any such offices, crippling the principal institutions of local governance under the Crown. The Privy Council had to act swiftly and decisively to safeguard not only the fenland drainage, but also the integrity and authority of the commissions of sewers and the royal power they represented.

On 8 November 1616, Attorney General Francis Bacon advised the Privy Council to reaffirm in the strongest terms the statutory powers and customary methods of commissions of sewers. The "sundry suits and vexations moved of late by certain obstinate and ill-disposed persons" had created a "manifest peril of destruction and inundation" in the Fens. Bacon had therefore "weighed and compared the said late undue proceedings, with the ancient laws of this realm . . . together with the continual concurring practice of ancient and late times, and also the opinion of the Lord Popham, late Chief Justice." There were, he wrote, four main issues at stake: (1) whether the commissions of sewers had the power to order new drainage works, as well as repair old ones; (2) whether they could levy taxes generally on whole communities or only on individuals; (3) whether they could commit to prison "persons refractory and disobedient to their orders, warrants, and decrees"; and (4) whether they could be sued at common law for carrying out their duties.

The Privy Council's determination in all four questions could not have been clearer. The councilors decreed that "it can neither stand with law, nor common sense and reason . . . in a cause of so great consequence" that a commission of sewers should be barred "from making new works to stop the fury of the waters . . . where necessity doth require it for the safety of the country." They declared that a commission could also "raise a charge upon the towns, or hundreds in general that are interested in the benefit or loss, without attending survey or admeasurement of acres," when the safety of the region required "speedy and sudden execution" of their decrees. They found it unthinkable that "a commission of so high a nature and of so great use to the commonwealth, and evident necessity, and of so ancient jurisdiction" should "want means of coercion for obedience to their orders, warrants and decrees," especially when "upon the performance of them the preservation of thousands of His Majesty's subjects, their lands, goods, and lives doth depend." Finally, they ruled that it would be "a direct frustrating and overthrow to the authority of the said commission" if they or their agents "shall be subject to every suit at the pleasure of the delinquent in His Majesty's courts of common law . . . and so to weary and discourage all men from doing their duties in that behalf." For all these reasons, "and for the supreme reason above all reasons, which is the salvation of the king's lands and people," the Privy Council ordered that those litigants committed to prison a month earlier were to remain there until they had released their lawsuits.[63]

The council then wrote to the commissions of sewers throughout the Fens to reassure them of continued Crown support. They pointed to the fate of "those disobedient persons" who had annoyed the commissioners "with unjust suits," expecting that the example made of them would "be a warning for others to take heed of the like contempts." Already most of those committed to prison had released their suits. "Only Hetley remains yet obstinate," they wrote, "whose behavior in this business being so insolent and without sufficient ground, either in justice or reason, we shall not hastily release him, until he humble himself, and release [his suit] as the rest have done."[64] Thus having laid to rest "all things . . . that may any way disturb this service of the sewers," they exhorted the commissioners to go about their customary duties "according as in former times you had wont to do, by virtue of your commission, and not to be discouraged by any new opinions or conceits of law, much less by the opposition of such common and mean persons, as easily spurn against all authority, when they feel any burden laid upon them, wherein you shall never want the assistance of this Table; *foreseeing always that there be no just cause of complaint given by any abuse of your said commission.*"[65]

That last clause in the council's letter is revealing. Even as they thunderously reaffirmed the broad powers of the commissions of sewers, tacitly endorsed an innovative and interventionist approach to fenland water management, and committed several men to prison until they dropped their pending legal challenges, the councilors also recognized the real potential for sewer commissioners to abuse their office. Coke's concerns about corrupt royal officers undermining faith in the king's justice were not unfounded: if the council could not tolerate his legal assault on the sewer commissions, neither could they allow self-interested and corrupt commissioners such as Miles Sandys to act in the king's name with no oversight or accountability. Since the common law courts could not provide that oversight without creating legal chaos in the Fens, the Crown would have to provide it instead. At the conclusion of their decrees, therefore, the councilors also stipulated that fenlanders with "any complaint or suit for any oppression or grievance" could appeal directly to the Privy Council for redress "if they receive not justice at the commissioners' hands."[66]

In spite of all the proposed innovations, bitter disputes, and legal wrangling, then, at the end of 1616 the situation in the Fens was almost exactly as it had been a generation earlier. The commissions of sewers had survived a serious legal challenge and had their broad powers resoundingly reaffirmed by the Crown, but all of the problems that had given rise to the crisis were still in play. The Fens continued to be troubled by worsening floods, and no one could agree on a proper course of action; accusations of greed and self-interest were ubiquitous; the Crown was still enmeshed in trying to settle protracted fenland drainage controversies, with little success; and the region was still no closer to having a comprehensive drainage solution than it had been on Lord Burghley's watch in the 1570s.

∽

By the time the Privy Council rendered its decision, Sir Edward Coke had fallen from royal favor. He had already been suspended from his own seat on the Privy Council the previous June; he was forbidden by the king from attending the summer assizes; and on 16 November (one week after the council overturned his sewer law opinions) he lost his post as lord chief justice.[67] Coke's perceived offenses were many, from his attacks on the Court of Chancery to his openly rebuking the king in council sessions; but his handling of the various sewer law cases was prominent among them. When Lord Chancellor Ellesmere, Coke's great rival, swore in his replacement as lord chief justice, he publicly berated Coke for his "new construction of laws against commissioners and judges of sewers." Coke's arrogance in law had created a real danger in practice, Ellesmere claimed, for while the commissioners were forced to appear in his court, "disputing of tricks and moot points concerning taxes, and making new gutters or walls," they could not attend to their sworn duties in the Fens. Hundreds of thousands of acres had been left to the mercy of the weather, "for the winds nor the sea could not be stayed with such new constructions and moot points."[68]

Legal opinion concerning sewer law after 1616 tended uniformly to confirm the decree of the Privy Council. By far the best-known and most influential example was the formal reading on the 1532 Act of Sewers delivered by Robert Callis (c. 1577–1642), a sergeant at law of Grey's Inn, in August 1622. Callis was a Lincolnshire man and had personal experience serving on the commission of sewers there.[69] In his renowned reading, he provided a highly detailed exegesis of the statute that was still the cornerstone of English sewer law.

The powers of commissions of sewers were derived from "the tenure and prerogative" of the Crown, Callis wrote, "and so this law is a prerogative law, and seems to be as ancient as any laws of this realm." It was of the utmost importance, "because by this statute safety was brought to the realm, and wealth and profit to the people thereof; greater and better fruits than which, no human law can produce."[70] The king had always had both the authority and the responsibility to protect the realm and its inhabitants from harm; this included securing the land from floods. Because flood control could be a matter of life and death in places like the Fens, the king's prerogative powers in this regard were extensive. When delegated to the sewer commissioners, they explicitly included the power to construct new drainage works as needed; to levy taxes generally on anyone who might receive benefit; and to use broad discretion in determining the best course of action to secure the realm from flooding.[71] Callis's reading thus confirmed the Privy Council's 1616 decree in every instance, and he even quoted the latter in its entirety.

A less influential but even more emphatic interpretation of the sewer statute was provided by John Herne in 1638. Like Callis, Herne also argued that the law of sewers in England was derived from the royal prerogative. As the ultimate master of both land and sea in his realm, the king was entitled to all benefits taken therefrom, including any new lands recovered from the sea. At the same time, he also had a responsibility to protect his subjects from threats, including floods, and to facilitate their navigation and commerce. Royal prerogative power was thus comprehensive in matters of water management. The 1532 Act of Sewers, he wrote, did not serve to restrict the royal prerogative, but rather *expanded* on it by giving the king power over a wider array of waterways (extending even to ditches) throughout the entire realm, not just along the coasts. Moreover, while prerogative power gave the king only the right to distrain the property of those who did not maintain the drainage works on their lands, the statute granted the Crown further powers to arrest, fine, imprison, and dispossess negligent landowners, even for small offenses.

Neither prerogative nor statute allowed the king to act arbitrarily. Herne stressed the preeminence of public good over private gain, gave careful attention to what punishments were most appropriate in which situations, and emphasized throughout that a king should act with discretion. Nevertheless, as long as the king's actions (or those of his appointed agents) were intended to improve either drainage or navigation, the Crown's powers in this regard were virtually without limit. Herne admitted that this made the king, in effect, a tyrant; yet far from seeing this as a problem, he claimed that royal tyranny was both necessary and wise in opposing the even crueler tyranny of disordered nature. "And yet herein appears the wisdom of the parliament in setting up one tyrant against another," he wrote, "for what [is] more tyrannous & more unmerciful than the sea & inundation of waters . . . ?"[72]

It was the practical implications and consequences of Herne's "tyranny" that had so troubled Sir Edward Coke in his own legal critique of the commissions of sewers; even when intended for the common good, unchecked royal power could do grave damage to the body politic. Coke was certainly aware of the very real threat to life and property posed by dangerous floods, and of the vital role played by sewer commissions in preventing them, but he did not trust them to act without judicial oversight. His reactionary reinterpretation of sewer law was part of his larger quest to prevent corrupt royal officials in all areas of government from abusing their powers and undermining the rule of law. The Privy Council, however, by placing all sewer law disputes beyond the jurisdiction of the common law courts, gave higher priority to the commissioners' statutory powers to innovate and act with broad discretion. As local officials formally appointed by the Crown, sewer commissioners acted *in loco regis*; they wielded royal prerogative powers to safeguard the king's lands and subjects, as the king was required by God and England's ancient constitution to do. It made no sense for the common law, which ostensibly also served the good of the king's subjects, to stand in the way of that mandate. What Coke viewed as self-interested abuses of power, others saw as the Crown's protection of the commonweal, and Herne, at least, was willing to accept the rational tyranny of the king over the much more dangerous tyranny of an irrational and unruly nature.

In shielding the sewer commissioners from challenges at common law, the Privy Council members also resumed their role as the arbiters of all drainage-related disputes, and they took this responsibility quite seriously after 1616. The councilors immediately launched an ambitious program for a detailed survey of the entire Great Level of the Fens, in order to settle outstanding disputes and lay the groundwork for the centralized, regional drainage that had long eluded them. Their efforts were not met with success, however; though the councilors sponsored multiple surveys of the region, they still could not broker resolutions for long-running disagreements where different communities' entrenched interests were in conflict. The king soon grew frustrated and sought to break the impasse a few years later by supporting another group of outside projectors who proposed to take on the drainage at their own charge in return for a sizable share of the drained lands. The ill-fated project further tarnished the Crown's reputation as an impartial arbiter in the region and provoked widespread opposition that finally managed to unite nearly all fenland stakeholders in common cause.

CHAPTER 4

The Struggle to Forge Consensus, 1617–1621

And if your lordships please to take it into your honorable care, and direct some real & effectual course for . . . the regaining of so many thousand acres that are now surrounded, [it] would prove a work of great honor to His Majesty, & good to the public, though there be many gentlemen of good worth in those parts that want neither judgment nor zeal to do service therein; yet it is conceived that it would be best effected by such as have no interest at all in the country, but standing indifferent betwixt each party, and having no other end but the common benefit, may proceed according to the rule of justice & reason. . . .
—Clement Edmonds's report to the Privy Council, 20 September 1618

For my own part I conclude with my own opinion as this. . . . That [the undertakers] will tire their bodies, empty their purses, weary their partners, be cumbersome to the country, and at the length without faithful performance of what they promise, return to the place from whence they came with more knowledge and less money.
—Anonymous commentator, c. 1620

In November 1616, the Privy Council affirmed that commissions of sewers could build new drainage works, levy taxes generally on entire communities, and imprison anyone who proved to be obstinately disobedient, all without fear of being sued in a common law court. The councilors reasoned that if the commissioners were to combat the worsening floods that threatened lives and property throughout the Fens, they must be free to act innovatively and aggressively to meet new challenges. However, they also recognized the potential for sewer commissioners to abuse the powers of their office in pursuit of private interests, and if the courts could not provide redress in such cases, then the Crown must do so. The council therefore reaffirmed as well its own role in the process, overseeing the commissions of sewers and acting as arbiters in settling conflicts when they arose. The Crown's priorities and methods in governing the Fens were thus little changed from what they had been forty years earlier.

The Crown still wished to promote a more robust regional drainage in the Fens, but despite their efforts in years past to learn more about the region and what ailed it, the Privy Council members still needed more and better information in order to act effectively. To resolve the festering controversies between various sewer commissions, they needed to come up with a unitary solution to perfect the drainage and then forge a regional consensus around it. This, in turn, required a much fuller picture of what the real problems were throughout the Great Level, how they might best be corrected, how much it was likely to cost, who would benefit from the work, and who therefore ought to contribute toward it—a much more precise and comprehensive understanding of the situation on the ground than anyone possessed in November 1616. The extensive local knowledge of the region's sewer commissioners had to be better coordinated and incorporated into the English monarchical state. Crown oversight of fenland drainage was thus a problem of state building: only by managing the Fens and their inhabitants more actively from the center could the Crown hope to improve matters there. This was not a new problem: just as it had with Humphrey Bradley in 1589, and Richard Atkyns in 1605, the council's first step was to commission a full and thorough survey of the Great Level.

Between 1617 and 1619 the Privy Council embarked on an ambitious program to collect as much information as they could get about the current state of the fenland drainage. They ordered not just one, but three major surveys to be carried out, each conducted under a different authority, which gave them a detailed and comprehensive picture of the pre-drainage Fens and their many problems. However, while the three surveys indicate the Crown's appreciation of the value of reliable local information, as well as the need for a powerful central authority to compile and control it, they also demonstrate how difficult such an undertaking was in the seventeenth century. To learn what they needed to know, the councilors had to rely primarily on the same local authorities who had always controlled access to such knowledge: the commissioners of sewers. Vital information about the Fens could not be acquired without its being filtered through the same local interests and bitter controversies that were at the heart of the problem. The resulting surveys were thus contradictory and pessimistic in tone; they compiled a great deal of information but did much more to highlight fenland disagreements than to settle them, and efforts to improve the drainage continued to stagnate.

Frustrated once again with the continued lack of progress, in 1619 the king attempted to break the stalemate by introducing outside projectors, this time in the person of two royal courtiers who saw the Fens as an opportunity to make a quick fortune. Their proposed project never got under way, however, in part because of the Crown's failure to secure an independent source of reliable information about the Fens. The very things the projectors most needed to know remained under the control of local authorities who opposed them. In promoting such an unpopular project, however, the Crown did stir up a great deal of resentment and ironically succeeded at last in uniting the landowners and sewer commissioners of the Great Level around a common cause: opposing any and all foreign drainage projectors, even those recommended by the king. The Crown, it seemed, could neither intervene effectively in the Fens to settle local feuds nor impose an outside solution in the face of united local opposition.

MASTERS OF ALL THEY SURVEYED?

The Privy Council had to hit the soggy ground running at the end of 1616, struggling to settle a long-standing regional conflict that had impeded any real progress in the Fens for more than twenty years. The particular issues of contention were many and complicated, but broadly speaking the commissioners of sewers from the Great Level counties had divided into two rival factions over the best means of draining the entire region. One group came from the more "upland" portions of the level, primarily in Huntingdonshire and Northamptonshire, which were located upriver and thus a bit further inland from the wettest parts of the Fens. The uplanders tended to favor drainage works that would vent as much water as possible out of their territory and carry it swiftly away toward the North Sea, and their top priority was to restore the River Nene and its lost outfall via Wisbech.

The uplanders were opposed by a smaller faction from the more "lowland" parts of the level, primarily in Cambridgeshire and the Isle of Ely. The lowland Fens were downriver, so any waters coming out of the upland Fens would obviously have to flow through their districts before they reached the sea. They were slightly lower in elevation and flatter in topography, the real sink of the fenland basin, so their lands were the most vulnerable to severe and sustained flooding. The uplanders' preferred approach, they believed, would place their own lands at much greater risk by conveying far too much water into their territory all at once. The lowlanders advocated instead building higher banks and sluices to try to control the existing rivers and prevent them from overflowing. These two competing views of how best to keep the Fens drained were rooted in differing topography and did not easily admit of compromise. Each faction refused to submit to sewer laws passed by their rivals that they considered detrimental to their own local interests or to pay any sewer taxes for building or repairing works of which they did not approve. The Crown had tried for decades to broker a resolution but had repeatedly failed.[1]

The long-simmering controversy flared up again in December 1616, when several lowland communities petitioned the Privy Council that they were being unjustly taxed to pay for drainage works that would do them much more harm than good. Believing that the petitioners had "great appearance of just cause of grievance," the council convened a hearing with representatives from each faction in May 1617, to consider not only the petition at hand but a whole host of outstanding fenland disputes with an eye to forging a regional drainage consensus at last.[2] But the councilors soon found that they lacked the evidence to make a sound determination on most matters and needed more information. "[I]t is conceived expedient," they wrote, ". . . before any resolution be taken therein an exact and true view [of fenland drains] be taken and presentments made by indifferent juries, of such things as are needful to be executed, for the drain and current of these waters, as shall be most behooveful for the public service of the country."[3]

A view and presentment by a jury of sewers was the normal procedure for gathering information about what drainage works needed maintenance, as had been the case since at least the 1532 Act of Sewers; the unusual factor in this instance was the sheer scale of the enterprise. Rather than looking only at particular trouble spots in a given area, this was to be a complete survey of the entire Great Level, an effort to harness the collected local knowledge of sewer commissioners from throughout the Fens to consider the problem from a centralized, regional perspective. The Privy Council began by calling for a new commission of sewers, a single large body that would include commissioners from all of the Great Level counties, to meet as soon as possible and empanel enough jurors to view each river, drain, and bank throughout the level. Once the jurors had returned their presentments, the commission would reconvene and decide on the best course of action for maintaining and improving the region's drainage, as well as who should pay for it. To discourage one faction from meeting and making decisions without consulting their rivals, as often happened in large commissions, the council explicitly ordered that at every session of sewers "as many commissioners of each county do meet as conveniently may," so that they might "proceed upon true and exact information."[4]

The new commission lost no time in responding to the Privy Council's mandate. At a joint session of sewers held at Wisbech on 12 June 1617, seventy-eight jurors were impaneled from among "the best experienced men" of Lincolnshire, Huntingdonshire, Northamptonshire, Norfolk, Cambridgeshire, and the Isle of Ely. One participant described the session as "so great an assembly of commissioners & inhabitants of all the counties . . . as hath scarcely been seen at any time before."[5] The jurors were divided up by county, and each group was ordered to survey the principal rivers and drains in its district, note any defects in them, determine who ought to be responsible for fixing them, and present their findings back to the commission.[6] The juries all returned their presentments three months later, on 16 September. Though the scale of the task was enormous, the large number of jurors involved allowed them to complete it with a surprising degree of precision, with individual landowners often being singled out for their failure to maintain specific drains and banks and for other transgressions.[7] The scope of the presentments varies, and in several cases they overlap, with two or more juries calling for the same repairs to be made, especially when a river bordered or flowed through more than one county.

The several jury presentments together constitute one of the most detailed and comprehensive views of the drainage situation in the Great Level ever compiled in the early modern period.[8] As a means to forge consensus around a coordinated, multilateral approach to improving the region, however, the exercise was a failure. The division by county allowed each jury to compile a wealth of detailed observations about territory the jurors knew best, but it also allowed them to give voice to festering grievances and opposing local interests that could not easily be reconciled, despite their claim to have reached "a wonderful & strange agreement & consent" in their views.[9] The Huntingdonshire jurors, for example, complained of a dam erected by the inhabitants of Willingham in Cambridgeshire that threatened to flood their own lands. Likewise, in their survey of the River Welland—which served as the boundary between Northamptonshire and Lincolnshire—jurors from the former county complained of numerous defects that ought to have been amended by landowners from the latter.[10] Such examples could be easily multiplied.

Predictably, the sewer commissioners had trouble translating the myriad conflicting jury presentments they had received into a coherent, regional plan of action. All of the most contentious issues were still very much in dispute, and the "variance of opinions therein being amongst the commissioners" was such that "voices did fall out equal on either side," so they "did proceed no further in the said business."[11] Subsequent sessions of sewers yielded only further stagnation and acrimony, with the most intractable divisions manifested between commissioners from the upland and lowland parts of the Great Level. The commission did eventually manage to pass a new, comprehensive sewer law in February 1618, but whatever temporary consensus that law may have represented did not last, and no actual work was done.[12] In seeking to compile the most comprehensive account of fenland drainage issues that could be had, the Privy Council had made sure to include all points of view and had uncovered an impressive catalog of local problems and grievances, but had failed to bring rival fenland factions closer together.

The Great Level commissioners petitioned the Privy Council for help in settling their ongoing differences in June 1618. The Crown had decreed that "one general commission of sewers should be granted for all these counties," precisely in order that "nothing should be done under a pretense of doing good to one part, which might annoy another." But the deliberate inclusion of all sides of the argument had only spotlighted the "variance of opinion amongst the commissioners, which distract their resolutions and judgments." Though each individual commissioner might "aim at the general good," yet as a body, ". . . all agree not in the manner of the effecting hereof, whereby it is come to pass that some to satisfy their own opinions, & some to prevent their fears of particular loss or charge, have made some oppositions and one in the absence of another have troubled this honorable Board [i.e., the Privy Council] with petitions & informations which hinder the work intended." In the meantime, badly needed maintenance was endlessly deferred, any funds expended thus far were wasted, and the Fens continued to "suffer, and undergo great damage." The commissioners therefore appealed to the council for assistance, as they were supposed to do whenever local disputes could not be resolved, according to the council's decree two years earlier. They asked the council to appoint one of its own clerks to attend their next session of sewers personally and to accompany them in a full survey of all the lands and drains in question. Such a man, "being indifferent to all parts and parties," could then advise the council "truly & without partiality," so that "one constant & resolute course may be directed."[13]

Sensing an opportunity to settle the matter of fenland drainage for good (or at least a chance to acquire some reliable, unbiased information), the Privy Council appointed one of its clerks, Sir Clement Edmonds (1567/8?–1622), to attend the next session of sewers at Huntingdon. Edmonds was an interesting choice; a scholar of military history and tactics, he had witnessed some military campaigns in the Low Countries with Sir Francis Vere during the early 1600s, and while he was there he may have observed some Dutch drainage works as well.[14] His instructions from the council were to make a personal view of all the principal rivers of the Great Level, consult with the commission of sewers, and then report back to London. The council also ordered that "a sufficient number of the commissioners of every of the said counties" should attend him throughout his survey.[15] At the Huntingdon session on 12 August 1618, three commissioners from each Great Level county and the Isle of Ely were appointed to accompany Edmonds in his travels and to provide any information or assistance he might require. The group made a sizable party for such a survey, including partisans from all sides of the various disputes that divided the commission.[16]

The party set out the following day (13 August) and spent a full week traversing the region. They surveyed the course of each of the three major river systems in the Great Level (the Nene, Welland, and Great Ouse), as well as a number of smaller drains and channels.[17] The survey was ambitious and sweeping, though rather general in nature. In contrast to the jury presentments of the previous year, which provided abundant local details, the surviving notes from Edmonds's survey make few references to specific landowners and infractions. Edmonds concentrated instead on the capacity, condition, and viability of each major river and its outfall, and how they might be improved. His goal was to compile a "big picture" view of the Fens and its problems, culminating in a series of suggestions about how best to proceed in draining the entire region, around which he hoped to build a consensus among the commissioners.

Toward that end, Edmonds made a sincere effort to give all points of view a fair hearing with respect to the many issues still in contention—this was the whole point of having such a large and unwieldy party to conduct the survey. The commissioners who accompanied Edmonds were each asked to endorse his notes and observations at the end of each day's travels, though in a few cases those who dissented from the rest of the group were permitted to record their reservations or caveats before signing their names.[18] For example, when the group surveyed Stretham Common (on Sir Miles Sandys's estate) to consider whether a new sewer was needed there, eleven commissioners endorsed the day's proceedings without comment, but Simeon Stewart and Richard Cockes noted that they "consent to all except the ground now to be cut up at Stretham," before signing their names. Likewise, after surveying the drain at Clough's Cross and another sewer called King's Creek, Leonard Bawtry and Richard Colvile signed their names only after writing, "I do not think that the King's Creek will be the best or safest drain, nor Clough's Drain to be any preservation for the north side of Wisbech, but rather dangerous; yet do agree that trial may be made thereof so as the countries adjoining be not charged or endangered."[19]

The opportunity for dissenting commissioners to put their objections in the record indicates Edmonds's desire to be seen as a fair and impartial observer.

The dissenters might not prevail in their argument, but they could not say their views were ignored.[20] On the other hand, the dissenters' willingness to record their objections over their signatures—a most unusual practice in early modern English governance—demonstrates just how intractable these disputes really were and how difficult it was to achieve a consensus around them. Indeed, few could have been more pessimistic in this regard than Edmonds himself. On his return to London, he reported his experiences and opinions to the Privy Council, with particular attention to the controversies he had been sent to try and resolve. He made a series of recommendations for much-needed repairs to the drainage network, but he acknowledged that there was still no real consensus among the sewer commissioners. He recounted numerous "questions & differences . . . concerning banks & sewers that had relation to *Meum & tuum*, and were insisted upon with more instance & siding than any other part of the business . . . wherewith your Lordships may chance hereafter to be troubled."

Edmonds assured the council that the situation in the Fens was far from hopeless. With cleverness, perseverance, and some guidance from the Crown, the drainage of the region *could* be improved and "the regaining of so many thousand acres that are now surrounded . . . would prove a work of great honor to His Majesty, & good to the public." But he questioned whether the sewer commissioners would ever be able to put aside their local interests, rivalries, and feuds to work together toward such a goal. "[T]hough there be many gentlemen of good worth in those parts that want neither judgment nor zeal to do service therein," he wrote, ". . . yet it is conceived that it would be best effected by such as have no interest at all in the country, but standing indifferent betwixt each party, and having no other end but the common benefit, may proceed according to the rule of justice & reason, as well in draining the waters to their true & ancient outfalls, which is the only means to do the country good, as levying the charge without favor or partiality." Draining the Fens would never be an easy task; Edmonds referred to a maxim common among the fenlanders that "he that will do any good in sowing must do it against the will of such as shall have profit by it." But if some competent projector from *outside* the region could be persuaded to undertake it, he felt, "yet the end will crown the work with honor & safety, and make large room for people & habitation, with as much advantage to the state as any other part of the kingdom can afford."[21]

The commissioners did manage to enact several new laws at their next session of sewers, held at Peterborough on 20 August, in the wake of Edmonds's survey, putting his recommendations into effect throughout the Great Level. The sudden flurry of activity did not reflect any newfound regional consensus, however, but only the numerical superiority of the upland commissioners, whose interests the new laws favored—a vocal lowland minority remained restive.[22] Nevertheless, the Privy Council was determined to treat Edmonds's survey and the laws resulting from it as the last word on the matter in spite of any remaining opposition. Over the following months, the council ordered that the new laws be "ratified and established and in no part impugned," and admonished the commissioners that "they do not proceed . . . to the repealing or altering of any law or ordnance already settled and established at the late general commission of sewers held in August last."[23]

But fenland resentments continued to trouble the council, just as Edmonds had feared. The next point of contention was the technical issue of duplicate commissions. A duplicate empowered two courts of sewers to operate within the jurisdiction of a single large commission, so that the commissioners resident in one area could take timely action to preserve and maintain their own drainage works without having to convene a full session with their more distant colleagues. In an area as vast as the Great Level, duplicate commissions were not just a matter of convenience; they could be vital in a flooding emergency when the need for immediate action was paramount and travel from the farther corners of the level was most difficult. In the midst of a bitter factional dispute, however, duplicate commissions tended to create more problems than they solved as each faction could assemble a quorum to convene a session of sewers on its own, repeal all of the laws their rivals had just approved, and enact their own to replace them.[24]

In November 1618, several upland commissioners asked the Crown to issue a new commission of sewers for all the Great Level counties and to grant only one formal copy of it to the Norfolk commissioners, with no duplicate for the lowland areas of Cambridgeshire and the Isle of Ely. The Privy Council approved the uplanders' request and issued only a single copy of the new commission, hoping perhaps to encourage both factions to convene and work together as one body. In practice, however, the upland commissioners were now empowered to legislate on behalf of the entire region, with no need even to consult their lowland rivals. They soon pressed their advantage, with a minimal quorum of just six commissioners scheduling the next general session of sewers at Stilton on 20 January 1619.

The lowland commissioners complained to the Privy Council that, although the time and place of the next session might not seem problematic to someone who did not know the Fens, the choice was in fact most prejudicial. They called it "an innovation not formerly known in our parts to have a general session of sewers summoned in this winter season of the year, the whole level of the surrounded grounds . . . lying usually as now it doth deep under water," so that making a proper survey of the land was impossible. Moreover, Stilton was located "in the remotest part of Huntingdonshire," a considerable distance from the Cambridgeshire Fens in the heart of the Great Level. The lowland commissioners could not travel so far in January "without great inconvenience and hazard . . . the ordinary passages being all drowned," and the uplanders surely knew it; the Stilton session was almost certainly intended to exclude them. Since the lowland counties contained by far the greater share of flood-prone lands in the level, the commissioners there had the biggest stake in any new drainage legislation. They therefore asked the council to stay any new laws enacted at Stilton without them and to grant a duplicate commission for Cambridgeshire and the Isle of Ely, "as in all former times we have had."[25]

The Privy Council initially decided that the lowlanders had a valid complaint. They drafted a letter to their upland counterparts requiring them to "respite the execution and penalty" for any new sewer laws passed at Stilton that impacted the lowland Fens. They also approved a duplicate commission for Cambridgeshire and the Isle of Ely, on the condition that they not use it to overturn any of the laws deriving from the Edmonds survey.[26] Believing they now had the council on their side, the lowlanders wrote to Nicholas Massey, clerk of the commission of sewers for the Great Level, informing him of the situation and commanding him to revoke any warrants recently issued by the upland commissioners. They also ordered him to hand over "all such laws, journals, orders, and decrees of sewers, together with all letters" from the Privy Council, which would effectively prevent the uplanders from conducting any further business. Finally, they took a cue from their rivals and scheduled their own session of sewers in Cambridge on 7 April 1619—the day after quarter sessions were to meet in all the Great Level counties. This would have prevented any commissioners of sewers who were also justices of the peace (as a great many were) from attending to both duties—unless they happened to be Cambridgeshire men.[27]

The upland commissioners soon gave the Privy Council their side of the story. They complained that their lowland counterparts had revoked all their warrants without so much as consulting them, stymying their efforts to carry out the recommendations of the Edmonds survey and leaving the Fens in grave danger of flooding. This was not what the council had in mind when they approved the duplicate commission; they had sought to facilitate the enacting of sewer laws appropriate to each locality and restore balance between the factions, not to enflame regional tensions and bring matters back into stalemate and stagnation. The council reacted sharply, rebuking the lowland commissioners for their disobedience and commanding that all of the laws approved by the uplanders at Stilton should be executed and enforced without delay. They also demanded the immediate return of all copies of their previous letter to the lowlanders, lest they give rise to further mischief.[28] The uplanders thanked the council for their continued support against "the causeless suggestions of some of our fellow commissioners," but warned that further conflict remained all too likely "unless your Lordships may be moved to call in the duplicate of the commission, late sued out for Cambridgeshire."[29]

The unseemly back-and-forth between the upland and lowland factions in the Great Level continued for months, with the Privy Council forced to act as unwilling arbiters. The upland commissioners still had numerical superiority on their side, and Edmonds's survey favored their point of view; but the concerns and interests of the lowlanders were both real and compelling, so that their cause could not simply be dismissed. The council, eager to reach some settlement but leery of imposing one unilaterally, finally decided to hold a hearing in the spring of 1619.[30] But when the hearing took place on 17 May, they once again found that nothing could be decided without more information. They asked the new bishop of Ely, Nicholas Felton, to make yet another survey of the area, particularly around the River Great Ouse, and to examine such of the inhabitants as he saw fit. Felton was invited either to render a verdict in the matter himself or else refer it back to the council with his recommendations.[31]

Felton was born in Great Yarmouth, Norfolk, and educated at Cambridge, where he was a fellow and master of Pembroke College; both communities lie just outside the Fens, in counties at the heart of the controversy.[32] As a native of the region, he probably had some general awareness of the situation but perhaps not much personal experience in the intricacies and politics of fenland drainage. He therefore sought assistance from someone with more understanding: Richard Atkyns, who had first surveyed the region in 1605 in connection with Sir John Popham's drainage project. Atkyns made another personal view of the entire Great Level for Felton in the summer of 1619 and compiled a report of his observations and recommendations. His report offers yet another clear and comprehensive description of early modern fenland drainage and its many problems. Atkyns had concentrated primarily on matters affecting the Isle of Ely, as Felton probably asked him to do, but he gave some attention to all of the Great Level's major river systems.[33] His report is more detailed and plainspoken than Edmonds's, and he made far less effort to present an impartial, balanced point of view—his sympathies clearly lay with the upland commissioners and their lowland ally, the maverick Sir Miles Sandys. Nevertheless, Atkyns's report serves to confirm and complement Edmonds's, and his conclusion was not a happy one: the Fens were poorly drained, the existing drainage network was badly decayed, and matters would most likely worsen. The most pressing material problems could all be addressed, Atkyns believed, but this was unlikely to happen because the fenlanders were congenitally unable to cooperate.

Controversy and contention stood in the way of progress everywhere Atkyns surveyed in the Fens. When someone tried to improve matters, "complaints and oppositions were made . . . which stayed the proceeding thereof." A principal issue of debate in several cases was who should pay for the proposed new works; Atkyns described more than one promising proposal that had languished only because "it sticketh upon the charge." Equally daunting was the sheer level of mistrust and acrimony among the principal inhabitants, especially in the Isle of Ely, where Sir Miles Sandys's various improvement schemes on his estates had been a source of local tension for more than twenty years. One drainage project in the Isle had been stymied, Atkyns reported, "more, as was thought, of mislike of the persons that pursued it, than of the work itself." Another was deadlocked in a dispute that had become more personal than pragmatic: "How this will stand I know not, but verily there were [those] that observed some defects on both parts; one over violently enforcing, the other too vehemently insisting . . . the quarrel, I doubt, is not yet ended, nor will be, I fear, till some passions be better qualified."[34]

But perhaps the single most discouraging aspect of Atkyns's 1619 report was the need for it, following so hard on the heels of Edmonds's report of 1618, and the thorough multicounty jury presentments of 1617. Compiling detailed information about fenland drainage problems was simply not enough to resolve the long-running disputes that paralyzed the region. The Privy Council had taken responsibility for overseeing fenland commissions of sewers three years earlier, in the wake of legal challenges that had threatened to render those institutions all but powerless. They still had not managed to build consensus among the commissioners for a regionwide drainage scheme and had little to show for their efforts beyond a growing pile of surveys and reports—surveying the Great Level was becoming an annual rite by 1619. New sewer laws had been enacted and stayed, warrants issued and revoked, and all the while the Fens continued to flood, and the inhabitants to suffer accordingly. "[O]nce pity it is," Atkyns wrote, "the country in general should be so much hurt as they complain they are for any private respect; and again as much pity it is to see grounds of that nature, and so well quality'd, to be through faction overthrown."[35]

Atkyns's frustration and pessimism come through in his report, with little suggestion that any change was likely in the coming year. Yet even as he was writing the words above, a new prospect had emerged. Clement Edmonds suggested in 1618 that the Fens could only be drained by an outside party with "no interest at all in the country," someone who could stand "indifferent betwixt each party, and having no other end but the common benefit, may proceed according to the rule of justice & reason."[36] In July 1619, two such figures presented themselves: Sir William Ayloffe (c. 1562–1627) and his son-in-law, Sir Anthony Thomas, proposed to the Crown that they would undertake to drain the Great Level at their own charge, in return for a large share of the newly drained lands. Ayloffe and Thomas could hardly be described as having "no interest at all in the country," and "the rule of justice & reason" was hardly their guiding principle, as will be seen. Nevertheless, James I and his Privy Council leapt at the chance to settle matters in the Fens once and for all by placing the whole problem into the (dubiously) capable hands of this pair of would-be projectors.

THE AYLOFFE-THOMAS PROJECT

Ayloffe and Thomas were an unlikely pair of drainage projectors. Neither was a native of the Fens, nor did they possess any known experience in, or special knowledge of, land drainage. Indeed, Ayloffe's only prior connection with drainage matters came in the 1590s, when he failed to maintain the banks on his principal estate in Essex, allowing the River Thames to flood Havering Marsh; the local commission of sewers fined him £500 for his negligence, and he was reprimanded by the Privy Council. The two men's introduction to the Great Level apparently came while accompanying the Earl of Arundel as part of a delegation sent by the Privy Council to tour the region in 1618, in the wake of Edmonds's report, to "prepare some opinion to be delivered to the Board, of what present course might be fit to be taken therein."[37] The reasons for their being included in the delegation are unknown.

Undaunted by their lack of drainage experience, Ayloffe and Thomas nonetheless pitched their project to the king and Privy Council with great confidence. They declared themselves to be "no common undertakers" and claimed to have consulted with "some rare engineers" (whom they never named) in making their own "especial observation" of the area. They proposed to drain the entire Great Level at their own expense, transforming the land into "good & profitable meadows and pastures." They were prepared to begin work immediately, and after some initial negotiation, they offered to accept whatever portion of the improved royal landholdings in the Great Level the king and council deemed appropriate after the project was completed. As for the many private landholdings and common wastes throughout the region that would also be improved, they

planned to negotiate suitable agreements with the owners and commoners, asking for as much as two-thirds of all the lands they drained. Nowhere in their
proposal did they indicate how they intended to achieve their design.[38]

The king was enthusiastic about the proposal, viewing it as a no-lose proposition for the Crown, and he declared himself "willing to give them all encouragement toward so worthy and public a work." He involved himself personally in
supporting the venture, writing to commend it to the Great Level commission of
sewers in September 1619.[39] He predicted that the project would bring "a great, &
inestimable profit" to the kingdom, and ". . . recommended the said undertakers,
& their designs to the especial care of all the lords & other commissioners of sewers . . . , requiring them to be aiding, & assisting to them in the contracts to be
made with the subjects according to law; and that all persons should be respective, & conformable to that His Majesty's royal pleasure."[40] Frustrated by the slow
pace of negotiations, the king also complained to his privy councilors about the
"long protraction and loss of time" by which the projectors were "hindered from
beginning their work" and ordered them to write to the sewer commissioners
speed things along.[41] The council did so on the following day, admonishing
the commissioners "for the readier, & more speedy execution" of the work, "to
observe well His Majesty's said gracious, & princely direction."[42]

The Great Level commission of sewers was still operating with a duplicate
commission for the lowland areas in Cambridgeshire and the Isle of Ely, so each
group met separately to consider the proposed project. The uplanders convened
a session at Peterborough on 8 September, with Ayloffe and Thomas in attendance, and resolved to give "all aid & assistance which lawfully we may by authority of our said commission." Two weeks later they ordered a formal copy of their
resolution to be printed and circulated throughout each upland county, in which
they scheduled specific times and places where landowners and commoners
could meet to negotiate contracts with the projectors. They also warned that if
terms could not be agreed on, they would take further steps to expedite the
proceedings at their next session.[43]

The lowland commissioners met likewise at Ely on 22–23 September, also
with Ayloffe and Thomas present, and declared themselves "assembled in all
humble obedience to his most excellent Majesty." Unlike their upland counterparts, however, the lowlanders required the two men to accept some conditions
before they would cooperate with the venture. Among other things, they agreed
that they would receive no recompense except where landowners and commoners
saw a clear improvement in their lands and that a local jury, impanelled by the
commissioners, would determine whether lands were improved. The commis-

sioners then resolved to "give them all aid & assistance which lawfully we may by authority of our said commission," and ordered their resolution to be printed and circulated widely, along with specific times and places for landowners and commoners to negotiate their own contracts. If any refused to participate, the commissioners pledged to "take further order for the expedition of the proceeding of the undertakers in their work according to His Majesty's pleasure."[44]

The projectors had good reason to be pleased with their progress thus far: they had secured agreements with both rival factions of sewer commissioners in the Great Level to support their project and facilitate the contract negotiations. Given the recent history of tensions between the upland and lowland camps, the fact that they appeared to be of one mind in promoting the undertaking was itself no small achievement. It appeared that Sir Clement Edmonds might be correct in his assessment: perhaps the Fens could only ever be drained by outsiders, acting with the explicit support of the Crown, supplying their own funding, and with no stake in the local controversies that had paralyzed the region for so long.

Fenland enthusiasm for the project was more ephemeral than real, however, as careful observers noted. Daniel Wigmore, archdeacon of Ely Cathedral, attended the Ely session on 22 September and reported his opinion that the project would face strong local opposition. He observed that the commissioners' support was ambivalent—despite the king's endorsement, they had debated the project all day before grudgingly approving it—and he predicted that the inhabitants would never willingly surrender a portion of their lands to the projectors. "To . . . bring them to a contract will be very hard," he wrote, "for a great part of their old drains they say are already opened and if they may have but the like time allotted them which these undertakers speak of, they would make all this part of the country dry themselves."[45]

Other reports were even more pessimistic. An anonymous observer of the session at Peterborough on 8 September contemptuously dismissed the two projectors for their "want of knowledge in the nature of the fens . . . [and] their weakness in the science of those things which they undertake." He disparaged their tendency to "confound all the fens in general . . . not making any distinction of the nature and quality of every several fen, which do much differ"—a criticism leveled against Popham's project as well. He described a revealing instance in which the two men very nearly accepted an offer from the commissioners to take their recompense out of lands in the salt marshes "surrounded at the low water mark"—in other words, the sea—and were saved only when a scrupulous commissioner pointed out their error. Given their ambitious claims and obvious ignorance of the Fens, he wrote, "these undertakers' large promises do give cause to fools to wonder at them, but to wise men and of experience to smile at them, and suspect their attempts."[46]

The observer also doubted the projectors' supposed intention to benefit the commonwealth, arguing that the entire affair was motivated solely by greed and would end in an expensive failure—they sought to make a fortune in the Fens but had no idea what they were doing. In any successful drainage project, he wrote, the *"bonum publicum* must carry the show," but it was clear in this case that *"privatum commodum* is the thing aimed at." He likened the two men to *"alchemists,* whom likewise the same *cupiditas habendi* hath possessed, and the arts much alike tempting, promising mountains, but in the end having consumed much silver to multiply gold, would be glad of brass." After discoursing at length on the projectors' incompetence, ignorance, and cupidity, the observer closed with a dismal prediction: "For my own part I conclude with my own opinion as this. . . . That they will tire their bodies, empty their purses, weary their partners, be cumbersome to the country, and at the length without faithful performance of what they promise, return to the place from whence they came with more knowledge and less money."[47]

Just as Daniel Wigmore had predicted, landowners and commoners through-out the Great Level were unwilling to surrender a portion of their lands to pay for a drainage project in which few had any confidence. Very few contracts were concluded, especially for the common wastes, and the commissioners of sewers were soon forced to intervene in an effort to move things along. The following weeks demonstrated just how much the commissions of sewers could do to ad-vance a drainage project if they chose to support it. The lowland commissioners ordered that every town and village within their jurisdiction should send represen-tatives to speak for them at their next session of sewers, held at Cambridge on 13–14 October, and to declare their intent to bargain with the projectors for draining their common wastes. According to the formal session minutes, this process went smoothly: twenty-four communities sent representatives who duly announced their obedient intent to negotiate an agreement in good faith.[48]

The uniform expression of assent was illusory, however. The minutes also rec-ord that representatives from at least three communities questioned whether their lands should even be included in the project, and several others stated that they lacked the authority to speak for all of their neighbors.[49] Moreover, eight communities had sent no representatives to the session despite the "so public notice given," whether "through ignorance, or negligence," and so had to be compelled to cooperate. The commissioners therefore levied a tax of twenty shil-lings per acre on all lands for which a contract had not yet been concluded—an amount substantially higher than the entire annual value of most of the lands in question, and thus not realistically payable. This was the same tactic used to advance John Popham's project in the face of local resistance, potentially empowering the commission to seize a portion of all lands in default and use them to conclude their own bargain.

If the formal session minutes indicate a certain degree of reluctance among the fenland commoners, another surviving version of events reveals a much more widespread and vocal opposition. According to a rough manuscript account of the same session of sewers, several representatives were actually bullied into giving their assent. The two from the city of Ely, Nicholas Cowper and William Morgayne, were recorded in the formal minutes as being "willing to contract with the undertakers, & they hoped to persuade the rest of the commoners of Ely to contract with them upon such offers & agreements as were ordered, & agreed upon."[50] In the alternative account, however, they had rather more to say for themselves. They initially "exhibited a petition" against the project, "which being read, it was much disliked." The commissioners reproved the men for their obstinacy and advised them "to think of a better answer and attend in the afternoon." At their second appearance, they still protested that "the grounds belonging to the City of Ely were not within the compass of the design of the undertakers, neither do they think they can receive any benefit by any works to be done by them." They asked for more time to confer with their neighbors about the terms to which they were authorized to agree, and only then did they indicate their own personal willingness to negotiate, and their hope of convincing the rest of Ely's commoners to do likewise.[51] And even this grudging compliance was belied by the Ely men's further actions: the official minutes record that Cowper was assessed a bond of £40 and ordered not to leave Cambridge because of some "offensive speeches which the undertakers offered in court to justify he had uttered, & spoken against them, & this their design." Not until he had "acknowledged his fault, & submitted himself to the said undertakers" was he given leave to depart and his bond forgiven.[52]

The Ely representatives were not the only ones whose truncated responses in the session minutes indicate the commission's suppression of local opposition. Although the men from Stretham and Thetford were formally recorded as agreeing to contract with the undertakers, in the alternative account they, too, presented a petition against the project, signed by several inhabitants of each town. Their petition "was redelivered to them & they admonished to make a direct answer whither they would contract or not in the afternoon," at which time they reluctantly consented to negotiate. Likewise, John Banks of Cottenham was formally recorded as being willing to bargain for his own lands "if it did them good & no hurt," though in the rough notes of the session he appears to have said the exact opposite.[53]

The projectors got no better reception from landowners and commoners in the upland fen counties, and the commissioners there were also forced to take action to get things moving. At a session held at St. Ives on 15–16 October, the commissioners recorded that "the lords, owners, commoners, and parties interested, have not come before this court . . . to make & conclude their contracts, according as was really expected at their hands." Rather than having the project "be further delayed or protracted . . . in such case of apparent neglect, or willful absence or omission," they also levied a tax of twenty shillings per acre on the lands of anyone who refused to negotiate a contract and ordered the county sheriffs to publicize their order. Yet they also felt it necessary to require the sheriffs to certify that they had in fact done so, on pain of a £10 fine—an indication that the project was equally unpopular in those counties and that local officials were offering at least some passive resistance to it.[54]

The heavy-handed and bullying tactics of the commissions of sewers illustrate the power of local officials to facilitate and advance a drainage project, even in the face of popular opposition. Such aggressive advocacy was vital for an unpopular project to have any chance of success, as James I perhaps acknowledged when he urged the commissioners to do all they could to assist the projectors. Yet while the commissioners might do much to overcome local resistance if they chose to support a project, the coming months would demonstrate their capacity to hamper any project they did not favor, even if it had the Crown's endorsement. For in spite of the commissioners' unified and active support of Ayloffe and Thomas to this point, their project made no progress at all. The large majority of fenlanders still refused to deal with them, and even more tellingly, the commissioners never made any pretense of forcing them to do so by threatening to collect the punitive taxes they had assessed. They may have hesitated to invest their local political capital in such a venture and face the popular furor that would surely result if they tried to enforce their decrees, especially as the greed and incompetence of the projectors became ever more apparent. Whatever the reasons, within just a few weeks the commissioners began to express their own doubts and frustrations, particularly over the issue of how much land the projectors felt they were entitled to receive in recompense.

The Privy Council soon noted the lack of progress in the Great Level. In December 1619, they wrote to the commissioners to complain that the work had been so long delayed by "the lingering, and slow proceeding of the meaner sort of people, not apprehending the benefit, and commodity intended unto them." Ayloffe and Thomas had by this time attended some fourteen sessions of sewers and had offered "such conditions as were expected would with all cheerfulness, and alacrity have been embraced," yet still the landowners and commoners refused to bargain with them. The council reminded the commissioners that the project had already "received the allowance, and approbation as well of His Majesty & the state, as of yourselves," and charged them with "removing . . . any difficulty that may hinder an undertaking of so great hope," since all the means necessary to overcome local resistance were "wholly in your own power."[55]

The council's letter provoked a defensive reply from the commissioners, particularly the implication that they had not acted vigorously enough to advance the project, "by which misinformation we take ourselves not a little wronged." Any delay, they argued, was rather "the fault of the undertakers themselves, [who] being not like to succeed so well as they expected, they would by such suggestions decline their own errors upon us." The main point of contention, it seemed, was the amount and manner of recompense they should receive. Ayloffe and Thomas insisted that large portions of land should be set aside for them even before construction work began, and thus "before it can possibly appear what proportion of benefit shall arise to the inhabitants." The commissioners dismissed their demands as "both unseasonable, as not made in any due time, & also most unreasonable in respect of the over large quantities of ground which they require to be allotted unto them." They countered that the projectors should receive only "a moiety of that clear profit which by their sole industry every private man should reap out of his surrounded grounds more than formerly he had done." This could only be determined *after* the work was fully completed, and would be decided by a jury of local men impanelled by the commissioners—terms the projectors had agreed to back in September.

The fenland commoners, meanwhile, would not cede any of their lands before they saw an improvement in them and refused to negotiate contracts on such terms. They "incessantly cried out unto us not to give away their common," according to the commissioners, and petitioned to be allowed to drain the land themselves instead. The commissioners forwarded the petition to the Privy Council with their endorsement, "the rather for that we hold so great a work may much more easily be performed, by the joint concurrency of so great a multitude, than by the particular charge of any private undertakers, though their abilities were much greater than those who have hitherto offered to undergo the same." Unable to break the impasse regarding the manner of recompense, both the upland and lowland commissioners suspended their respective negotiations with the projectors, and from that point onward the two rival factions were united in opposing the venture.[56]

Open resistance was risky, to be sure, given the king's explicit endorsement of the project and the Privy Council's long-standing frustration with the commissioners' endless infighting. Indeed, the fenlanders' petition to be allowed to drain the land themselves must have seemed disingenuous, since their proven inability to agree on a means for doing so was a key factor in the Crown's move to employ outside projectors in the first place, as Edmonds's report had advised. At least one fenland observer was wary of the possible consequences, noting that the Crown had backed Ayloffe and Thomas's project "for the taking away of quarrels about draining, whereof the [council] is weary." The commissioners might reasonably object to the terms of a given proposal, citing "sinister intents only of private gain to the country's wrong," but rejecting the entire project outright was dangerous because of the Crown's direct involvement. "If there were no greater men engaged in the business, than these visible undertakers," he wrote, "we might stand more stiffly both upon the conditions & against the work itself." But since it obviously had the support of "greater persons," including the king, the wiser course would be "to contract as warily as we can . . . that yet we may do it with the least disadvantage." Otherwise, when the council grew impatient they might "take a new & perhaps (or rather without all doubt) a harder course for the country."[57]

Yet despite the observer's prudent warning not to antagonize the Crown, both the upland and lowland commissioners persisted in their opposition. Ironically, Ayloffe and Thomas had managed what the Privy Council had long failed to accomplish: they had kindled a sort of consensus among all of the Great Level commissioners of sewers, albeit against their own proposed drainage project. Moreover, the commissioners' unity proved a political force to be reckoned with, one that had not been seen in the Fens in a long time. With their exclusive control over critical local knowledge and the political authority to wield it, commissioners of sewers were still strong enough collectively to resist the Crown's attempt to impose an ill-advised and unpopular drainage project on them.

LOCAL KNOWLEDGE, LOCAL CONTROL

In February 1620, with the project at a standstill, both sides appealed to the Crown for support. The king summoned the projectors and representatives from the commissions of sewers to appear before the Privy Council so that "the whole matter shall be thoroughly debated to give every man satisfaction so far as may be." He still endorsed the venture, deeming it "so worthy an end as His Majesty will vouchsafe to honor the consultation with his own presence."[58] The meeting took place on 11 April and opened with a full reading of Clement Edmonds's report, "which was much commended by His Majesty then present," including his conclusion that outside parties would be "best effected" to drain the region.[59]

The projectors then presented the terms of their proposal to the council, both orally and in writing: they pledged "to drain & make dry the whole level, & latitude of the Fens" at their own charge—a total of some 330,000 acres in the Great Level—and to complete the work within three years and maintain it in perpetuity. In return, they sought "assurance of competent land" to reward their efforts and pay for future maintenance, "wherein we desire all plain, and honest construction hereof by the country." They provided a long list of various plots of land they aimed to drain and asked for between one-quarter and one-half of each, an amount suitably reflecting "our merit, & industry, pains, & trial, charge, & adventure in so great a business." The lands in question were to be surveyed and set aside *before* the work was begun, as security for the projectors' investment, though they would not actually take possession until the drainage was success-fully completed. They also acknowledged that their information about some plots of land might be uncertain and incomplete, but lamented that they had received "little instruction, or information from the owners, or dwellers there, who rather endeavor wholly to conceal the same from us." As for how they planned to effect the drainage, they offered few details and asked that "we be not pressed to dis-cover until our agreements be made perfect, and assured."[60]

The commissioners of sewers responded, objecting to the proposal on several counts. They argued that several thousand acres slated for "improvement" were already valuable in their present state, worth between eight and twenty shillings annually, and thus comparable with good pasture outside the Fens. Since lands of such quality were unlikely to be significantly improved by any drainage proj-ect, they asked that the projectors "desist from meddling with grounds of that value."[61] Regarding those lands that might see some improvement, the commis-sioners insisted that the terms already agreed on were for "the moiety of the clear gain, which they added to the former benefit of the said grounds," the amount of which was to be determined after the fact by a local jury.[62] Setting out any quan-tity of land in recompense before the work was even started not only made no sense but was illegal according to the terms of their commission:

[W]e have no power in due form of law to take any man's land from the proper owner thereof without his voluntary consent. But the authority, which the commissioners have by their commission (unto the observing whereof they are strictly sworn) is only to rate the charge of every particular person toward any such general work according to the profit, which such person shall receive by the same. And forasmuch as it is impossible to be discerned before the work be finished, either who shall have profit by their intended work, or how much, the commissioners cannot legally procure any such preposterous assurance of lands beforehand unto the undertakers as they demand.

They concluded by suggesting that if the projectors would not abide by the reasonable terms they had agreed to in September, the fenlanders "may lose no more time by entertaining any further treaty with them, but may be left to the speedy draining of ourselves."[63]

The commissioners' hypocrisy in this case is stunning. For more than a decade they had insisted in court that they did have the power to levy taxes on whole communities generally, before any works were under way, based on a potential benefit they assumed everyone would enjoy equally. Now suddenly, it seemed, a commission could only assess sewer taxes case-by-case, and only after the work was completed so that each individual landowner's gains could be carefully measured. This shrewd departure from their earlier legal arguments illustrates the capacity of fenland political culture to adapt to shifting circumstances and to manipulate the law to suit their needs. Commissioners' powers could be broad and innovative when they wished, limited and conservative when they didn't. Their crafted response also indicates a newfound unity emerging within the fractured and dysfunctional political culture of the Fens. The session of sewers at which the above response was drafted included commissioners from all of the lowland and upland counties—the first such meeting since Edmonds submitted his report nearly two years earlier.[64] Finally, the projectors' complaint that they could get "little instruction, or information" from the fenlanders, and their plea for "all plain, and honest construction . . . by the country," suggests that they realized the weakness of their position. The commissioners' ability to work at last in common cause, manipulate the law, and keep strict control over important local knowledge, gave them a clear advantage.

The members of the Privy Council handed down their decision at the end of May. While they continued to support Ayloffe and Thomas in their project, they acknowledged the merits of the fenlanders' case. The drainage would go forward, but on the sewer commissioners' terms. The projectors' recompense would be determined only after the work was complete, taken from "a moiety of the clear profit, above the rate that now the same [lands] are valued at," and all improvements would be "rated by the commissioners of the same county." Moreover, no lands already worth eight shillings or more per acre annually would be partitioned, whatever their improvement might be. The council hoped this determination would allow the drainage to "proceed to execution without further protraction, or loss of time," but in fact it handed the fenlanders a decisive victory.[65]

The sewer commissioners held the power to determine which lands were improved, by how much, and how many acres would be ceded to the projectors as a result. But more immediately, they had the power to ensure that the project never advanced to such a point; for the next step was to make a full valuation of all lands in the Great Level to determine which plots were already worth at least eight shillings per acre and were thus excluded from contributing the project. Such a task would require a thorough knowledge of the "lay of the land," quite literally—its productivity, annual value, and the degree of flooding to which it was prone, often stretching back over several years. And the only people in authority with the immensely detailed local knowledge required for such a survey were the commissioners of sewers; the council's ruling, in other words, enabled them to act as judges in their own cause.

Surveying flooded lands to determine their condition and value had been a routine function of the commissions since the Middle Ages, and even Clement Edmonds had relied heavily on their assistance. The obvious conflict of interest in this case was recognized, but no one else was so well equipped to undertake the task at hand. The king reminded the commissioners that Ayloffe and Thomas had shown "their zeal to us, and the weal public" by taking on "so matchless, and hazardous an enterprise," and he admonished them to be honest and fair in making their valuations: "we cannot but require, and expect square, and plain dealing, such and so legal as against which they may not make just exception."[66] The Privy Council added their own admonition, writing that "His Majesty doth expect that you entertain their readiness herein with all due, and correspondent assistance," and urging them to complete their survey as quickly as possible.[67] But there was little the Crown could do to compel a just, fair, and speedy valuation from the commissioners, who understood their advantage only too well.

The lowland commissioners soon met to organize the survey in their district. They decided that every parish in Cambridgeshire and the Isle of Ely should nominate "four able, & sufficient men," who would "truly view and particularly set down in a bill under their hands all the fens & surrounded grounds within their several parishes . . . & by what names they are known, & what quantity of acres they contain *& how much of every of them* (comunibus annis) *for these ten years past are and have been under the yearly value of viij^s the acre,* & how much under, truly rating them at the just worth as they will justify the same upon their oaths if they be required."[68] Along with whatever verbal cues may have been passed along, the commissioners embedded within their written instructions the key criterion that would render a parish's lands immune from depredation by the projectors: an annual value of at least eight shillings per acre. Nor was this helpful hint wasted on the parishes. The many surveys returned to the commission of sewers vary greatly in style; some are just a few lines long, while others delineate every plot of land in the parish over many pages. One curious trait many share in common, however, is a lower limit of eight shillings for the value of most of their lands. Of the nineteen parishes whose written responses survive, ten claimed that all or nearly all of their lands were already worth eight shillings or better.[69] While some plots would naturally be worth more than others, such uniformly high valuation strains credulity, especially when the remaining nine parishes reported a wide range of land values. The majority of land in Tydd St. Giles, for instance, was valued at two shillings or less; and Leverington (who submitted the longest and most detailed report) listed seven full pages of individual plots worth less than eight shillings per acre.[70]

Perhaps the most interesting survey is the one submitted by the village of Sutton, insofar as it suggests a local conflict having to do not so much with land values as with personal scruples. The original draft of the survey listed every single plot of land in the parish as being worth at least eight shillings. However, someone else apparently edited the document later on—in a different hand and in different ink—sharply lowering the values reported. Several entries were amended to put them below the eight-shilling threshold, and while many were still rated above that mark, they were reduced from even higher figures.[71] The original scribe may simply have committed an honest error of overvaluation, which a better-informed colleague later corrected. But it is far more likely that the editor was just a bit more conscientious regarding the information he was prepared to submit to the commission, over his signature and potentially under oath. If other parishes similarly overvalued their lands and were less troubled by the potential legal and moral consequences, it could help to explain the returns from villages such as Haddenham, whose surveyors simply reported, "All the fens belonging to Haddenham is worth 8s. an acre and better," with their signatures.[72]

For their part, the lowland commissioners of sewers accepted the valuations they received from each parish without question and made them the basis for all further dealing with Ayloffe and Thomas. On 28 September, they made their final offer to the projectors. More than half the land in Cambridgeshire had officially been valued at eight shillings per acre or more and would therefore be exempt from contributing to the project. The remaining lands would yield only "a moiety of the clear profit . . . above that they are now esteemed at." The commissioners of sewers would determine how much improvement had been realized, and only after a period of seven years, "for that it cannot be adjudged until wet years happen—the grounds in dry years being in as good cause as can be expected by any draining." If the projectors should refuse these terms, they were required to "desist from any further undertaking, & to hold the country no longer in suspense."[73]

Ayloffe and Thomas recognized that the proffered terms were "not such as they could imagine we would accept, but whereby they might compel us to desist," and they knew that they had been outmaneuvered. The commissioners' unrivaled local knowledge and experience, together with their new unity of purpose, made them more than a match for two inexperienced outsiders, even with the backing of the Crown. The projectors protested to the Privy Council that they had not received the "square, and plain dealing" the king had commanded but had been dealt with in bad faith at every turn. They knew very well from their own surveys and prior negotiations that a lot of land was being blatantly overvalued and alleged that the surveyors in this case had been willing to swear to "they cared not what." Indeed, the commissioners of sewers may also have been troubled by the moral implications of the surveys they received, fearing they might imperil the surveyors' immortal souls. But instead of demanding greater honesty, they simply "gave over to take any more oaths because (as some of themselves said) they would send no more to the devil." But be that as it may, the projectors still lacked the information they needed to contest specific land values, and the commissioners would not share what they knew: "Against those valuations we cannot make so particular exceptions because we have no copy of them . . . we are utterly out of hope to do any thing by their help."[74]

Ayloffe and Thomas did have one more, desperate gambit to try: they petitioned the Crown to ask that a new commission of sewers be issued "to the most indifferent commissioners of each county, to make just, true and particular valuations of all their fen grounds." These valuations, they suggested, should then be returned to the Court of Exchequer, to remain on public record. If the lands in question turned out to be of little value after all, then the drainage might go forward and "a competent proportion" might be set aside for the projectors in draining it. But if the lands really were "of such value as the commissioners now pretend," then Exchequer might see to it that the inhabitants were taxed accordingly thenceforth, to pay their fair share of taxes "which hitherto out of the Fens hath not been had."[75] In the meantime, they asked to have their case referred to the High Court of Parliament, where they anticipated a more indifferent and equitable judgment than they had received in the Fens.

The projectors' ploy to turn the fenlanders' exaggerated land values against them was ingenious but unsuccessful. The Crown did not grant their petition, and their case was never heard before Parliament. After the fall of 1620, the proposal of Sir William Ayloffe and Sir Anthony Thomas fades from the historical record, with not a single construction work ever begun. The two men departed the Fens deeply in debt, a burden from which Ayloffe, at least, never recovered.[76]

King James I's desire to see the Fens drained did not fade, however, and one year later he declared that "for the honor of this kingdom . . . [he] would not suffer any longer that said land to be abandoned to the will of the waters, nor to let it lie waste and unprofitable."[77] In July 1621, he issued another new commission of sewers for the Great Level, seeking to reignite the project with a more tractable group of commissioners.[78] The Privy Council informed the new commission that the king planned to take over the project as the sole undertaker and ordered them to assist in securing agreements with all of the landowners and commoners in the affected region. If they met with any "opposition and disturbance from such as are ill affected or otherwise value their own private occasions, above all public respects," they should either punish them to the limits of their own powers or report them to the council for further discipline.[79] James was determined to succeed where Ayloffe and Thomas had failed.

Yet even the king's personal investment in the project did not overawe the fenlanders. The new commission of sewers impanelled juries in August 1621 to conduct yet another detailed survey of lands to be drained throughout the Great Level, but the presentments they returned made it clear that fenland political culture would not yield to pressure from the Crown. The Northamptonshire jury, for example, denied any need for new drainage works. They insisted that if local landowners would only live up to their customary responsibilities, then "every man shall be undertaker of his own work. . . . And this great business shall be brought to a speedy and happy end, without one penny cost unto his gracious Majesty."[80] Other juries claimed that "we have no fenny or surrounded grounds," or that "the said grounds receive more benefit that hurt" from the flooding and submitted annual land values ranging between twelve and twenty shillings per acre annually. Some jurors went so far as to suggest that their lands had actually decreased in value in recent years because various drainage efforts had diminished their grass yields.[81]

The king was undaunted; he remained determined to bring about a full, regional drainage of the Great Level by whatever means necessary. "We undertake [the drainage] wholly, and not by fractions of little particulars," he wrote to the commissioners, "neither [do we] intend to tie ourselves to old drains only, but to make use either of new cuts, bankings, or any other means, or engines, whatsoever, for the effecting of the work, where it shall require."[82] He laid claim to a total of 120,000 acres in recompense for his efforts, roughly one-third of the entire Great Level, and he warned the commissioners that they had better deal with him honestly in allocating the lands to be transferred: "For the particulars . . . you will remember, that it is your king, whom you treat with . . . so he expects that . . . you shall proceed in such manner, as may be worthy of his acceptation. . . . [W]e shall observe, if any thing should be done to the contrary, though obliquely."[83] James obviously had no intention of overseeing the drainage himself but sought out yet another group of projectors who would subcontract for a share of the royal recompense. In February 1622, the Crown reached a preliminary agreement with two Dutch drainage engineers, Cornelius Liens and Cornelius Vermuyden, who proposed to drain the Great Level at their own expense in return for seventy thousand acres out of the king's total share.[84] The bargain came to nothing, however; as Vermuyden recounted twenty years later, "His Majesty's great occasions and the time would not permit so great and good a resolution but it was deferred."[85]

The condition of the Fens continued to deteriorate during this time, owing in part to the king's inaction. The commission of sewers admitted that little had been done even to maintain the existing drains, "every man neglecting as well their public as private duties therein, occasioned (as it is conceived) principally through the expectation of some happy success of His Majesty's undertaking." The region's many rivers, sewers, and drains were normally cleansed at least three times every summer, but no one wanted to invest money or effort to repair insufficient drainage works that would soon be rendered obsolete, and the ongoing neglect caused them to become choked up with vegetation. The result was a severe flood in the summer of 1622 that covered both the lowland and upland fens, with the waters remaining on the land well into the summer and causing great damage. The commissioners wanted to take "a stricter course" in enforcing existing sewer laws, but their efforts could only be provisional and temporary "until His Majesty shall be pleased to go in hand with this great work, and for the better advancement of the same."[86] As late as May 1623, the Privy Council wrote that the king still intended "to undergo the said work with all convenient speed" and charged the commissioners with making sure their surveys were complete and up to date; but the letter itself is a tacit admission that yet another year had passed with no progress made.[87]

James died in 1625 with the drainage still not begun. William Dugdale speculated that the king's project may have been a casualty of "the great disturbance he had about that time, and after till the end of his reign, for regaining the Palatinate," but whatever the reason, "Certain it is, that no farther progress was made therein; nor anything else conducing thereto." The first (partially) successful attempt to drain the Great Level had to wait until the 1630s, years after the succession of James's son, Charles I.[88]

✑

After the Crown took stern action in 1616 to reaffirm the broad powers of the commissions of sewers and shield them from suits at common law, the Privy Council was determined to engage more actively with the Fens, hoping to see an improved regional drainage network created there at last. The councillors' approach reflected the needs and methods of a burgeoning early modern state. They began by reorganizing the many local commissions of sewers into a single, regional body representing the entire Great Level and then sought to resolve the lingering disputes between them by compiling more and better information about the Fens and the drainage problems that plagued them. They got their information from a variety of surveys: from the comprehensive, coordinated jury presentments undertaken in 1617; from the personal views taken by various commissioners of sewers; and from the firsthand reports of Richard Atkins and even their own secretary, Clement Edmonds. But however hard it was to survey hundreds of square miles of flooded fenland, noting the precise condition and value of thousands of acres and the many defects of each river and drain, gathering information was really the easy part, and the council could not progress beyond this initial step. For each survey told them much the same thing: there was no agreement in the Fens about how to drain the land, who should pay for it, or even whether it ought to be attempted. Compiling a detailed catalog of the problems did not generate solutions for them, and those with personal knowledge of the Great Level, such as Edmonds, despaired of ever reaching a consensus there.

It was Edmonds who suggested, as Humphrey Bradley had almost thirty years earlier, that if the Fens were ever to be drained, it would have to be done not by the fenlanders themselves, but by an outsider. The undertaking would require someone with "no interest at all in the country, [who] standing indifferent betwixt each party, and having no other end but the common benefit, may proceed according to the rule of justice & reason." But what could that possibly mean, in the real world of the Great Level? Such an enormous and capital-intensive project could be no work of public-minded philanthropy. Edmonds probably had the Crown in mind as chief undertaker, a large landowner in the region that in principle could not but act in the best interests of the commonwealth, and he therefore urged the council to "take it into your honorable care" so that it would "prove a work of great honor to His Majesty, & good to the public."[89] But the Crown was also in dire financial straits by the late 1610s and in no position to take on such a massive project—or to do so without an eye toward maximizing the revenues to be gained by it.

Enter the projectors, Sir William Ayloffe and Sir Anthony Thomas, offering an ideal solution from the Crown's perspective. The drainage would be undertaken by (allegedly) capable outsiders who were indifferent to the many ongoing local disputes besetting the Fens and who would do the work at their own charge, with any reward to come only from the drained lands themselves. Such a project would cost the Crown nothing except a reasonable share of the improved royal landholdings, and if successful would yield a tidy profit not only for the projectors, but for the fenland inhabitants and the royal treasury as well. Moreover, the drainage would finally put a stop to the endless squabbling and infighting among the commissioners of sewers that had so taxed the patience of the Privy Council—it would mean "the taking away of quarrels about draining, whereof the board is weary."[90] In this, at least, Ayloffe and Thomas were ironically successful; the projectors did manage to forge a fenland consensus at last, but only by uniting the Great Level commissioners in opposition to their project, a political obstacle they were unable to surmount.

The drainage project of Ayloffe and Thomas is certainly not worthy of note based on the works they accomplished or the increased value they created in the Great Level—given that ground was never broken, they had even less to show for their time and trouble than Sir John Popham had. The importance of the affair lies in the reasons for their failure to get their project under way, despite having the explicit support of the Crown. One reason was certainly their inexperience and manifest incompetence, their "want of knowledge in the nature of the fens . . . [and] their weakness in the science of those things which they undertake," as the anonymous observer at Peterborough put it. But the projectors were also hampered by the same shortcoming that had bedeviled the Privy Council for so long: their inability to acquire and control vital local information. The same writer who criticized the projectors' ignorance was quick to contrast it with his own hard-won knowledge and experience: ". . . [F]or such fen grounds as lie in our parts . . . I dare profess knowledge as being both native near unto them, and all my lifetime I have traveled through them both observing the nature of every soil, what effects the inundation of fresh waters from time to time do work, what the sea doth in her courses, besides mine own experience in those kind of works, which did tend to the making, and maintaining of those grounds aforesaid depasturable for the summer time (for more cannot be expected, or desired)."[91]

Maintaining an accurate and comprehensive local knowledge of fenland drainage matters was a principal raison d'être for the commissions of sewers. The ability to command that knowledge was closely tied to the question of authority in the Fens. Who had the right, the duty, and the means to survey the lands in question? To determine not simply whether they were flooded, but *hurt-fully* flooded? To determine what the land was worth, and what if anything could be done to better it? The capacity to answer these questions with authority carried with it the power to define the very status of the land and to determine whether its inhabitants stood to gain anything by surrendering a portion of it to outsiders who aimed to "improve" it. In 1620, this authority lay not with the king, the Privy Council, or the projectors they recommended, but with the fenland commissioners of sewers. Their local knowledge was still indispensable in managing the Fens, and the Crown was forced to rely on their judgment. With so much at stake, for themselves and so many of their neighbors, the commissioners carried no small burden: "We look that all public matters which concern our country should be wielded by our own commissioners, and on their wisdom and care we rely. It behooveth therefore that they be very circumspect in this action for upon them in this matter will lie either the blessing, or the curse of the people because in their hands is managed the authority of this work."[92]

The fact that the sewer commissioners were also interested parties in this case made little difference. They were suffered to act as judges in their own cause because no one else had the knowledge and experience to render a definitive verdict, or to gainsay theirs. For all of the Crown's efforts to collect information in and about the Fens, it still did not *control* that information. The king might command the commissioners to use "square, and plain dealing" with his chosen projectors, but the early modern state still relied on the willing cooperation of local officials and elites, who served their own communities at least as much as they served the Crown.

The project of Ayloffe and Thomas thus represents a failed attempt at state building in Jacobean England. It serves as an example of the ways in which early modern authority and power came not only from royal favor or from one's social and economic status, but also from a command of valuable local knowledge and experience. The king's drainage projectors were outmaneuvered by his commissioners of sewers, leaving the Fens undrained and entirely in the possession of the fenlanders. No large-scale drainage project could proceed there until the Crown either won the cooperation of the commissioners or secured an independent source of expertise by which their authority might be superseded. By the end of the 1620s, both of these changes had taken place; how this dramatic shift in fenland affairs came about is the subject of the next chapter.

PART II

Drainage Projects, Violent Resistance, and State Building

CHAPTER 5

Draining the Hatfield Level, 1625–1636

But whilst this great projector [Cornelius Vermuyden] thought
himself secure of His Majesty's favor . . . he found himself mightily
annoyed by the gnats and flies, that is the common sort of the
inhabitants that set upon him when he should rest. . . . [T]hey
disturb'd him in his works, and when that would not do, in great
numbers they burnt his carts, and barrows, and working instruments,
in great heaps by night.

—Anonymous, *The State of That Part of Yorkshire*, 1701

In the early afternoon on Wednesday, 13 August 1628, an English laborer named
John Kitchen was walking to the village of Haxey in Lincolnshire. Kitchen
worked for Cornelius Vermuyden (1590–1677), the Dutch land drainer and projector commissioned by King Charles I to drain the fenlands that spanned several royal estates in the area, including the manor of Epworth where Haxey was
located. Vermuyden had pledged to transform the fens there, known collectively
as the Hatfield Level, into land "fit for tillage or pasture."[1] Though ostensibly
undertaken for their benefit, the drainage project had never been popular among
the local inhabitants and had lately provoked considerable unrest. Kitchen later
testified under oath to Robert Portington, a justice of the peace in neighboring
Yorkshire, that as he approached Haxey, he met an acquaintance named Vincent
Taylor, who warned him that if he dared to proceed any further "the women of
the town would stone him to death." Kitchen did not heed Taylor's warning, but
continued on into Haxey where he was "then & there cruelly stoned by a company of women, he the said Taylor animating them thereunto."[2]

The violent assault on Kitchen was not an isolated incident. More than a
dozen English and Dutch laborers involved with the Hatfield Level drainage
testified to Portington that over the previous few weeks they had been "cruelly
beaten and sore wounded," "sore beaten & hurt by throwing of stones & other

things," "[thrown] into the River of Vickars dyke [Bickersdike]," and "cruelly threatened to be burnt by the multitude."[3] Most of the assaults took place during a series of riots on Wednesday and Friday, 13 and 15 August, in which at least three hundred of the local inhabitants took part, according to the victims' testimony. The rioters included both men and women; the men favored clubs and pitchforks as their weapons of choice, while the women mostly threw stones at the victims.[4] The crowd was socially diverse as well: William Thompson, who was beaten and stoned by the rioters, testified that there were "in the said company both horsemen & footmen to the number of 5 hundred at the least," implying that the crowd may have included some of the "better sort" among their ranks.[5]

Given that Vermuyden's drainage project had provoked the riots, it is not surprising that some rioters were heard to exclaim that "if Mr. Vermuyden had been there they would tear him in pieces," and even a local bailiff allegedly declared that "if Mr. Vermuyden came to the work, he would desire but to have one blow, if he lost his life for the same."[6] But it is perhaps more surprising that so many explicitly identified King Charles as the main source and target of their discontent. The Epworth rioters were well aware that the king had personally concluded the agreement with Vermuyden to drain their lands, and they aggressively challenged his right to do so, "shaking their pitchforks & other weapons." Several were heard to declare that "they cared not for the king for their lands were their own & he had nothing to do with them"; "if the king did send 1000 men he should have no common there, and if the king himself came thither he should have nothing to do there"; "they neither cared for God nor the king"; and "if the king was there they would kill him."[7]

Vermuyden and his fellow Dutchmen were not intimidated by the violence. After their beating on Wednesday afternoon, they took Thursday off to regroup, lick their wounds, and better prepare themselves for subsequent confrontations. Vermuyden gave firm orders that the crew should continue their work regardless of local unrest and sent a contingent of reinforcements and a variety of arms to protect them. When the Dutch and English laborers and their overseers returned to Haxey Carr on Friday morning to begin damming the River Idle, they were well armed, carrying between them three muskets, one firelock, two pistols, a quarter-pike staff, two pitchforks, and seven swords. Their work proceeded apace until they were once again confronted by a large group of inhabitants from Haxey. As the crowd approached, the overseers ordered all of their men who were working on the far side of the river to cross over to the near side, to meet the mob and present a united front. They were concerned lest any of their party should be left isolated and beaten once again, or worse; the leader of the work crew, a

"Mr. Laynes," told his men to "Come, come, make haste & let us get over the river to save our men, or else they will be killed."[8]

What happened next is difficult to determine, as the laborers' and inhabitants' testimonies contradict one another. According to the laborers' account, three of the local inhabitants (Hezekiah Browne, Richard Taylor, and John Newland) approached the Dutch overseers for a parley as they crossed the river; the Dutch fired one shot in response, to warn them not to come any closer. The three men stood their ground, however, and began to parley with Mr. Laynes once he and his crew reached the shore. The rest of the armed Dutchmen then moved toward the main body of the Haxey crowd, who were still several hundred yards away. Their intent was "to dissuade the people from interrupting of their work or beating of their men (as they had done formerly) & to keep them back until Mr. Laynes & the said Browne & the other two of his side had talked together."[9] The inhabitants' account describes a much more belligerent encounter; they denied that any parley ever took place and claimed that the Dutch fired on them in earnest even as they approached, causing the crowd to flee. The Dutchmen, they alleged, "did pursue them vehemently" and continued to fire, "saying they would send them home singing with bullets in their tails," though the Dutch denied this allegation.[10] However it started, both sides agreed that when "the affray began betwixt them," several shots were fired by the Dutch party, and an Englishman, one Robert Coggin of Haxey, was killed.[11]

News of the incident soon reached the royal court. On 21 August, Sir Ralph Hansby, a prominent local man, sent the king's chief adviser, the Duke of Buckingham, copies of all of the witnesses' testimonies recorded by Robert Portington up to that point. "[I]t is clear," Hansby wrote, "that great riots hath been committed by the people, and a man slain by the Dutch party, the killing of whom (as I conceive) is in all that were present or gave direction for them to go so armed . . . that day for that purpose, murder."[12] Yet when the Privy Council wrote to the Lincolnshire justices of the peace, their priority was not to investigate the circumstances of Coggin's death but only to reestablish order so that the drainage work might proceed as rapidly as possible. The councilors expressed their dismay at "the great disorders and riots committed lately in that country . . . in grievous wounding and beating of the workmen," and admonished the justices to "have a care that His Majesty's peace be kept there, and to deal with the country, that *no interruption be given to the proceeding of the works.*" The justices were to reassure the restive inhabitants that "free way shall be given to any just or legal complaint on their side, but *in the meantime the works, which only tend to the general good, may not be interrupted.*" The letter made no mention of Coggin's killing but commanded instead that the Dutchmen's progress *"may not be interrupted by vexatious warrants procured against them* upon pretense of these riots." They also sent a warrant for eighteen of the inhabitants who had participated in the riots to be bound to appear in person before the council, brought up on charges, or committed to jail, "as in your discretions you shall find most fitting for the peace to be kept *and the works to be proceeded in.*"[13] If only through sheer repetition, the Crown's fervent interest in promoting the drainage project above all else could not be more clearly expressed.

The Hatfield Level drainage, undertaken during the late 1620s, was at that time the largest such project yet attempted in England. The Hatfield Level was a peat fen like the much larger Great Level to the southeast, though the two areas were not contiguous, they were not drained at the same time, and the circumstances of each project (e.g., land tenure and rights of common) were quite different from one another. Both levels were similar in their geography, ecology, and economy, however, and the drainage projects undertaken in each shared important connections, so that the Great Level drainage cannot be fully understood without reference to the earlier project at Hatfield.

One key link between the two projects was the individual ultimately responsible for directing each of them. Cornelius Vermuyden, the Dutch drainage engineer, was not only the architect of the Hatfield drainage, he also designed and oversaw the Great Level project during the early 1650s.[14] His approach to fen drainage was consistent, and much of what he later set out to do in the Great Level, he attempted first at Hatfield. Vermuyden's involvement in the Hatfield project was pivotal, because his prior knowledge, experience, and skill in land drainage gave the Crown an independent source of expertise outside the local commissions of sewers. Queen Elizabeth and King James I, and their respective privy councilors, had sought to work through local officials whose interests, as fen landowners, did not always align with the Crown's. James had also tried to introduce outside projectors, but they had lacked the knowledge of drainage affairs needed to circumvent the commissioners' authority or overcome their opposition. Vermuyden, and the Dutch investors who backed his scheme, knew more about large-scale drainage projects than anyone in England, and they had the experience, the confidence, and the liquidity to carry them out. Vermuyden gave Charles access to the expertise he needed to make draining the Fens a reality, and his service to the Crown helped to undermine the power of the sewer commissions.

King Charles I's ascension to the throne in 1625 was another important factor in driving large-scale drainage projects forward at last, as opposed to the stillborn efforts of his father's reign. While James and his privy councilors had tried to mediate resolutions to local disputes and to forge consensus around a new, regional drainage scheme, Charles's style of governance was markedly different. During his reign, the Crown was much more determined to proceed with drainage projects despite any local opposition, and the quest for consensus was soon overborne by the royal demand for progress. Charles viewed drainage projects as an undoubted public benefit, as well as a way to enhance revenues from royal landholdings, and anyone who resisted them was liable to be treated as a self-interested adversary of the good of both Crown and commonwealth.[15] It was no coincidence that the Hatfield Level drainage and all of the other large-scale projects that soon followed it were undertaken after 1625.

But Charles's willingness to pursue fen drainage with little concern for securing local approval or consensus also provoked a much more heated level of fenland opposition, expressed through both legal and violent means. Seventeenth-century resistance to drainage projects followed a pattern common to other popular protests against enclosure movements in early modern England. Protestors used petitions, lawsuits, and riots (usually, but not always, attacking property rather than people) as complementary means of engaging with both local elites and the institutions of the state to seek some redress of their grievances. Historians have sometimes characterized such protests as "nonideological" or "prepolitical" behavior, arguing that they sought only to correct local injustices and did not engage with the broader political or socioeconomic crises that roiled England during the seventeenth century.[16] Keith Lindley, in particular, has written that anti-drainage riots were "essentially defensive, conservative and restrained in character; the fenmen were defending their traditional economy against innovation and . . . were not striving to turn the world upside-down." Fenland protesters and rioters "did not give expression to political feelings, but contented themselves with drawing attention to specific grievances of immediate concern while in most other respects observing their traditional place and obedience."[17]

This apolitical understanding of fenland anti-drainage protests, both legal and violent, is shortsighted. Other social and political historians have lately taken a broader and more fruitful view of what "politics" might mean in early modern England, giving much greater attention to local power relations at the level of the village, manor, parish, and household.[18] Viewed in this light, the myriad disputes and power struggles among commissioners of sewers, landowners, and commoners must be understood as part of a multifaceted and dynamic fenland political culture. Moreover, drainage projects gave rise to extended and often contentious interactions between fenland inhabitants and officials on the one hand, and the central governing and legal institutions of the state on the other. Tim Harris has argued that "[t]he pre-eminent instrument of politicization . . . is the government and what it is doing to people," and this was certainly true with respect to Crown-sponsored fenland drainage projects.[19] Insofar as the Crown promoted, pursued, and justified such projects as being for the good of both the Fens and the commonwealth (and also as a means to boost royal revenues), they were fundamentally political in nature, and so was the opposition they provoked.

Whenever anti-drainage protestors petitioned the Crown or Parliament, landowners brought anti-drainage suits in the central law courts, or anti-drainage rioters were made to appear before the Court of Star Chamber, fenland inhabitants engaged directly with the state. Their concerns were undeniably provincial and their goals conservative, usually expressed in terms of their customary rights and practices—they were, admittedly, not trying to turn the world upside down. But as Andy Wood has shown with respect to free miners in the Peak District, this did not mean that they were unaware of the broader political context or that they lacked a political culture of their own through which they could articulate their interests as against those of the state. Like the free miners, fenland commoners used their understanding of customary practice and established sewer law to oppose an alternative legal framework that emphasized private property and the landlord's right to improve, enclose, and exclusively exploit it. They were also capable of interpreting their customary rights conservatively or dynamically as necessary, to build a more favorable case for themselves.[20]

Moreover, while potentially hostile magistrates' descriptions of seditious language and coordinated attacks on the part of fenland rioters should not be accepted without question, neither should they be dismissed that way. It is not unreasonable to believe that anti-drainage rioters might aggressively challenge the king's legal right to drain and enclose their commons, especially when the king was himself the lead drainage projector, as at Hatfield.[21] In many cases the local magistrates were not even hostile to the rioters; some riots were allegedly instigated by the "better sort," and parish constables and even justices of the peace were known to have taken part in the violence. Finally, the anti-drainage protests at Hatfield Level must be understood in the context of an increasingly fraught political situation at the national level, as king and Parliament were continually at odds over a variety of issues having to do with Charles's arbitrary use of royal prerogative powers.

Fenland drainage projects and the resistance they provoked must therefore be viewed as manifestations of early modern state building in England. In some cases, anti-drainage protestors acted as voluntary participants in state formation, pursuing their claims against drainage projectors through petitions and lawsuits and thereby reinforcing the authority of state institutions. At other times, fenlanders violently resisted the Crown's interference in their affairs through anti-drainage riots, and in such cases the Crown did not hesitate to use the full coercive power of the state against them. Charles I and his Privy Council could be sympathetic to fenlanders' complaints of ill treatment, as long as they were expressed through acceptable channels. When confronting violent riots or explicit challenges to the king's rights as lord of the soil, however, the Crown responded aggressively. This sometimes put local officials in a bind, forced to choose between protecting the interests of their families, friends, and neighbors against greedy foreign projectors and obeying the Crown's demands that they support the drainage projects at all costs for the putative good of the commonwealth.

THE HATFIELD LEVEL AND CORNELIUS VERMUYDEN

The Hatfield Level was a peat fen of some seventy thousand acres (about twenty-eight thousand hectares). It consisted of two main parts: Hatfield Chase in the West Riding of Yorkshire, and the Isle of Axholme in northwestern Lincolnshire, together with a small strip of fenland at the northern tip of Nottinghamshire. In 1625, it lay within the overlapping floodplain of several interconnected rivers, including the Don, Torne, Idle, Bickersdike, Trent, Aire, and Ouse, all of which eventually drained into the Humber and thence past Hull to the North Sea.[22] As in the Great Level, the topography is very flat, so the rivers meander slowly through it and tend to silt up, a trait exacerbated by the fact that the Trent, Ouse, and Humber are all tidal rivers (fig. 5.1). The region was prone to serious flooding each winter, and during an especially wet year, in the summer as well. William Dugdale recorded that even heavily laden boats could sail right across the countryside during serious floods, when the land might be under several feet of water.[23] Geographically the region was also quite similar to the Great Level, and the inhabitants exploited the land in much the same way: grazing livestock on the vast common wastes, harvesting the natural wetland produce, and planting cereal crops on the few elevated "isles." In Hatfield Chase, they might also take advantage of the little-used royal hunting ground to poach deer and other game. Inhabitants made full use of their right to the common wastes, and there is evidence that the land available for pasture was becoming insufficient for the growing population of the area, as poorer cottagers migrated to the region precisely because of the easy access to common.[24]

The Hatfield Level differed from the Great Level in one critical circumstance: the four principal manors that made up the region—Hatfield, Epworth,

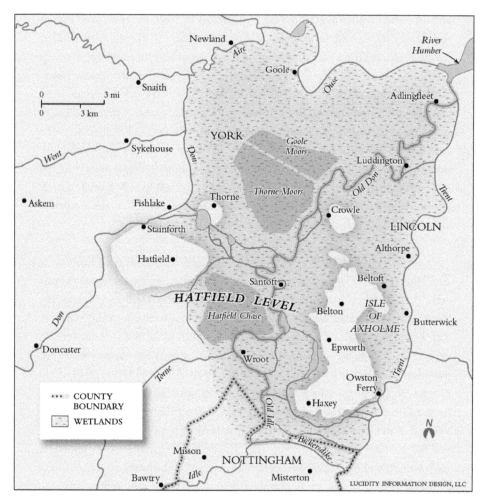

Figure 5.1. A map of the pre-drainage Hatfield Level.
Adapted from a map in Korthals-Altes, *Sir Cornelius Vermuyden.*

Crowle, and Misterton—together with another thirteen adjacent manors, were all royal landholdings and so directly subject to the king as landlord.[25] This greatly simplified matters for a would-be drainage projector; while customary common use rights were still a potential complicating factor, a projector at Hatfield had only to reach an agreement with a single landlord, the king, in order to proceed. This eliminated the need to work through a commission of sewers, whose main responsibility in promoting fenland drainage projects to this point had been to use their broad powers to persuade or compel landowners and commoners to cooperate with the projectors. Sewer commissioners in the Hatfield

Level, by contrast, played only a minimal role in the drainage and had far less opportunity to impact the proceedings, either positively or negatively.

Royal interest in draining the Hatfield Level probably originated with James I. In 1617, the Crown ordered "an exact survey" to be made of Hatfield Chase, to determine the quantity and quality of land there.[26] In 1622, according to the eighteenth-century antiquary Abraham de la Pryme, a group of local gentry were commissioned "to consider about the drainage, improving, & disafforestation" of the area. The commissioners responded that "considering how great the levels were, & how continually deep with water, how many rivers ran thereinto, & such like they did humbly conceive that it was impossible to drain & improve them."[27]

This was not the last word on the matter, however. The king's need for revenue was ever increasing, and after James's experience in the Great Level, with Ayloffe and Thomas, he had perhaps learned to regard the opinions of fenland inhabitants concerning drainage matters with some skepticism. In any case, he continued to insist on "disafforesting the said Chase, that by the improvement thereof, an increase might be made to his highness's yearly revenue." By 1624, the commoners there had expressed "their loyal submission and approbation to His Majesty's pleasure" and asked that "those parts of the said Chase lying subject to surrounding . . . be drained & freed from the annoyance of water, at his highness's charge." The lord treasurer and the chancellor of the Exchequer then "appointed one Mr. Vermuyden, a man well experienced in draining, to view the said annoyance." Vermuyden deemed the drainage "to be feasible," and "did propound to undertake the said work at his own charge," in return for a portion of the drained lands.[28] This was not long after James had reached a preliminary agreement with Cornelius Vermuyden and Cornelius Liens to drain the entire Great Level in return for seventy thousand acres; but as in the latter case, James's interest was not sustained, and the matter was deferred until after his death.[29]

At this point it is appropriate to examine the origins and early career of the individual whose credentials had so impressed King James I that he had already awarded him the two largest drainage projects in early modern England. Cornelius Vermuyden was born in or around the year 1590 in the town of Sint-Maartensdijk, located on the Isle of Tholen in the province of Zeeland, near Bergen op Zoom in what is now the Netherlands, the son of Gillis and Sara Vermuyden.[30] The Vermuydens were a prominent family in Sint-Maartensdijk: Bartel Vermuyden (grandfather of Cornelius) was a city alderman in 1570, while Gillis served as high sheriff in 1591. They were also drainage engineers and had been involved in land drainage projects in the region since at least the early fourteenth century. Cornelius's mother, Sara, was the daughter of Cornelius

Werckendet, a local magistrate and drainage engineer in nearby Zierikzee; both he and Sara's brother Lieven Werckendet were active in local land drainage projects and in the construction of the Zierikzee harbor.[31] Vermuyden was thus connected on both sides of his family with generations of Dutch land drainers and was certainly exposed to the rudiments of drainage practice and theory through his relations. He is believed to have worked on drainage projects himself in Brabant, Beveland, and Flanders during the 1610s, and he wrote a book in Dutch on the subject in 1615.

Vermuyden was also related through his sister's marriage to the prominent Liens family, whose background was similar to his own. It was probably through his brother-in-law Joachim Liens that Vermuyden was first brought to England in 1621 or 1622. Liens was part of a Dutch diplomatic delegation sent to England in 1618 to negotiate several issues of contention between England and the Low Countries. He and his colleagues made a good impression at court and were knighted by the king in July 1619. Liens would have been at the English royal court when Sir Clement Edmonds made his report to the Privy Council on the Great Level and throughout the contentious undertaking of Ayloffe and Thomas. When James declared his intent to take over the Great Level drainage himself, in July 1621, he may well have consulted with Liens about it; certainly it can be no coincidence that in February 1622 he commissioned Joachim's brother Cornelius Liens and his brother-in-law Cornelius Vermuyden to design and undertake the work.[32]

When James's pledge to drain the Great Level came to naught, Vermuyden found himself in England with nothing to do, though he soon found other employment. He worked for a time with his father-in-law, Joos Croppenburgh, a Dutch merchant and drainage projector residing in London who had contracted to reclaim some 490 acres of flooded land in the Erith Marshes, beside the River Thames, in the summer of 1622. At roughly the same time, Vermuyden also managed to secure another drainage commission, this time in his own name. The levels of Havering and Dagenham, located along the River Thames in Essex, had been badly flooded when an "outrageous tide" broke through their protective banks in September 1621. The commissioners of sewers there had already contracted with an Englishman named Standon to repair them, but Vermuyden submitted his own proposal offering more favorable terms. The commissioners dismissed Standon (absorbing a significant loss on the funds already disbursed to him) and contracted with Vermuyden to repair the breached riverbanks on 23 March 1622. But Vermuyden badly underestimated both the scale and the cost of the work that needed to be done at Havering and Dagenham, and he seems to have botched the entire affair. Though he claimed to have spent at least £3,600 of his own money on the project, the works he built were inadequate, and his workmen complained that their wages were unpaid.[33]

The Essex sewer commissioners complained to the Privy Council that Vermuyden's works were "discovered to be neither permanent nor durable." They also castigated his approach in effecting the repairs: rather than merely fixing the original breached banks, Vermuyden tore them down instead and proposed to build new, artificial ones several meters distant from the river to replace them. This strategy, which Vermuyden would later employ both at Hatfield and in the Great Level, was meant to give the river room to swell before flooding the country, but it sacrificed hundreds of acres of once-valued land to the flood zone. Essex landowners resented Vermuyden's "leaving out to the Thames great quantity of good land," as well as his expensive construction of an entirely new riverbank. In short, the Essex commissioners complained bitterly of Vermuyden's "not performance of his promise and agreement," pointed out that he still owed great sums of money to local laborers and tradesmen, and lamented that "by his ignorance in workmanship the levels are now in worse state by many hundred pounds than when Vermuyden first undertook them."[34] The Privy Council urged some sort of compromise between Vermuyden and the aggrieved landowners, but the commissioners reported that Vermuyden's demands were "so unreasonable," and the landowners' losses "so great and grievous to them" that no resolution was possible.[35]

Despite his apparently disastrous failure in Essex, however, James continued to look with favor on Vermuyden. He was commissioned to drain the royal park at Windsor in 1623, and in the following year a contract to drain the Hatfield Level was awarded to "one Mr. Vermuyden, a man well experienced in draining."[36] Nor had the Essex landowners heard the last of him. In August 1625, just a few months after Charles's ascension to the throne, the Dutchman received royal letters patent granting him tracts of land in Havering and Dagenham in recompense for his efforts to reclaim them from the Thames. The patent said that Vermuyden had completed the work as contracted in 1623 but had not received his due reward. The land in question was obtained in the usual manner: an unpayable sewer tax was levied, and the land was then confiscated by the sewer commission and granted to Vermuyden in perpetuity.[37]

The abrupt shift in the Crown's handling of the dispute between Vermuyden and the Essex landowners is indicative of the broader change in the style of royal governance after Charles came to the throne. James's Privy Council had encouraged negotiation and sought to broker a compromise resolution, and when this effort failed, the king compensated Vermuyden instead through his own continued patronage. Under Charles, however, the Essex men were brought to heel; they were made to reward Vermuyden for his (dubious) service to their country, and to do so not with a one-time cash payment but by sacrificing still more of their lands to him forever. While James preferred in practice to rule through mediation and consensus, despite the absolutist rhetoric he sometimes favored, Charles was much more aggressive in promoting the interests of his favorites, and he made sure that his subjects understood that the royal will would brook no opposition.

THE HATFIELD LEVEL PROJECT

Not long after the resolution of the Essex dispute in Vermuyden's favor, Charles gave him a further sign of his esteem by awarding him a new contract to drain the Hatfield Level. The project would take place on several contiguous manors in the area, most of which were owned by the Crown and so subject to the king's feudal rights as landlord. The preamble of the agreement characterized the region as "diverse wastes, waste grounds and commons . . . subject to be surrounded and drowned with water in such manner that little or no benefit is or can be made thereof." It also emphasized the king's concern for "the good and welfare of his subjects inhabiting near or about the places aforesaid," and his desire that the land should be "laid dry and made useful" for their benefit.[38] William Dugdale echoed the agreement in his account of the Hatfield project, lauding Charles's "royal and princely care for the public good, in regaining so great a proportion of surrounded land; which, at the best, yielded little or no profit to the common wealth, but contrariwise nourished beggars and idle persons."[39]

Charles may well have had a genuine wish to see his fenland subjects prosper on improved lands, but the Crown's primary interest in the project was to enhance the rent rolls of royal estates at a time when the treasury was in dire need of new revenues. De la Pryme stated flatly that the king, "being reduced to great straits for want of money, was resolved to endeavor the drainage & improvement thereof."[40] And even Dugdale, who was an ardent royalist during the Civil War, acknowledged that Charles undertook the project "partly for the easing of his charge, and increase of his revenue."[41] The potential profits were substantial; attorney general Robert Heath, the project's greatest advocate at court, advised that the drainage should yield the Crown at least £6,000 annually in increased rents, in addition to "the great good to the kingdom" that would result.[42]

Vermuyden agreed that he and his partners would "drain and lay dry the said drowned and surrounded grounds in such manner as to make the same fit for tillage or pasture," and would maintain the drainage in perpetuity. He was entitled to make use of any existing waterways and to make new ones wherever he saw fit, even through lands not previously subject to flooding. He was also permitted to reserve a corridor along each waterway, "for receptacles of the sudden downfalls of waters," so long as these encompassed no more than three thousand acres and landowners were compensated for their lands lost. In other words, he was allowed to employ his preferred tactic of building his river banks well back from the rivers themselves, creating "washes" to absorb surging floodwaters. Once the drainage was completed, a corporation was to be created (members of which would be nominated by Vermuyden), with sufficient lands set aside to pay for the continued maintenance of the drainage works. In return for his service, Vermuyden and his partners would receive one-third of all the lands drained—a total of just more than 24,400 acres—to be held by them and their heirs in free and common soccage. Compared with subsequent drainage projects, the terms of the Hatfield Level agreement were rather vague and permissive: in particular, there was no specified time limit for completing the work nor any penalties for failing to do so.[43]

Charles agreed to sponsor the project "in his own name," and that he would "from time to time direct that assistance shall be given to the said Cornelius Vermuyden and his partners . . . as need shall require." In return for his role in promoting such a massive and expensive improvement, the king planned to enclose another one-third share of the drained lands for his own use and profit. This would leave only the remaining one-third share for all of the local inhabitants to use in common (though it would in theory be a much-improved common). The contract explicitly acknowledged that some inhabitants of the level "do claim common of pasture in some part of the said lands," rights that the king as landlord was bound to acknowledge. Charles therefore pledged to create a special commission "to treat, deal, agree, and conclude" bargains with all of those whose common use rights would be infringed on, and construction work was not supposed to begin until these claims were resolved.[44]

During the negotiations over compensating the Hatfield Level commoners for the loss of two-thirds of their commons, Vermuyden returned to the Low Countries to procure the laborers, materials, and (most important) the investors he would need to get the project under way.[45] He had little trouble finding prospective investors among his countrymen; within a few months he had recruited thirty-four partners into the venture. The original investors, who were generally referred to as "the Participants," were mostly Dutchmen from wealthy and noble families; besides Charles I, only a small handful of Englishmen were initially involved in the scheme (though many bought into it afterward). A few of the Dutch Participants even came to England to assist Vermuyden and take up residence on their allotment of land, but most never laid eyes on the Hatfield Level. They remained in the Low Countries and trusted Vermuyden to design and oversee the works.[46]

In addition to Dutch money, Vermuyden returned to England with so many Dutch laborers and such a store of supplies that one commentator referred to the assemblage as "a Navy of Tarshish."[47] Some of the laborers and overseers chose to settle permanently in the area after their work was complete, as tenants of the Dutch investors who had funded the venture. They were soon joined by more than "two hundred families of French and Walloon Protestants (fled out of their native country for fear of the Inquisition, only to enjoy the free exercise of their religion here)."[48] Together these newcomers, most of whom were already familiar with various crops and farming techniques suitable for newly drained lands, would eventually settle numerous farms and build a sizable community at Sandtoft in the Isle of Axholme. They even had their own reformed church, with permission to hold services in Dutch or French.[49] Given the high degree of Dutch and Flemish involvement in the project, as investors, overseers, laborers, and tenant farmers, L. E. Harris has suggested that the Hatfield Level drainage is best understood not as an English venture, but as a Dutch *onderneming* (undertaking) on English soil, modeled after so many similar projects in the Low Countries.[50]

No original design proposal or plat survives for the drainage, so the works actually carried out can only be surmised from contemporary descriptions and complaints, and from the rivers that still exist in the area. The project was complicated, with several new-built drains, dams, sasses, and sluices, but the overall strategy was fairly straightforward: Vermuyden identified the rivers that were the main source of flooding and sought either to restrict their flow through the area or bar them altogether. The three rivers on which he focused most of his attention were the Don, the Idle, and the Torne (fig. 5.2). To the north, the River Don originally split into two branches near Thorne, with the lower branch flowing through Hatfield Chase in Yorkshire and causing frequent flooding. Vermuyden planned to dam the lower branch and reroute its flow into the upper, which skirted the northern boundary of the level and emptied via the River Aire into the Humber. He also planned to build high banks along the southern (Hatfield) side of the newly combined Don, set back from the river itself, to prevent floodwaters from spilling over into the level. He took a similar approach with the River Idle to the south, intercepting it before it flowed into the Isle of Axholme in Lincolnshire and shunting part of it instead into the nearby river of Bickersdike, which skirted the level's southern boundary and discharged into the River Trent. The remainder would flow through the level via a new, straighter channel. The

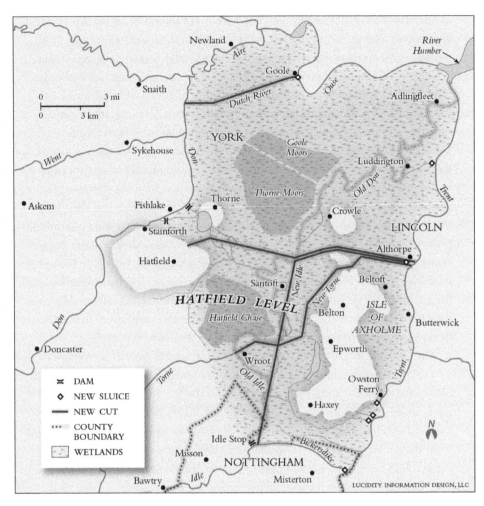

Figure 5.2. A map of the Hatfield Level, including new drainage works. Because no contemporary plan of the drainage project survives, the location of new drains is based on conjecture from manuscript sources, and the location of rivers in the region during the modern era.
Adapted from a map in Korthals-Altes, *Sir Cornelius Vermuyden.*

River Torne would continue to flow through the level, but Vermuyden planned to dig a new channel for that river as well, eventually merging it with the New Idle and giving both a new outfall into the Trent at Althorpe. Finally, he planned to create a network of smaller drains to carry off surface water during rainy periods, as well as some sluices to prevent the tides from silting up his redesigned rivers.[51]

Construction of all the new drainage works went smoothly, for the most part, and was completed in a surprisingly short time given the scale and complexity of the job. By the end of 1627, just eighteen months after signing the agreement (though with significant construction still yet to be done), Vermuyden already felt confident enough to appeal for a commission of adjudication to survey the level, declare it drained, and award him his allotment. All of the originally designed works were completed by 1631, at a total cost (according to Dugdale) of £55,825.[52] The work also provoked only scattered protests early on; the majority of the fenlanders either endorsed it or offered little outright resistance. Especially in the manors of Hatfield and Crowle, most inhabitants were copyholders; their customary use rights in the common wastes were more limited, and their land tenure was less secure. They soon reached agreements with the royal commissioners for the partition and enclosure of their commons, despite some occasional grumbling, if only because they lacked a strong legal basis on which to challenge it. Some also embraced the drainage as a real improvement, even after the enclosures. The village of Crowle petitioned the House of Commons to express their gratitude for the project, and to ask for speedy passage of a bill to settle lingering disputes and reward the Participants for their noble efforts.[53]

Not everyone was so acquiescent, however, and several protests soon erupted. The drainage system devised by Vermuyden had already given rise to new troubles in the region by 1627, even as it improved matters in most of the Hatfield Level itself. The River Don to the north and Bickersdike to the south were especially problematic. Vermuyden had redirected large volumes of water into rivers that could not accommodate them and then constructed high banks along the Hatfield Level sides of each river to prevent the floodwaters from entering. This had the unintended consequence of forcing excess water to overtop the banks on the far side of each river, causing severe flooding in neighboring communities in Yorkshire and Nottinghamshire that had been relatively dry before. It aroused considerable resentment in the newly flooded villages and sparked a flurry of petitions, lawsuits, and riots. Within the level itself, the new channel constructed for the River Torne was not deep enough—a substratum of rock had actually forced Vermuyden to raise his new riverbed *above* the surrounding lands. When it overflowed, it flooded the Isle of Axholme with water that, before the drainage, would never have entered the area. The inhabitants there complained that Vermuyden had drained Hatfield Chase to the north only by shunting the floodwaters onto their lands instead.[54]

The most serious and sustained challenge to the project came from the manor of Epworth in the Isle of Axholme. Epworth had a large population of freeholders with absolute tenure over their lands and extensive, legally protected use rights to the common wastes, and they did not recognize the king's right as lord of the soil to "improve" their commons and enclose a portion of them. They based their claim on a 1359 indenture, granted to the commoners and their descendants by the lord of the manor at that time, John de Mowbray. Mowbray had pledged that in return for allowing him to enclose a portion of the common waste for his own use, he would guarantee the commoners' use rights on the remainder forever after.[55] The original of this indenture was kept in a specially made wooden chest in the parish church of Haxey, and the inhabitants of Epworth manor were well aware of its existence and its import. One local yeoman testified that he had seen the original and been told of its contents, and at least one member of the local gentry had a copy in his possession.[56] Epworth had since passed to the Crown, but as lord of the soil, the king was still bound by the terms of Mowbray's indenture. The commoners there were convinced the king had no legal right to drain their commons and enclose two-thirds of it, and they resisted all compromise or negotiation. Since the king's contract with Vermuyden obliged him to "treat, deal, agree, and conclude" bargains with those who could demonstrate their common use rights in the level before work could even begin, the Epworth commoners' claims threatened the entire project.

Opposition to the Hatfield Level drainage was thus every bit as complicated as the drainage work itself. Different communities resisted the project at different times, in different ways, for different reasons, and with varying degrees of success. Likewise, the Crown's response to anti-drainage resistance depended on the nature of the complaint and the means used to pursue it. Those fenlanders who argued persuasively that the project had done them real harm and petitioned for some sort of relief met with a reasonably sympathetic response, as long as they confined their protests to legal channels and eschewed violence. But those who rioted or who challenged the king's right to drain and enclose the land in the first place found themselves facing the full coercive power of the state. In order to understand the multifaceted opposition and the divergent royal responses to it, the cases must be examined independently.

UNINTENDED CONSEQUENCES: YORKSHIRE AND NOTTINGHAMSHIRE

Communities located just outside the northern and southern peripheries of the Hatfield Level experienced very similar difficulties as a result of Vermuyden's drainage scheme, and they reacted in similar ways. In both cases the root of the problem lay in Vermuyden's plan to prevent a troublesome river from flowing through the heart of the area he was trying to drain by diverting the water into another riverbed that skirted the bounds of the level. To the north, in the West Riding of Yorkshire, the southern branch of the River Don was dammed off and rerouted to join the northern branch instead; to the south, the River Idle was diverted into Bickersdike near the village of Misterton in Nottinghamshire. Along each river, Vermuyden also constructed an elevated bank on the Hatfield Level side, to allow some room for the waters to rise before they flooded the neighboring level.

The new works performed as intended with respect to the Hatfield Level, but two fundamental design flaws created new problems elsewhere. Neither Bickersdike nor the northern branch of the Don was wide or deep enough to accommodate the much greater volume of water it was now expected to carry, leaving each river more flood-prone than ever. And while Vermuyden's new banks protected the level, he did not bother to build similar banks along the far side of each river, so surging floodwaters now spilled into communities that had not been much troubled by flooding before his intervention. By 1630, nearly two dozen villages in Yorkshire and Nottinghamshire were complaining to the Privy Council and the Council of the North that Vermuyden had only managed to drain the Hatfield Level by drowning them instead.

Richard Bridges, "a woeful spectator of the lamentable destruction of my native soil and country," wrote in September 1630 of the devastation in the village of Sykehouse in Yorkshire, where he saw "[a]ll sorts of people in pitiful distress." He described the many farms that were ruined and abandoned, the inhabitants having fled to higher ground, their fields full of corn ready for harvest but now under water. Nor was he in any confusion about where to lay the blame for the catastrophe: "Thus have strangers prevailed to destroy our inheritance, and to convert that waste ground to their profit and our subversion, which our ancestors left for a sink and receptacle of inundant waters for our future safety."[57] Likewise, the inhabitants of Misterton in Nottinghamshire complained to the Privy Council of "the great damage & prejudice" Vermuyden had caused them "in drowning their meadows, pastures, & arable lands under pretense of draining their surrounded commons & waste grounds, which he hath likewise made worse, by diverting the ancient course of some waters and bringing other new waters into the said wastes."[58]

The Yorkshire men were more violent in their protests, which began while construction of the new drainage works was still under way. An anonymous chronicler of the area wrote long afterward that "whilst this great projector [Vermuyden] thought himself secure of His Majesty's favor . . . he found himself mightily annoyed by the gnats and flies, that is the common sort of the inhabitants that set upon him when he should rest. . . . [T]hey disturb'd him in his works, and when that would not do, in great numbers they burnt his carts, and barrows, and working instruments, in great heaps by night."[59] As early as June 1627, the Privy Council complained to Sir Thomas Wentworth (the future Earl of Strafford, and soon to be lord president of the Council of the North) that "diverse loose and dissolute persons" living near Hatfield had "in a mutinous and tumultuous manner assaulted the workmen and set fire on the carts and timber employed for the draining of some parts thereof." They called the attacks "an insolence not to be suffered in a peaceable and well governed state" and told Wentworth to raise "a competent number of trained forces, as need shall require to aid and assist . . . those that are employed in that work, and to inflict punishments upon those that shall willfully offend in this kind."[60] They likewise charged the local commission of sewers to "prevent all future disturbances and outrages" and issued a warrant to arrest four of the ringleaders to answer for their crimes before the council, after the king had given order "for the punishing of some of the principal offenders for their mutinous and seditious behavior."[61]

Further rioting erupted in December 1628, however, when floodwaters from the River Don drowned more than 1,200 acres, after which "some ill disposed persons averse unto His Majesty's general work of draining . . . maliciously cut some gaps and holes in and through the said new erected wall" that Vermuyden had built to keep the Don from flooding the Hatfield Level. The vandals were trying to drain their lands by letting the waters back into Hatfield, where they would have gone in the first place before Vermuyden's intervention. Their actions were therefore logical and defensible, after a fashion, and they had the support even of the local justice of the peace and steward of the royal manor of Hatfield, Robert Portington. It was Portington who, just a few months earlier, had recorded the numerous depositions concerning the Haxey riots discussed at the start of this chapter; but in this case, he was accused of joining with the rioters. Portington's sympathies with the commoners may have been awakened by the Haxey riots, or perhaps he was simply more sensitive to the suffering of his fellow Yorkshire men than he was to that of the rioters in Lincolnshire. In any case, he was described in an affidavit as "a great encourager of these disorders, who by his place and employments ought in his duty to have been a principal furtherer of the king's service." The council had him arrested, together with a few other ringleaders who were ordered to attend the council in London until discharged.[62]

When the Privy Council finally held a hearing on the matter in April 1629, with the king in attendance, they continued to support Vermuyden and his drainage scheme, but they also acknowledged that the Yorkshire men had some legitimate concerns regarding the deterioration of their lands. Vermuyden, for his part, insisted that the fault lay not with his drainage, but with the inhabitants for neglecting to maintain the riverbanks on their own side of the Don. The councilors decreed that the banks along the north side of the river should be restored by the landowners and commoners traditionally responsible for them and offered Vermuyden's "best advice and direction if the same be desired." If this proved inadequate, then Vermuyden would be held responsible for augmenting the banks to whatever additional height was needed. The alleged ringleaders in the riots were required to give security that they would appear again if called on, and were then "discharged and set at liberty" to return home. The humbled Robert Portington was even permitted to continue on the bench as a magistrate, "so long as he behaveth himself well in the furtherance of this work, which he hath now undertaken to His Majesty to perform with his best endeavor."[63] From this point onward, the Yorkshire fenlanders remained peaceful in their protests.

Vermuyden, in the meantime, continued to enjoy the full and active support of the king. Charles knighted him on 6 January 1629 and the following month bestowed on him the lordship of Hatfield, along with the West Riding manors of Thorne, Fishlake, Stainforth, and Dowcethorpe, in return for a cash payment of £10,000 and an annual rent of £425 (a figure increased from just over £195, presumably reflecting the improvement in the drained land). This gave Vermuyden direct control over both his own and the king's former shares of the commons there and enhanced his legal and social standing in dealing with the more recalcitrant inhabitants of his new demesne.[64]

Conditions in the West Riding did not improve. Despite the Privy Council's decree, the northern bank of the Don was still "greatly in decay through the neglect of the repair thereof," and by the spring of 1630, the inhabitants were again complaining that they had "lately sustained infinite losses by the inundation of waters" and were "much impoverished." They backed their claims with a certificate from several justices of the peace, attesting to their destroyed crops and drowned livestock, and pleaded that "that part of the country would in short time become uninhabitable if some means of prevention were not speedily used." They blamed their sorry state on "the banks raised by Sir Cornelius Vermuyden," though the projector continued to insist that the Yorkshire men were at fault for failing to mend their own banks. The council held a second hearing on the situation in May 1630 and once again ordered that the northern banks of the Don be restored and augmented, but this time they divided the responsibility a bit more equitably. Vermuyden and his partners were to perform the work and maintain it ever afterward, and the affected communities were to contribute £200 toward the work plus an annual sum for maintenance. The council ruled further that "no demands should be made on either side for damages or losses already sustained."[65]

The council's decree that all losses should simply be forgotten was hard on the fenlanders, who claimed that the unprecedented flooding had already done as much as £20,000 in damage to their property. Even worse, however, was Vermuyden's vindictive attempt to arrest and prosecute anyone suspected of vandalizing his banks in Hatfield. Just days after the Crown's even-handed attempt to resolve the dispute, Vermuyden took out warrants to have several West Riding inhabitants committed to prison at York. Those not immediately apprehended were forced to flee, not knowing how many warrants Vermuyden had or his authority to execute them. For his part, the new lord of Hatfield boasted of his power to intimidate the country; he allegedly "threatened to hang divers of the petitioners, His Majesty's native and freeborn subjects," and to "set up a pair of gallows to have terrified the people."[66]

Vermuyden's aggression in dealing with his Yorkshire commoners overstepped the bounds laid down by the Crown, however. Alarmed at the situation, the Privy Council referred the matter to a special commission headed by Thomas Wentworth, now lord president of the Council of the North, in June 1630. They charged the commission to investigate the matter and to "settle the differences between them (if they can) as that His Majesty and this board be no further troubled therewithall."[67] The commissioners met over the following months and finally ruled that "all suits depending in any court of justice" were to be remitted, suppressing all of Vermuyden's recent efforts to prosecute his tenants. They also upheld the Crown's decree of the previous May, that Vermuyden should repair and maintain the riverbanks on the north side of the Don, with the inhabitants contributing £200 toward the cost of the work.

But the commission went further, adding a few new provisions to give greater security to the fenlanders: Vermuyden was to be held liable for any flooding that occurred after the banks were completed, and he was required to make good any future losses. Moreover, if the augmented banks should continue to prove insufficient, Vermuyden and his investors would be required to build an entirely new drain through Hatfield Level at their own charge, to vent the River Don's excess waters. They also placed restrictions on Vermuyden's feudal rights as the new lord of the manor, requiring him to confirm the tenants' copyhold leases with the traditional customs and certainty of fines. They advised that their orders should be decreed in both the Court of Exchequer and the Council of the North, to ensure that Vermuyden and the rest of the Participants complied.[68]

Although most of the Participants accepted the commission's ruling, Vermuyden stubbornly resisted it for as long as possible. He refused to pay his share of the costs in building the required banks, and in order to escape the jurisdiction of Wentworth and the Council of the North, he even conveyed his allotment of land to trustees and left Hatfield altogether. Wentworth complained to the Privy Council that the much-needed banks were being deliberately neglected, and the land remained in a pitiful state. He also warned that "he found it a work of some difficulty to contain the people from sudden violence, unless they find the state hath some regard toward them . . . and that they might soon expect the work to be done," and he asked for the council's assistance in compelling Vermuyden to comply with their judgment.[69] The council summoned Vermuyden to answer for his disobedience in May 1631, threatening that if he continued to resist "the board will take notice thereof and direct such further course for the performance of the same as shall be fit."[70] Even so, he continued to stonewall through the summer, relenting only when the Privy Council summoned him once again in July and charged him with "neglect, and disobedience, to the order of the board." The special commission's orders were finally decreed in the Court of Exchequer in November 1631.[71]

After 1631, the Participants finally took action to alleviate the plight of the West Riding inhabitants. They started construction on a new drain across the Hatfield Level to relieve the overcharged northern branch of the River Don in the summer of 1632, but the work was held up for years by a legal battle between Vermuyden and the other Participants over who should bear financial responsibility for it. The other Participants blamed Vermuyden's poor design and argued that he should pay to fix his mistakes, while he insisted that he should not be required to bear the burden alone.[72] In the meantime, neither side contributed their stipulated share of the construction costs—Vermuyden was even committed to debtor's prison for a few months in 1633—and flooding in the West Riding continued. The fenlanders petitioned the Privy Council in 1634 regarding "their utter ruin and confusion especially by their last inundations," and in 1635 they complained again that "our dwelling houses & barns & corn fields have been drowned & our seed lost though sown 3 times in one year, & no dry ground left us for ourselves or our cattle to tread upon, our lives endangered, our estates wasted heightening our banks, our towns likely to become desolate & we & our families utterly undone."[73]

The Participants eventually prevailed in their case against Vermuyden, whose remaining lands in the region were placed in receivership.[74] The construction work proceeded slowly, but once it was finally completed in 1635 or 1636, the new drain (still known as the Dutch River) did much to mitigate flooding along the north side of the Don. However, it also added at least £20,000 to the overall cost of the drainage project, all but eliminating any hope that the original Participants would realize a profit from the venture, so that most disposed of their shares at a significant loss. They were bought out by eager English investors who sought to take advantage of the newly improved lands at little cost to themselves, since the expensive drainage works were all but completed.[75] Despite the judgment against him, Vermuyden insisted throughout that his original drainage scheme was sound, and he was entangled in further lawsuits with his former partners for years afterward.

∽

The issues faced by the fenlanders of northern Nottinghamshire were similar to those in the West Riding of Yorkshire. Vermuyden's decision to divert the River Idle into the drainage channel of Bickersdike had caused unprecedented flooding all along the southern periphery of the level. The inhabitants of the Nottinghamshire town of Misterton, located just outside the Isle of Axholme on the south side of Bickersdike, were particularly hard hit. They complained to the Privy Council in June 1630 of "the great damage & prejudice" they had suffered as a result of Vermuyden's drainage scheme, and the council responded by appointing a special commission to investigate Misterton's claims. The commissioners were instructed to make a personal survey of the area and broker a resolution if they could, or else to report their findings back to the council.[76] This was just two days after the councilors had commissioned another group to investigate complaints of flooding in the West Riding, at the other end of the level.

The commissioners were forced to postpone their visit to Misterton for nearly three years, because the town "hath been visited with the sickness." When they finally arrived in June 1633, their survey quickly confirmed the inhabitants' complaints: "we find that by the late proceedings of Sir Cornelius Vermuyden and the Dutchmen much harm is done within Misterton in arable grounds, meadows, and pastures, and also in their common wastes. All which are much worse drowned by the late works than ever heretofore they were." The problem was caused, they judged, by diverting the River Idle into Bickersdike, which was at least three feet shallower than the old Idle and thus could never accommodate all the water it now carried. They could not broker a resolution, however, because Vermuyden refused to attend their sessions, and his agents would not agree to anything in his absence. The commissioners therefore recommended to the Privy Council "that the River Idle may have its ancient course restored," and that Vermuyden and the other Participants should make full restitution for all the damage caused by their drainage scheme. The Misterton inhabitants especially deserved fair consideration, they told the council, because they were utterly dependent on their right of common for their livelihood and had "long even full five years, to their great impoverishment, patiently suffered much loss without any tumultuous or riotous courses"—a degree of forbearance even the Participants were forced to acknowledge.[77]

The commission's recommendations had no immediate effect, however, and northern Nottinghamshire continued to suffer. In November 1633, eight more villages petitioned the king and Privy Council for relief, claiming that unprecedented flooding in their communities had caused more than £1,500 in damages over the previous five years. They asked that the River Idle be restored to its old course, and that the Participants be made to pay them restitution.[78] The inhabitants of Misterton also petitioned for relief once again, complaining that they were "greatly impoverished" by the damage done to their lands, and "many of them constrained to forsake their habitations & seek elsewhere to live by some new and other course of life."[79] This time the council referred the matter to the attorney general, William Noye, for investigation.

Noye soon met with both the Participants (including Vermuyden) and several of the Nottinghamshire inhabitants. In his report back to the council, he emphasized the areas in which there was some measure of consensus. Both sides, for example, agreed that the lands in question had flooded occasionally even before the Hatfield Level project was begun, but that some areas had clearly suffered more in recent years because of the changes to the old drainage network. They also agreed that building some new sluices, sasses, and drainage ditches would probably help to alleviate the worst of the flooding and that paying for them ought to be the responsibility of the Participants. The real sticking point, as in the West Riding, was that the Participants could not agree among themselves who should bear the full financial burden for the new works. Vermuyden argued that the cost should be borne by all of the Participants equally, while his erstwhile partners insisted that he alone should be responsible since his flawed plan had created the problem.

Noye had no patience for the Participants' internecine squabbling, especially when he considered the ongoing damage it would likely cause: "This work must be done by all those whose interests occasioned the drowning," he wrote, "& it is not convenient that the petitioners should be delayed until [the Participants] and Sir Cornelius Vermuyden be agreed, but that they forthwith make it."[80] He sided with Vermuyden in arguing that the burden should fall proportionally on all the Participants who had benefitted from damming the River Idle, including even those whose allotted lands were on the other side of the level in Yorkshire.[81] The Participants, however, bitterly objected to the king that they should be expected "to make endless and fruitless works within the level to satisfy idle humors & causeless clamors of every discontented town or person although it tend to their utter undoing."[82]

The Privy Council accepted Noye's recommendation. In 1634, the Crown issued a new commission of sewers covering Nottinghamshire, Lincolnshire, and the West Riding of Yorkshire, with the power to levy sewer taxes on the Participants' lands throughout the Hatfield Level to pay for additional drainage works and repairs wherever they were needed.[83] The new commission assessed the Participants several shillings per acre, though they had continual trouble in collecting it and were eventually forced to distrain and sell the lands of those who refused to pay what they owed.[84] Though the Nottinghamshire fenlanders received little, if any recompense for the losses they had suffered, by 1636 they finally got some relief from the flooding that had plagued them for several years, thanks to the additional drainage works constructed at the Participants' expense.

The disputes in Nottinghamshire and the West Riding of Yorkshire demonstrate that the Crown was not invariably on the side of drainage projectors in their disputes with fenland inhabitants. The wheels of justice ground slowly, to be sure, but they did grind in the fenlanders' favor. Their complaints of unprecedented flooding were investigated and validated, and the Participants were made to pay for additional drainage works to relieve their plight, even though many were consequently forced to sell their shares in the project at a loss. One reason for the fenlanders' success in appealing to the Crown for relief in each case was their general forbearance from rioting and vandalism. The Yorkshire men confined their protests to petitions after a few violent incidents early on, while the Nottinghamshire men were explicitly commended for eschewing violence throughout.

Another important factor in each case was the manner in which the fenlanders framed their petitions. Rather than attacking the drainage itself and challenging the legitimacy of the Crown's right to pursue it, they presented specific, demonstrable grievances that could be addressed without threatening the entire project. The Crown could grant their petitions a full and sympathetic hearing without risking the Hatfield Level drainage or undermining royal authority in the region. Indeed, in expressing their complaints mostly through legal and political channels, rather than violence, the fenlanders' actions reinforced the power of the monarchy by voluntarily engaging the king and his council in settling fenland affairs—a fine example of cooperative state formation. In the Isle of

Axholme, however, the anti-drainage protests were of a very different character, and they met with very different results.

AN EXISTENTIAL THREAT: THE ISLE OF AXHOLME

Anti-drainage sentiment in the Isle of Axholme posed a much more serious threat to the entire project because of the topography of the Hatfield Level and the particular legal circumstances of Epworth, the chief manor of the Isle. Epworth was initially of little concern to the Participants, except insofar as it was located directly in their way. The land there was already comparatively dry, but based on the scheme Vermuyden had devised, some of his newly constructed drains had to flow through Epworth to empty into the River Trent. A later critic of the Hatfield Level project, Daniel Noddel, commented in 1653 that the Participants "did but desire to cut through the manor of Epworth, to drain Hatfield Chase, and many other manors. . . . And that without cutting through the manor of Epworth, Hatfield Chase, and those other manors could not be drained."[85] John Lilburne likewise claimed that "they were forced to drain all the waste grounds of the other 14 manors, through the manor of Epworth," and that "not one drain" was cut in order to improve Epworth itself.[86]

Epworth thus had to be included in the project, but unfortunately for the Participants, the inhabitants there were staunchly opposed to the drainage, partition, and enclosure of their commons, and they had a strong legal case to stand on. Unlike most other Hatfield Level manors, Epworth was home to a large number of freehold farmers who believed their extensive rights of common were supported not only in custom, but also in property law. The Mowbray indenture of 1359, preserved in the parish church, limited the powers of the lord of the manor by guaranteeing that no further improvements and enclosures would be made in the common there without the commoners' consent.[87] Epworth, so crucial to the success of the entire project, was also the only manor in which the king did not have a clear right as lord of the soil to do as he pleased with the common. This fact was implicitly recognized in Charles's original agreement with Vermuyden, which stipulated that some inhabitants "do claim common of pasture in some part of the said lands," and that construction could not begin until three months "after . . . the King's Majesty shall have agreed and concluded with such person or persons as have or claim to have any estate, interest, or common of or in the said grounds."[88]

The king never secured the necessary agreements, yet the drainage works proceeded just the same. This was what provoked the riots that greeted the Dutch laborers when they entered Haxey Carr in August 1628, described at the start of this chapter. It was also the reason for the rioters' seditious declarations that "they cared not for the king for their lands were their own & he had nothing to do with them," "if the king did send 1000 men he should have no common there," and "if the king was there they would kill him."[89] The Privy Council, taking note of the "extraordinary circumstances, both of numbers [of rioters] and of violence, in grievous wounding and beating of the workmen," issued a warrant for eighteen of the principal offenders to be apprehended. At the same time, however, they acknowledged the legal claims of the Epworth commoners and attempted to reassure them that "as nothing shall be done against them but by legal course, so free way shall be given to any just or legal complaint on their side."[90]

Yet the violence continued in Epworth; in September 1628, the Misterton magistrate Francis Thornhill wrote to Vermuyden to inform him that some three hundred rioters had "cast down your new bank builded over the River of Idle, broke some of your planks & barrows, and drive your workmen from your work in riotous manner." Two of the perpetrators had since been caught, he reported, but "the greatest part being women, boys, servants & poor people whose names cannot be learned," Thornhill worried that most would "escape unpunished, hinder the proceeding of your work, & make a scoff of all authority & government unless some speedy severe course be forthwith taken against them." He also observed that while most of the rioters were of "the baser sort," he believed they were being encouraged by "the better sort," who remained in the background.[91]

In response to the September riots, the Privy Council adopted a more coercive stance in behalf of the Participants. They commanded local magistrates to "apprehend and commit to the common gaol of the county" all the "principal animators" of the riots, with the "chief offenders" to "answer this their outrage in the Court of Star Chamber." They also asked the attorney general to prepare a royal proclamation forbidding the Axholme inhabitants from interfering in the drainage works, "upon such penalties and punishments to be inflicted upon the offenders as in such cases are usual." Finally, having been informed that the "better sort" might well be in league with the rioters, the council further commanded "the substantialest inhabitants of those places, their sons and servants" to assist local constables in arresting the rioters. If they should "refuse or negligently perform" their duties, the magistrates were to notify the council immediately so that "further course may be taken for the exemplary and severe punishment of the offenders of all sorts." Nor did the council fully trust the magistrates to whom they wrote, as they concluded with a stern warning: "hereof you may not fail as you tender His Majesty's service and would avoid the danger to yourselves which will follow your neglect therein."[92]

The royal proclamation was delivered to the Axholme town of Haxey by a sergeant-at-arms, the sheriff of Lincolnshire, and (according to one report) thirty armed horsemen who threatened to sack and burn the town if the rioting continued.[93] This show of force had the desired effect—the violence abated for a time, and the fenlanders even helped to restore the damaged works. But the summer of 1629 saw further rioting erupt throughout the Isle of Axholme, with large numbers of rioters destroying the newly repaired works and assaulting the workmen. The Privy Council once again ordered the sheriff to quell the riots and commit the offenders to jail, using troops if necessary, though they also felt the need to admonish the sheriff in less than cordial terms: "Hereof you must not fail as you tender His Majesty's service and will avoid that just reprehension and punishment which will fall upon you for your neglect herein."[94]

Beyond the persistent threat of violence, the battle over the fate of the Isle of Axholme was also being waged more quietly through legal channels. In October 1628, during a lull in the rioting, the Axholme commoners appealed to the Lincolnshire commission of sewers "with intent . . . to cross and interrupt the proceeding" of the drainage project. When the Privy Council was made aware of these plans, they quickly put a stop to it. Writing to the commissioners, the council reminded them that "the general good and His Majesty's profit and service is not a little concerned" in the drainage and that the project was not "unadvisedly undertaken, but carefully digested as well by the great officers of His Majesty's revenue as by select commissioners of good experience and integrity." They ordered the sewer commissioners "by His Majesty's special command (who hath been likewise made acquainted therewithal) . . . not to suffer any proceedings to be had before you which may hinder or cross the work already done or to be done," without first giving notice to the king and council.[95]

Although the tone of the council's letter was cordial, its implication certainly was not. The commissioners of sewers would not have missed the references to the king's personal involvement in the scheme or the enhanced Crown revenues that were at stake in it. To cross the project, the letter made clear, would be to offend the king and undermine the well being of the royal treasury. These were risks that few commissioners of sewers were willing to take during Charles's reign, in contrast to the bolder behavior of the Great Level commissioners under James toward the proposals of Ayloffe and Thomas. When Charles commanded the cooperation and obedience of local officers, he expected to get it, and any hope the fenlanders had of blocking the project through the Lincolnshire commission of sewers quickly dissipated.

The Crown also worked to press its legal case. In late 1628, attorney general Robert Heath exhibited a bill in the Court of Exchequer asserting the king's right as landlord to improve and enclose a portion of Epworth common. The king wrote personally to the barons of the court to inform them "what opposition hath been against so profitable and so public a work," and he reminded them "how dangerous it would be to give way to the willfulness of any to oppose, and how much discouragement it would be to the undertakers to admit the least delay to finish all things." The Epworth commoners opposed the bill, arguing that the land was already productive and valuable as pasture and required no draining, but the linchpin of their case was the Mowbray indenture, which they presented as evidence against the king's right to enclose any of their common.[96] In April 1629, the court appointed a special commission (the first of many) to investigate the details of the case, including the current condition and value of the land and the king's rights as lord of the soil.[97]

The commission soon determined that the king was entitled to improve and enclose a portion of the Epworth common, the Mowbray indenture notwithstanding. A second special commission, appointed to negotiate a settlement with the commoners, proposed that from a total of 13,400 acres of common waste, 6,000 acres should remain in common for the inhabitants. Of the remaining 7,400 acres, roughly 4,500 would go to Vermuyden as his agreed-on one-third share, leaving the other 2,900 acres for the Crown. This offer would have left the inhabitants with almost 45 percent of their original commons intact—considerably more than the one-third share left to the other Hatfield Level manors, with the difference taken entirely from the king's own portion. It was an unusually generous offer from the Crown's point of view, granted perhaps in acknowledgement of the commoners' legal claim through the Mowbray indenture. The Epworth commoners, however, still insisted on their legal right to the entire 13,400 acres, and they "absolutely refused to yield themselves in any sort conformable" to the king's wishes, so that "no full and absolute agreement could be made."[98]

This was far from the end of the case. Over the next two years, the Court of Exchequer appointed a series of commissions to investigate whether the Isle of Axholme was really drained, how much the land had increased in value, and what portion the Epworth commoners might be willing to part with as a result. The commoners' opposition remained firm; they were "resolved that His Majesty had no interest or power to make any improvement there . . . and [they] would not assent to any allotment to be made."[99] But whereas King James might have backed down eventually in the face of such stout opposition, Charles's will was likewise unshaken, and he did not hesitate to use his full prerogative powers in pressing his claim. When one of the many Exchequer commissions doubted whether the land really had been "made fit for tillage or pasture," the king accused them of returning "a foul certificate against the truth," threatened to punish them for "so signal a breach of public trust," and had them replaced. The new commissioners, predictably, returned a much more favorable verdict, declaring that "the said grounds are dry and pastureable and so likely to continue."[100]

The legal impasse was broken at last in February 1632, when twenty-two Epworth commoners suddenly and unexpectedly acknowledged in court that the land in question was well drained and that the king and Vermuyden were entitled to a fair portion of it. They announced their willingness to consent to the enclosure and subscribed to a bill submitted by Vermuyden to the Court of Exchequer for the purpose.[101] Why did the commoners' resolve, which had been unshakeable for years, abruptly collapse? The answer lay not in the Court of Exchequer proceedings, but in another legal venue: Vermuyden and his allies on the Privy Council had sought to subdue some of the more rebellious fenlanders by hauling them before the Court of Star Chamber to answer for their riotous behavior. Under Charles's reign, Star Chamber was increasingly used as a means to silence and discipline those who resisted unpopular royal policies, such as the collection of prerogative taxation not sanctioned by Parliament. In 1631, fourteen Axholme rioters—five men and nine women—were brought before the court and punished with a crushing series of fines, ranging from 500 marks to £1,000 apiece. The court also awarded Vermuyden an additional 2,000 marks in damages.[102]

The fines were probably enough to ruin each defendant and their families several times over. But the Crown was less interested in collecting the fines than in using the threat of them to secure the defendants' cooperation. Very shortly after the Star Chamber verdict, William Torksey, Hezekiah Browne, James Moody, and Henry Scott—each of whom had been fined £1,000 and committed to Fleet prison for good measure—privately agreed to support Vermuyden's bill in Exchequer. In return Vermuyden assured them "that they should not be troubled or questioned for the said damages and that he would use means to save them from the danger of the said fines." As if the threat alone was not enough, the fenlanders' solicitor, John Newland, was later alleged to have received a bribe of £80 from the Participants for securing his clients' cooperation. Thus when Vermuyden submitted his Exchequer bill in February 1632, allowing him to enclose more than half of Epworth common, all four men subscribed to it, along with several other Star Chamber defendants.[103]

The commoners who subscribed to Vermuyden's bill claimed to do so "on behalf of themselves and the rest of the tenants who claimed common," though there is no evidence that their neighbors had authorized them to act in any representative capacity. In fact, the agreement was immediately challenged in a variety of legal venues, but the Participants managed to block the Axholme commoners at every turn and deny them a full hearing in any court. In the spring of 1633, Edmund, Lord Sheffield, and several other prominent landowners brought suit in the Court of Exchequer "for and on behalf of themselves and the rest of the freeholders, tenants, and commoners within the manor of Epworth . . . to cross all the said former proceedings in this court." The suit alleged that those who had subscribed to the bill did not represent the rest of the manor's common- ers. The bill should therefore be null and void, since most of those whose rights were implicated in it had not consented to it, nor were they even informed of it until after the fact. The plaintiffs also described the underhanded means by which Vermuyden had used the crushing fines issued in Star Chamber to com- pel the subscribers' compliance. As they were preparing to examine witnesses, however, proceedings in the suit were mysteriously stopped.[104]

Sheffield's suit in Exchequer was matched by a contemporaneous case brought in Star Chamber against Torksey, Browne, Moody, Scott, and Newland for committing their fellow Epworth commoners to an agreement without their knowledge or consent. Surprisingly, the Star Chamber judges ruled for the plain- tiffs in this case and sentenced the defendants to stand in the pillory for their offense. Yet the defendants "nevertheless escaped the shame and punishment due to them (by tampering with . . . the prosecutors) by reason whereof the sentence was omitted to be entered," and the Star Chamber verdict did nothing to nullify the Exchequer bill.[105]

Finally, in 1634, the Epworth commoners tried to obtain a hearing in the Court of King's Bench through a writ of replevin, a legal procedure used to re- cover property unlawfully seized. They drove some of their cattle into the lands lately enclosed by the Participants, and when the latter impounded the cattle for trespass, the commoners took out a writ of replevin ordering the Participants to release their cattle in return for their pledge to have the matter decided at trial. The Participants would thus be forced to present their case in open court and abide by whatever judgment was handed down, or else suffer the commoners to continue grazing their livestock on the land without further challenge. Realizing that they might well lose in any trial, however, the Participants appealed to the Court of Exchequer instead and succeeded in obtaining a stay on all writs of re- plevin in the Isle of Axholme.[106]

Despite their best efforts, the Epworth commoners discovered they could get no legal remedy in the courts of Exchequer, Star Chamber, or King's Bench and found themselves "without hope of any relief in law, to recover their right." In their frustration, many of the "desperate" commoners "strived to defend force with force, and so to keep their possession" by extralegal means.[107] Rioting erupted once again in the region in July 1633, in the midst of the various trials, and the Participants complained somewhat hyperbolically to the Crown that the fenlanders had "risen in troops & cut & thrash their rape, trod the writs under their feet, threatened to kill the servants of the Dutch, rip up their bellies & throw their hearts in their faces." The sheriff, magistrates, and lords lieutenant of Lincolnshire were ordered by the Court of Star Chamber to use whatever troops were necessary to quell the violence. According to later accounts, some rioters were allegedly "murdered and shot to death, and many wounded," while others were arrested and imprisoned.[108]

This time the Participants were able to use the riots as a pretext to suppress all local opposition through sheer intimidation. They obtained warrants to arrest the rioters and added to them the names of several prominent Axholme landowners, even those who had not taken part in the riots, to have them jailed and bound over—a tactic Vermuyden had also used in the West Riding. Many were arrested, and others were "forced to fly out of the kingdom at that time, having their goods illegally distrained." On top of all this, the Participants brought their own suit in the Court of King's Bench, alleging that the riots had caused damage to their new drainage works and enclosures amounting to thousands of pounds. A jury eventually ruled in their favor and awarded them damages.[109]

The Participants had brought to bear in their behalf the monarchical state's full powers of suppression by the mid-1630s. The Epworth commoners had been denied any means of legal redress, punished with crushing fines and damages, put under threat of arrest and imprisonment, and subdued by armed force. "[B]eing thus persecuted, and under such cruel vexation and tyranny," some of them had finally had enough. In February 1636, a total of 370 commoners "did submit themselves to such order and award as His Majesty's attorney general should be pleased to make therein." The attorney general's order confirmed the previous decrees given in the Participants' favor: the latter were to continue their possession and enjoyment of 7,400 acres that had formerly been part of the Epworth common, while the remaining 6,000 acres would remain in common use. A few small concessions were granted to the commoners, the most important of which was their being "left at liberty to implore His Majesty's grace and favor for pardoning the said issues and that all suits betwixt the said parties shall cease and no further persecution be made therein." The award was read and decreed in the Court of Exchequer, in the presence of several of the commoners.[110]

The 1636 Exchequer decree was deeply unpopular in the Isle of Axholme, despite the public approbation of 370 commoners. The fenlanders argued in subsequent legal proceedings that those who had consented were but a small minority, with no power to bind the majority to such an agreement; that they had done so only to escape the threat of fines and imprisonment; that many had not understood what they were endorsing; that others did not live in the Isle or have right of common there; and that several names in the list were included three or four times, included only a surname, or were simply fictitious.[111] But they could not hope to prevail against the Participants while the full legal and military power of the state was arrayed against them, and the 1636 decree finally brought a fragile peace to the Isle of Axholme. The Participants, seemingly secure in their title, began leasing out their 7,400 acres to their Dutch, French, and Flemish settlers, who "plowed and tilled . . . to the great benefit of the common wealth."[112] The Isle of Axholme remained restive, but quiet, until the outbreak of the Civil War in 1642.

FENLANDERS, PROJECTORS, AND THE STATE

The events and issues surrounding the Hatfield Level drainage project are complicated, encompassing a wide array of conflicting interests and understandings. As would become typical in subsequent ventures, there was never any real consensus about how effective the drainage was or whether it was a good idea, even long after it was purportedly completed. William Dugdale wrote in 1662 that in the years following the Hatfield Level drainage "a great part thereof hath been sowed with rape and other corn, for three years together, and born plentiful crops." He described increased land values, an abundance of produce harvested on lands that had once yielded next to nothing, and "several houses [that] have since been built and inhabited in sundry places . . . which formerly was drowned land." He noted approvingly that "before the draining, the country thereabouts was full of wandering beggars, but very few afterwards; being set on work in weeding of corn, burning of ground, thrashing, ditching, harvest work and other husbandry: all wages of labourers, by reason of this great use of them, being then doubled." The drainage, for Dugdale, was a manifestation of the king's "royal and princely care for the public good, in regaining so great a proportion of surrounded land"—a transformation of waste into an extension of the prosperous commonwealth, and of its slothful and beggarly inhabitants into productive English subjects.[113]

Dugdale's verdict was far from universal, however. Daniel Noddel, a local inhabitant who later served as solicitor to the Epworth commoners, viewed the drainage as a disaster for the district. Much of the "improved" land, he wrote in 1653, was already worth a substantial amount as pasture before the Dutch drainage projectors interfered with it. Turning away the annual floodwaters had done more harm than good by reducing the yield of grass and thus the number of livestock that could be fed on it, forcing the freeholders there to sell their herds and their lands and move elsewhere.[114] John Lilburne, in his 1651 apologia for the inhabitants' riotous behavior, offered a slightly more balanced assessment. He conceded that the drainage had "made some of the commons of Epworth manor, fitter for corn than formerly," but when he also considered the diminished grass for grazing and the loss of fishing and fowling as profitable by-employments, he found it "very doubtful . . . whether there hath been any improvement made of those grounds to a greater profit than formerly."[115]

Besides its disputed efficacy and value, another important question in the history of the Hatfield Level drainage is how it ever went forward in the first place, when so many similar proposals for drainage projects had come to nothing over the previous half-century. The Crown's ownership of all four principal manors in the region was one key factor: Hatfield Level was the king's own land, to be drained and improved at his pleasure (for the most part). This presented a very different set of circumstances from the Great Level, a far larger area where a multitude of contentious landowners and commoners with conflicting interests had frustrated every effort to forge consensus around a single drainage policy. Forging consensus was not necessary with only a single landowner in the mix, one who already favored the project. Moreover, the district's numerous copyholders had little legal basis for challenging the king's right to improve and enclose the land, and local opposition to the project was comparatively slight. Only the residents of Epworth proved to be implacably hostile, and their resistance constitutes a special case because of the number of freeholders living there and the existence of the Mowbray indenture.

Another factor in King Charles's success in promoting England's first large-scale drainage project was his access to experienced Dutch drainage engineers and confident Dutch investors. Early modern drainage projects were labor intensive, requiring a large capital outlay to get started and with little hope of any return on investment until after the work was completed—would-be investors had to have a committed belief that real value and profit would ultimately arise out of common waste. Nearly all of the original Hatfield Level Participants were Vermuyden's countrymen; they had ample experience with such investments in the Low Countries and a greater willingness to risk their finances in the English Fens than their English counterparts yet possessed. As for Vermuyden himself, engineers and historians have debated his skills and qualifications, and second-guessed virtually all of his decisions, from before his works were completed down to the present day. His original design for the Hatfield Level drainage certainly had some costly flaws. Nonetheless, he managed to present himself as a credible drainage expert to English and Dutch investors alike. His family background had given him considerable knowledge of, and experience with, various Dutch land drainage projects, and he offered to bring his valuable knowledge to England in the king's service. Whatever his shortcomings, Vermuyden had enough expertise and confidence to promote the project and carry it through to completion.

Yet the initial impetus for draining the Hatfield Level had come not from Charles but from his father, King James, who had also been lord of all the principal manors of the region and likewise had access to Vermuyden's expertise. Why, then, did the project not commence in the early 1620s, when James first took an interest in it? Some of the answer must lie with each king's priorities and style of governance. For all his absolutist rhetoric, in practice James preferred to resolve conflicts by building consensus. He dispatched commissions into the Fens to investigate and held hearings before his Privy Council to broker compromises, but when these measures failed, they resulted in paralysis.

Charles, in contrast, cared far less about achieving consensus among the various fenland officials, landowners, commoners, and other stakeholders than he did about promoting progress even in the face of opposition. Draining the Fens, Charles believed, was an intrinsic good that would benefit the commonwealth and boost royal revenues through improvement, and anyone who opposed it was understood to be motivated by selfish, private interests. This could put local officials under considerable strain when they had to choose between obedience to the king and protecting the interests of their neighbors, friends, family, and themselves. The Privy Council was aware of the tension and sought to overcome it by admonishing fenland sheriffs and magistrates to carry out their duties with diligence and alacrity, or face the king's wrath—it thus became an issue of loyalty. This approach proved reasonably effective; though it alienated many of his subjects in the long term, it was also a main reason why virtually all of the major English drainage projects undertaken in the seventeenth century were begun after Charles came to the throne in 1625. The king's willingness to frame things in terms of the royal will versus selfish local interests, with the good of the commonwealth at stake, compelled local officials to fall in line or face the consequences—as Robert Portington discovered to his detriment in 1628.[116]

The combination of Crown ownership of the land, access to Dutch investment and expertise, and a more aggressive style of direct royal governance enabled the Hatfield Level drainage to overcome opposition and go forward where
earlier projects had foundered. The most striking indication of the altered circumstances, besides the progress of the drainage itself, is the virtual disappear-
ance of commissions of sewers as principal actors in the story. Where James had
sought to promote the undertaking of Ayloffe and Thomas exclusively through
the Great Level commissions of sewers, whose united opposition had ultimately
doomed the project, Charles and his advisers negotiated directly with the projectors and barely consulted the Hatfield Level commissioners. In the single instance
when the latter tried to exert some influence in the project, the Privy Council ordered them to stand down.[117] The commissioners' input was neither needed nor
wanted because the king owned the entire level himself, eliminating the need for
them to broker myriad contracts; Charles's access to Dutch expertise diminished
his reliance on their local knowledge and experience; and his preferred style of
governance did not prioritize negotiation and building consensus in any case.

Yet another important question concerning the Hatfield Level drainage is
why some communities were more successful in their protests, receiving a sympathetic hearing from the Crown and redress for their grievances, while others
felt the full weight of royal suppression. How did the commoners of Misterton
and the West Riding of Yorkshire eventually succeed in forcing the Participants to
construct costly new drainage works, even to the point of bankrupting them, while
the Epworth commoners could not even get a proper hearing of their cause?
While the means of opposition were outwardly similar in each case—petitions,
lawsuits, occasional riots—the differences between Epworth on the one hand
and Misterton and the West Riding on the other were rooted in their different
perceptions of the project's very legitimacy.

The complaints from Misterton and the West Riding were earnest and persistent, to be sure, but were really incidental in nature. They were sparked by the
unintended consequences of particular drainage works that had caused unprecedented flooding around the periphery of the level, and they were remediable
by constructing additional works at the cost of the Participants. The inhabitants petitioned the Crown for relief, but they did not challenge the legitimacy
of the drainage itself, nor did they question the king's right as lord of the soil
to undertake it. Indeed, by acknowledging the Crown's rightful role in overseeing drainage matters and its agency in granting relief, the protesters actually
served to reinforce royal power. The Crown thus had every incentive to give
them a sympathetic hearing, even if the means of redress should impose a severe

financial burden on the Participants. Charles's chief priority was not to guarantee a profit for Dutch investors but to enhance Crown revenues while promoting the good of the commonwealth. He had no interest in seeing his own loyal subjects ruined through the poor design and execution of a project that was supposed to benefit all concerned. The Crown's handling of protests from Misterton and the West Riding are thus an example of cooperative state formation across several strata in early modern English society, a case in which royal and local interests could be reconciled and brought into alignment for their mutual benefit.

Matters were very different in the Isle of Axholme, where the challenge to the drainage project was not merely incidental, but existential. Epworth was vital to the success of the project because of its location—some of Vermuyden's new rivers had to flow through the manor to reach their intended outfalls at the River Trent. Unfortunately, it was also the only manor in which the king's rights as landlord were limited by the terms of the Mowbray indenture. When the Epworth commoners refused to consent to the drainage and enclosure of their common wastes, they posed a much more serious threat to the whole enterprise than their neighbors in Misterton and the West Riding had. The king's unique offer to allow them to keep a larger portion of their commons was an acknowledgment of the strength of their claim, though it did not win their approbation.

But the Epworth commoners' opposition not only threatened the drainage, it also challenged the legitimacy of arbitrary royal governance more generally. The Epworth protesters argued that their customary use rights in the common wastes were guaranteed in law, and that even the king could not simply ignore them or set them aside—there were, in short, clear legal limits to the king's powers. Rather than reinforcing royal authority by petitioning the Crown for just relief, they subverted that authority by insisting the king had no business meddling in their common. Their goals may have been conservative and provincial, but politically their protest was incendiary; it had broad ramifications at a sensitive time. The king's arbitrary use of prerogative powers was already under widespread attack at the national level in the late 1620s, culminating with Parliament's adoption of the Petition of Right in 1628. King and Parliament clashed repeatedly over a series of contentious issues such as nonparliamentary taxation, forced loans, arbitrary arrest, forced billeting of troops, and the governing of the established church. Charles's unwillingness to negotiate with Parliament over such matters would lead to his resolving to rule England without it throughout the 1630s.[118]

Apart from the fact that the Hatfield Level drainage could not succeed without including Epworth, Charles could ill afford yet another challenge to his royal prerogative powers. This, too, may explain why the Epworth commoners could not be suffered to present their case in any court and why such harsh and even underhanded means were used to thwart them. The Crown's ruthlessness in suppressing opposition in the Hatfield Level was consistent with other contemporary cases in which critics of arbitrary royal power were silenced. Driving the drainage project forward was a means to demonstrate the king's right to govern as he thought best, despite the opposition of a few disaffected subjects. The Crown's handling of the Epworth commoners' protests should therefore be understood as an example of coercive royal state building, at a time when Charles and his ministers worked to expand and consolidate the king's powers on many fronts.

Long before the drainage works were completed in the Hatfield Level, another project of vastly greater scale was already getting under way: the first serious attempt at draining the Great Level took place during the early 1630s, and it soon eclipsed the Hatfield undertaking. In many ways, Hatfield Level was a sort of dress rehearsal for the much larger scheme to follow—the two projects shared a number of similarities, from the technical means ultimately used to effect the drainage, to the state's handling of popular resistance. As in previous attempts, however, the sheer complexity of local, regional, and Crown interests within the Great Level ensured that the path of the project would be as winding and difficult to control as the unruly rivers themselves.

CHAPTER 6

The First Great Level Drainage, 1630–1642

[W]e still hold our resolution, to have all the fens drained. . . . [W]e
advise you not to stop in your proceeding upon unnecessary difficulties,
nor to give ear to froward and ignorant men, who had rather live a
poor and lazy life than a rich and industrious, for in works tending to
a general good, your considerations are not so much to reflect upon
the loss or benefit of this, or that particular, as thereby to hinder the
good of the public.

　　　　　　　　　—Charles I to the Lincolnshire commission
　　　　　　　　　　　　　　　of sewers, 20 February 1630

In 1620 or so, an anonymous fenland commentator offered a scathing critique of
the would-be Great Level drainage projectors, Sir William Ayloffe and Sir Anthony Thomas. Rather than criticizing the details of their proposed project,
which they had refused to reveal, he attacked the very notion of "the keeping dry
of all fens winter and summer," a goal he characterized as "utterly impossible"
and "beyond reason." The author claimed great knowledge in fenland drainage
matters, acquired not only from "mine own experience in those kinds of works,"
but also from having learned "the state of these countries in ancient times by
reading the records of religious houses herein seated, and other authors' writing
of the subject." He contrasted his own lifetime of experience with the projectors'
"want of knowledge in the nature of the fens . . . [and] their weakness in the science of those things which they undertake." While admitting that he might yet
be proven wrong, he was nonetheless confident in his assessment: "if I err, I err
with the multitude of those whose judgment and long experience is (I acknowledge) far beyond mine."

　　The author explained that the Fens were not all of one type or quality but
were highly variable and that the key to managing and exploiting them properly
was a sound understanding of each local variation and the particular possibilities

it allowed. The best quality fenlands, which he termed *"cenosa palludes,"* were "exceedingly profitable" as summer pasture in a good year, but only if they were permitted to flood in their turn: "experience hath taught that except they be in the winter drowned they are the worse in the summer next following." If such lands could be improved so as to make every year a good year, then "the undertakers will deserve reward, if by their labors it prove so. But to think by any means to make them better than in good years we have found them, let no man once nourish such a concept in his mind." The wetter types of fenland, *"profunda palludes"* and *"profundissime palludes,"* were incapable of much improvement, yet not without value. They yielded a diverse wetland produce, including reeds, sedge, fish, fowl, and peat, from harvesting all of which "a multitude of people get their living."

To support his claims, the writer cited the manor of Thorney Abbey as an example, a fenland estate of several thousand acres near Peterborough in the Great Level. Thorney was once a wealthy monastery, but after the Reformation it was seized by the Crown and was then acquired by John Russell, 1st Earl of Bedford, in the mid-sixteenth century. The Russells, like the abbots before them, had long sought to improve the land, though they had never succeeded in permanently draining it. But the anonymous commentator denied that such sound and sober men had ever been ignorant enough to aim at such a thing in the first place: "I doubt whether ever any men of experience and judgment did desire it, knowing it to be impossible to keep it dry winter and summer." Indeed, on hearing the projectors' claims that they could do what was patently impossible, those living on the estate believed they meant to accomplish it "by conjuration . . . men not finding any reason how they should perform their large promises, do attribute to the devil what they think man cannot do."

The Fens could never be permanently drained, wise men understood, because it was simply in their nature to be flooded. "[N]otwithstanding the most industrious care of the abbots in maintaining the drains and watercourses," the author wrote, ". . . to find a time when they were not fens, I think cannot be found within the compass of time. And for my own part I do think, that they were even so ordained by God." Nor did the Fens' watery nature indicate any flaw in God's divine plan for creation; they were a manifestation of the "beautiful order of nature . . . in disposing of rivers, meres, plains, &c. so as every thing helpeth other in his kind." This natural harmony and abundance was "like to be destroyed" if the projectors had their way, in their shortsighted belief that only dry, solid land was of any use to the commonwealth. Such a limited point of view had no place in God's great, diverse, and miraculous creation: "If this little world of our bodies were all arms, or all legs, it would prove very unfit for use, and the same I do conclude to betide in the bigger world, I mean this fenny region." Ayloffe and Thomas, then, were narrow-minded, ignorant, foolish, and irreverent in their desire to drain the Fens and would certainly come to grief if they pursued it. "We have always observed," the writer concluded, "that no man having true knowledge of the fens did ever undertake them; neither did any undertake them but with repentance."[1]

A decade later, in 1629, another knowledgeable, anonymous commentator offered a vigorous endorsement of a new proposal to drain the Great Level. The author, identified only as "H. C.," considered not only whether such a project was feasible but "whether it would be honorable and profitable to the king, and common-wealth in general, and to those countries in particular, if it might be effected."[2] As to its feasibility, he believed the land was clearly "well disposed to a draining, both because it hath a sufficient fall to the sea, and because the sewers . . . may be in a short time sufficiently eased." In most years, the region's rivers already drained the land by summertime; the question was whether those rivers might be improved, "sufficient to empty them sooner and with a stronger current than now they have," something he deemed eminently possible.

The author was hardly alone is his view; indeed, he wondered how anyone could doubt the feasibility of draining the Fens when so many august and knowledgeable people believed it to be possible, and he invoked several notable figures who had shared his optimism. He cited "my Lord Popham of worthy memory . . . a wise, experienced gentleman . . . very desirous to have adventured upon the draining." He praised the careful and comprehensive survey of Sir Clement Edmonds, who "in the return of his report to the state . . . resolveth that it may be done." And no less an authority than King James I had declared his intent to achieve it, "and would certainly have performed the work royally and really, saving that he was at that time and unto his death, taken and kept off by his weighty affairs of state." To pessimists who might argue that "many and great men have attempted it and failed," H. C. offered a more progressive view: "it may be their misses have taught other men wisdom: there is no greater advantage than to learn by other men's losses."[3]

The honor and profit of draining the Fens were equally obvious. The flooded Fens were unhealthy and unpleasant for those forced to live in them. "What should I speak of the health of men's bodies," the author wrote, "where there is no element good? The air nebulous, gross and full of rotten harres; the water putrid and muddy, yea full of loathsome vermin; the earth spewing, unfast and boggy; the fire noisome turf and hassocks: such are the inconveniences of the drownings." The land itself he likened to a man with failing kidneys: "the disease of the Fen" resembled "the stopping of the urine . . . it first filleth up all the veins of the body . . . and then drowns the patient in his own water." Moreover, the winter floods isolated the inhabitants for weeks, leaving them with "no help of food, no comfort for body or soul, no woman aid in her travail, no means to baptize a child or to administer the communion, no supply of any necessity, saving what those poor desolate places can afford." The remoteness of the flooded Fens also provided a natural haven for outlaws and rebels, "a fastness" from which they might hope to escape the king's justice.[4]

Once purged of all the impediments that prevented the land from draining efficiently, H. C. believed the Fens would be transformed into a healthier landscape and a veritable cornucopia. He saw potential for the Fens to become "a goodly garden of a kingdom; yea a little kingdom itself: as much and as good ground, it is supposed, as the states of the Low Countries enjoy in the Netherlands." His catalog of the likely benefits of drainage was every bit as comprehensive as his criticisms of the flooded Fens:

> So great a quantity therefore of rich land being gained, would marvelously increase & support the multitude of His Majesty's subjects, wherein consisteth the glory and strength of a kingdom. . . . What should I remember the profits which would accrue to the common-wealth? the abundance of provision for victual, flesh, fish, and whit-meats, the breed of horses serviceable both in peace and for war, the rich and necessary merchandizes of wool, hides, tallow, hemp, rape, and such like; the transportation of the commodities of the country from place to place for the use of the neighbor-parts; the ease of travellers, who now are fain to make compass journeys to avoid the overflowings; the convoy of His Majesty's armies if occasion should require, are all public profits, and of excellent consequence to the king and kingdom.

While some skeptics might claim that the flooded fens produced more fodder for livestock than they would if drained, the author countered that this only applied to "such rank trash" as reeds and sedge; it was certainly not the case for good, wholesome grass. And to those who believed the Fens should be left to the divine will, the author replied that the Lord had given mankind—including the fenlanders—the capacity and the mandate to improve their condition through labor: "God Almighty hath taught them by the experience of some dry years . . . the difference between a wilderness of water & a goodly green meadow: to lead them by sense (who are hardly governed by reason) to discern what may be best for themselves and their posterity."[5] Draining the Fens, then, was not only feasible, honorable, and profitable, it was also God's manifest will; to fail to heed God's lessons in husbanding His creation was to ignore reason and shirk one's divine duty.

⌒

These two contrasting portrayals of the pre-drainage Great Level—from "the beautiful order of nature" to "the disease of the Fen" where "there is no element good"—illustrate the debate that had been going on for at least a generation by 1620, in which one's understanding of the Fens determined one's approach to managing and exploiting them. The first text represents the traditional, local, conservative understanding of the Fens as a diverse and productive landscape, pleasant and valuable even in its perennially flooded state, a view long held among the fenland commoners and many landowners as well. Though occasionally troublesome and even dangerous, the floods were nevertheless understood to be the source of an abundant produce that in a good year enabled the fenlanders to make a decent living. Proper management of the Fens thus consisted in maintaining the region's drainage network such as it was, in order to maximize the number, length, and quality of good years.

The second text reflects the more interventionist and centralized understanding of the Fens. The projectors, improvers, and speculators who favored this view saw the land as an unpredictable and irrational waste, ugly, sickly, broken, and dysfunctional. Its true productivity and value existed only potentially, in its capacity to be drained. Proper management of the Fens, in this view, meant taking whatever radical steps were necessary to transform them into dry, healthy land that resembled the surrounding uplands as much as possible. The floods were to be banished; new crops, new techniques, and new tenants were to be introduced. The Fens, both the land and its inhabitants, would then be fully integrated at last into the wider English state and economy.

I have dwelt on these two texts because both were written in the 1620s, on the eve of the first (partially) successful effort to drain the Great Level. The long debate about the true nature of the Fens, begun in the previous century, shifted markedly toward the interventionist side after the mid-1620s. Partly as a result of the apparent success of the Hatfield Level drainage, both the Crown and a circle of prominent landowners within the Great Level became convinced of the possibility and the desirability of draining the region and transforming its economy. The shift did not take place all at once, and it was shaped as much by the complex political and economic circumstances of the time as it was by the lived experience of the Fens, which could be interpreted to suit different stakeholders' needs. Much depended on conflicting understandings of terms such as "drained," "valuable," and "productive." The power to define the nature of the Fens in the present was the power to determine their future—and that of the inhabitants as well. By 1640, the interventionist position had utterly prevailed among the political elites; the English Civil War had little impact in this regard, as fenland drainage was eagerly promoted by Crown and Parliament alike. The imperative to drain the land and keep it dry would continue to dominate fenland discourse throughout the seventeenth century, and almost to the present day.

THE EARL OF BEDFORD AND THE FIRST DRAINAGE PROJECT

In the summer of 1629, as construction on the original plan for the Hatfield Level drainage was nearing completion and the many problems with it were not yet fully apparent, Charles I took an interest in the much more elaborate project to drain the Great Level, an area roughly five times as large. The king wrote to the commissions of sewers for all of the Great Level counties in June and ordered them to take up the drainage project there once again. He noted his predecessors' long-standing desire to drain the region, though "for want of perfecting a general bargain with the country, this worthy design hath been foreslowed," and he declared his resolution to see the project through to completion this time. To get things under way, he recommended to the commissioners none other than Sir Anthony Thomas and his partners—the same Thomas who had failed to drain the region a decade earlier.

Charles commanded the commissioners to convene two sessions of sewers— one at Huntingdon "for the hithermost Great Level," and the other at Boston "for the other level" in northern Lincolnshire, an area comprising the East, West, and Wildmore Fens. At each session they were to "aid and assist [the projectors] in expediting of their contracts," and "to further them with your authority in all reasonable and lawful things . . . to the uttermost of your power and extent of your commission." No one other than Thomas and his partners should be permitted to negotiate contracts with the region's landowners or to interfere with the undertaking in any way. Charles pledged to support the project "with our kingly favor and Council of State," and concluded by "charging and commanding all persons whatsoever to be respective and conformable to this our royal pleasure."[6]

The king's will met with some resistance. The commissioners in northern Lincolnshire expressed their willingness "to assist and aid the undertakers . . . in all dutiful and obedient manner," but they made no move to advance the project.[7] Charles wrote again in December 1629 to prod them back into action. He warned that "the public good is often hindered or retarded by the interest of particular men" and advised the commissioners not to "give ear to any . . . who out of a froward and crafty nature . . . shall endeavor with unnecessary disputes or unseasonable averseness to retard the conclusion of a general bargain and agreement with the said undertakers."[8] After receiving the king's second letter, the commissioners finally impanelled a jury to examine the state of their fens, but the jurors returned a presentment concluding that "[w]e have no surrounded grounds, neither with fresh nor salt waters . . . but such as are sufficiently drained and in convenient time by our own works of sewers . . . made fit and good pasturing for our cattle."[9]

When the Lincolnshire commissioners reported this finding back to the king, he was apoplectic. "[W]e still hold our resolution, to have all the fens drained," he thundered to the commissioners, and commanded them "not to stop in your proceeding upon unnecessary difficulties, nor to give ear to froward and ignorant men, who had rather live a poor and lazy life than a rich and industrious, for in works tending to a general good, your considerations are not so much to reflect upon the loss or benefit of this, or that particular, as thereby to hinder the good of the public." He dismissed the offending jury presentment, noting "how partial and unsafe verdicts are," and ordered the commissioners to proceed "by your own view or knowledge . . . [and] comply with our pleasure herein." Otherwise, the king would be "constrained to interpose our royal power and prerogative . . . to force froward and adverse men to give way to that which is for the public good."[10] Unlike his father, Charles was not interested in building a fenland consensus before proceeding with his intended drainage. He threatened and bullied the local commissioners into obedience, just as he had in the Isle of Axholme. This approach had the desired effect in Lincolnshire; by the following May, the commissioners had obediently contracted with Sir Anthony Thomas and his partners to drain the East, West, and Wildmore Fens.[11]

Charles also encountered opposition in the Great Level. Just as they had a decade earlier, the commissioners there informed the king that without knowing the precise details of Thomas's plan for the drainage and who was likely to receive benefit from it, they had "no legal power to contract with them, nor to bind the country unto so unsupportable a charge."[12] After several weeks of fruitless negotiation, Charles was determined to break the impasse. The Great Level commissioners had shown "no respect nor conformity to our pleasure," he railed, "but rather such a proceeding as could not but induce distraction, and in the end the overthrow of the whole business." Rather than allowing the project to stagnate any longer, the king declared in February 1630 that "we have taken into our own hands the care of draining the said level of the six counties [the Great Level]."[13] He believed the commission of sewers would not be bold enough to obstruct the proceedings if they were dealing directly with the king as projector—a conclusion James I had also reached in 1621, after Thomas's previous failure to secure an agreement there. It had been comparatively easy for Charles to bend the sewer commissioners to his will in northern Lincolnshire, as he had in the Hatfield Level, because in each of those regions the Crown was by far the largest landowner and thus had unilateral power to settle such matters. But circumstances were different in the Great Level, a much larger territory in which the king would have to contend with a multitude of other landowners, the wealthiest and most powerful of whom was the Earl of Bedford.

Francis Russell (c. 1587–1641), 4th Earl of Bedford and lord of Thorney Abbey, had inherited his title and extensive properties in 1627, upon the death of his cousin Edward (fig. 6.1). The Russells had owned lands in the Fens since the mid-sixteenth century and had long desired to improve them.[14] Francis's father, William Russell, Baron Russell of Thornhaugh, had served as a military commander in Ireland and the Low Countries during the mid-1580s and was appointed governor of Flushing from 1587 to 1589.[15] He took an interest in drainage projects while in the Low Countries, presumably with an eye toward his fenland estates, and on his return to England in 1589, he sought to put his new knowledge to good use. Within a year he had become a leading figure in the new commission of sewers for the Great Level counties, and he soon sought permission from the Crown to bring over from Holland some men "of good ability & skill in draining marsh grounds" to drain Thorney Abbey, "which is now desolate & unprofitable." He intended for these "men skillful and industrious, that are experienced and brought up in such affairs," to become his tenants and farm the land as they did in the Low Countries, providing a profitable example to the English fenlanders "who are now doubtful or careless."[16] William Russell's attempt to drain his Thorney estate came to nothing, and his interest in drainage projects made no direct impact on the English Fens, but it may well have inspired his son and heir to pursue a similar course.[17]

In addition to his ongoing political responsibilities as a highly active and respected member of the House of Lords, Francis Russell proved himself to be an unusually attentive and capable manager of his vast estates and a visionary investor in improvement and urban development projects. As an investor, he was willing to sustain financial losses for years together in order to see a project through to a successful, and ultimately profitable, conclusion. Among his most notable endeavors was the development of the Covent Garden neighborhood in London during the 1630s, an expensive and trouble-prone venture that eventually came to fruition and added considerably to the family fortune.[18] He also shared his

Figure 6.1. Francis Russell, 4th Earl of Bedford, by George Glover and Peter Stent, c. 1643–67.
Portrait D19959, © National Portrait Gallery, London.

father's interest in land drainage. Long before he became the Earl of Bedford in 1627, he was already a prominent and active member of the Great Level commission of sewers. He had personally accompanied Clement Edmonds's survey in 1618, and Edmonds noted in his report that he "gave great assistance to the business."[19] He had also been involved in the commission's successful effort to scuttle the proposals of Ayloffe and Thomas the following year. By 1629, as the largest and most powerful landowner in the Great Level after the king, Bedford was able to dominate the commission of sewers, and he used that body to block Thomas's efforts once again. But like his father, Bedford was not opposed to draining the Fens per se—he simply wanted to control the terms of the undertaking himself and not be forced to cede a large portion of his estate to an outside projector or to the Crown.

After Charles announced his intention to take over the Great Level project in his own right, Bedford emerged as the chief proponent of fenland drainage among the commissioners of sewers and the leading alternative to the king as projector. In July 1630, he wrote to Sir Henry Vane, English ambassador to The Hague, telling him of his enthusiasm for the Great Level drainage and encouraging Vane to invest in it. "And as touching the fen business," he wrote, "I do assure your Lordship it is so feasible and may be so profitable, the quantity and extent of the drowned grounds by fresh waters being three hundred and three score thousand acres of ground, makes me very willing to force you into such an adventure." He shared his intention "to venture some part of my own shrunken fortune" in the scheme and mentioned that he was negotiating with Sir Cornelius Vermuyden to undertake the actual drainage work.[20] The king, apparently content to have the project advancing at last, gave Bedford at least the tacit blessing of the Crown and had the Privy Council help facilitate the negotiations with Vermuyden.[21]

Under Bedford's direction, the commission of sewers reached an agreement in principle with Vermuyden to drain the Great Level in September 1630. Given the great difficulty and expense of the project, they agreed on the need for someone "who as well for his art & skill in such business as for his other abilities in securing the country from loss, might best undertake the work and in most likelihood to perform the same," and they declared Vermuyden to be "a man every way sufficient." Vermuyden was to drain the Great Level together with neighboring Deeping Fen, a total of roughly 360,000 acres, in return for which he and his financial partners would receive 90,000 acres. In acknowledgment of his approbation and support for the venture, the king would receive 30,000 acres. The commission then ordered that a survey be performed and announced plans to draw up a formal contract at their next session of sewers in November.[22]

Before the deal was completed, however, something derailed it. The traditional account, based on the preamble of the sewer law that finally got the project under way the following year, is that the agreement with Vermuyden broke down when the latter insisted he should receive an additional five thousand acres in recompense and the commissioners rejected his demand. Vermuyden also allegedly faced strident opposition from the fenland commoners because he was a foreigner: "the country by their several petitions . . . [showed] much unwillingness that any contract should be made with an alien born, or any other stranger."[23] William Dugdale, in his account of the affair, also stressed the fenlanders' resis-

tance to having a foreigner lead the project: "But the country being not satisfied
to deal with Sir *Cornelius*, in regard he was an alien, they intimated their dislike
to the commissioners; and withal, became humble suitors to *Francis* then Earl of
Bedford . . . to undertake the work."[24]

A far more likely explanation is the theory advanced by Margaret Knittl, a
historian of early modern fen drainage schemes, that Vermuyden was ousted
from the Great Level project because he was unable to raise capital for the work
from the Dutch investors who had backed the Hatfield Level undertaking. In a
letter written by the cleric Thomas Blechynden to Sir Henry Vane, his patron,
Blechynden mentions that "the business of the fens" had stalled "because the
College of Drainers on that side will not trust Sir Cornelius Vermuyden with the
oversight of the work but from some misunderstanding of his former proceedings
would question him before them."[25] By the end of 1630, as the Great Level negotiations were nearing an end, Vermuyden's relations with his Dutch partners in
the Hatfield Level project were increasingly strained; they had come to distrust
his business practices and to doubt his technical competence in designing and
executing a drainage plan that was obviously problematic. Under the circumstances, Dutch investors were unlikely to finance the much larger and more complicated Great Level project as long as Vermuyden was in charge of it.[26]

In light of Vermuyden's unacceptability—not in the eyes of the fenlanders for
being a foreigner, but in the eyes of Dutch financiers for being untrustworthy
and incompetent—Bedford was forced to alter his plans. Vermuyden was ousted,
but that apparently was not enough to reassure the Dutch of the project's likely
profitability after so many problems had arisen at Hatfield. Bedford, undaunted,
turned instead to his own countrymen to secure the requisite funding. Blechynden informed Vane that the "business & great work of draining [is] now conceived
so feasible & full of hope, that he [Bedford] shall have bearers & sharers sufficient
in England to carry him through that vast undertaking."[27] Unable to find a suitable replacement for Vermuyden, in the end Bedford agreed to take over leading
the project himself.

On 13 January 1631, forty-one commissioners of sewers for the Great Level assembled at King's Lynn and formally named the Earl of Bedford as chief undertaker
in the Great Level drainage project. According to the formal agreement, commonly known as the "Lynn Law," Bedford was to complete the drainage within a
period of six years from its inception. He was authorized to construct whatever
drainage works he deemed necessary for the project, wherever he saw fit, only paying compensation to landowners for any damage or losses to their severals—
commoners would receive no compensation for losses to their commons. In

return for draining roughly 312,000 acres in the Great Level, Bedford and his fellow investors would receive 95,000 acres, to be held in free and common soccage, paying only a nominal rent to the king of £10 annually.[28] Bedford also agreed to form a corporation for the purpose of maintaining the drainage in perpetuity, and the annual revenue from forty thousand acres of his total allotment would be allocated to pay for the necessary maintenance. For his generous approval and agreement to release any royal claims based on his father's abortive 1621 project, King Charles was to receive twelve thousand acres out of Bedford's allotment as well.[29]

The Lynn Law gave the drainage project a sound legal footing, and any violent resistance to it would not be tolerated. In the event that "any riotous or unlawful act shall be committed either openly or secretly, to the destruction of any part of the said works, or to the hindrance or impediment thereof," the commission of sewers was empowered to "repress and suppress all such insolences and disturbances, and . . . do their best endeavors to discover and severely punish the offenders."[30] The king gave his royal assent to the law by letters patent in July 1631, upon which the Privy Council declared it to be "a matter of state, not to be altered or impeached," and ordered "all those whom it may any way concern, to . . . maintain the aforesaid Act, without permitting any alteration." They further warned that if anyone should "presume to stop or hinder the prosecution & execution of the said Act," the Crown would proceed against them for their contempt "in such sort that others may be thereby deterred from committing the like offense."[31]

In February 1632, a little more than a year after the enactment of the Lynn Law, Bedford and his partners agreed to what became known as the "Indenture of Fourteen Parts," which articulated the terms for investing in the project.[32] The land allotted to Bedford for performing the drainage was to be divided into twenty equal shares, which were ultimately purchased by a total of fourteen original investors (hence the "fourteen parts" of the indenture). All but one of the shareholders was English, including Bedford himself (two shares), his son Sir William Russell, the future 5th Earl (two shares), Sir Miles Sandys (two shares), and Sir Robert Heath, the attorney general (one share). The only non-English investor was Sir Philibert Vernatti (originally one share, later two), a Dutchman of Italian heritage who was also one of the leading investors and overseers of the Hatfield Level drainage.[33] The holder of each share was responsible for contributing one-twentieth of the cost of the drainage works, to be paid in as needed, and was entitled to an equivalent share of all profits and land allotments. The shares could also be subdivided, with all costs and allotments to be divided proportionally.[34]

Work on the project began even before all the investors were lined up for it. Bedford informed the king in December 1631 that they had already "made a fair and hopeful beginning of the said work, with the expense of many thousands of pounds."[35] Construction of all the new rivers, ditches, banks, and sluices continued over the next few years. In August 1634, three men traveling through the region described the "new made sluices & devices for turning of the natural course of the waters," and the "little army of artificers" who were building them, "which they need not be long about, having 600 men daily at work in it."[36] The following summer, on a subsequent journey, one of the same party recorded seeing "a numerous company of lusty, stout sweating pioneers hard at work, digging, delving, casting up, and quartering out, new streams, and rivers, to gain ground, and to make that large continent of vast, foggy, miry, rotten, and unfruitful soil, useful, fruitful, & beneficial, & for the advantage of the commonwealth."[37]

Most of the original sources detailing precisely what the "little army of artificers" were building in the Fens during the 1630s have not survived; they were most likely destroyed along with the Fen Office in London during the Great Fire in 1666. However, William Dugdale would have had access to those documents as he conducted research for his 1662 history of the drainage, which contains a list of the major works constructed at this time (fig. 6.2). The most ambitious of these was an entirely new river, seventy feet wide and twenty-one miles long, running from Earith in Cambridgeshire to Salters Lode, near Downham Market in Norfolk. This channel, known as the Bedford River, redirected some of the water from the River Great Ouse, cutting a straight path across the countryside in order to take full advantage of the land's limited gradient, allowing the water to flow more efficiently toward its outfall at King's Lynn. The Nene and Welland Rivers were also deepened, widened, and redirected in places. This effort included the construction of Bevill's Leam, a new river forty feet wide and ten miles long, from Wittlesey Mere just south of Peterborough to Guyhirn in Cambridgeshire; the construction of Peakirk Drain, a new river seventeen feet wide and ten miles long, from Peterborough Great Fen just north of the city to Guyhirn; and the rebuilding of Morton's Leam, a twelve-mile-long artificial channel from Peterborough to Guyhirn, originally commissioned in the 1470s by Bishop Morton of Ely. Several smaller works were either repaired or newly built, including a number of navigable sluices intended to regulate water flow and prevent tidal silt from being deposited in the drains. The largest of these were located at

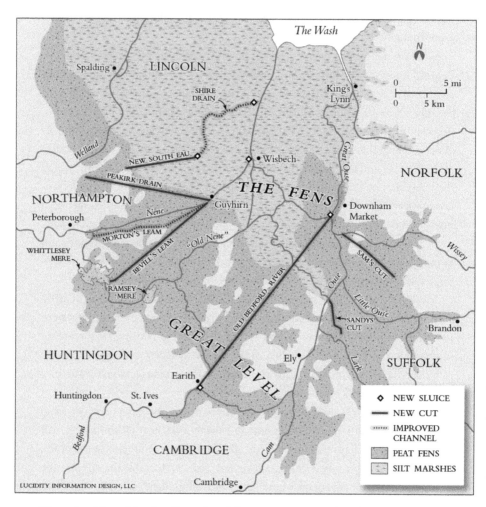

Figure 6.2. Map of the first Great Level drainage project, undertaken 1631–36.
Adapted from a map in Darby, *Changing Fenland*, 66.

either end of the Bedford River at Earith and Salters Lode, and the Horseshoe
sluice just downriver from Wisbech.[38]

The source of the plan for Bedford's drainage project has been the subject of
some controversy. It was long assumed by historians that Bedford engaged Cornelius Vermuyden as the chief overseer of the work in 1631 and that Vermuyden implemented his own plan as approved by the commission of sewers in late 1630.
However, Margaret Knittl has persuasively argued that Vermuyden had little or
nothing to do with the Great Level drainage at this stage. Vermuyden himself

made no mention of his involvement after 1631, nor does Dugdale mention him in his 1662 history.[39] Knittl has shown that throughout the years in question Vermuyden was involved in business ventures elsewhere, embroiled in the myriad Hatfield Level disputes, and even committed to the Fleet Prison for the better part of 1633, including the whole working season that year, for not paying his debts.[40] Nor is it likely that Vermuyden designed the plan carried out by Bedford and his partners, given that so much of what they did differed from Vermuyden's known practices, both before and after the 1630s. Knittl has argued that the plan for the 1630s drainage project came not from Vermuyden but from Bedford's fellow commissioners of sewers, and it consisted of a series of works that several of them had been advocating for a generation or more. All of the principal works constructed in the 1630s project can be traced to various pre-1630 drainage proposals, from John Hunt's design for Popham's venture in 1604–5 to Sir Clement Edmonds's survey in 1618.[41]

If Knittl's analysis is correct, and there is much to suggest that it is, then the series of works constructed by the Earl of Bedford and his partners should be seen not as a foreign project imposed by outsiders, but as "the collective wisdom of the fen community," or at least "its collective wish-list" in redesigning the region's drainage network.[42] I wish to argue that it should also be understood as *transitional* rather than transformative, exhibiting both interventionist and conservative elements. Insofar as the undertakers aimed "to drain the said marsh, fenny, waste, and surrounded grounds" of the Great Level and render them "fit for meadow or pasture, or arable," they certainly viewed the land as flawed, of inferior value, and in need of deliberate human intervention to improve it.[43] Some of the investors, such as Sir Miles Sandys, had sought for years to drain and enclose their estates but had been stymied by the opposition of their neighbors and commoners, who saw such projects as promoting private interests at the expense of the common good. With Charles I on the throne, and especially after the (perceived) success of the Hatfield Level drainage, such resistance was no longer tolerated, and the more interventionist commissioners finally had their way.

Yet the goals, methods, and management of Bedford's project all suggest a lingering conservatism. As local landowners and experienced commissioners of sewers, they had confidence in their ability to manage their own drainage affairs, and the drainage plan they implemented was the product of many years' deliberation among themselves, not a new scheme designed by a foreign projector. The Englishmen's preference for implementing local solutions to local problems is expressed in H. C.'s 1629 pro-drainage pamphlet. He opposed entrusting the drainage of the Fens to Dutch projectors, in spite of their impressive achievements in the Low Countries, partly out of English national pride, but also on the grounds that their preferred methods were inappropriate for the English context. He particularly objected to the Dutch reliance on "mills and other devices to heave out the water," arguing that such "costly and ingenious devices . . . are not proper for our business" because, unlike the Low Countries, the Fens were entirely above sea level.[44] Dutch engines and expertise were all well and good in their place, in other words, but that place was not the English Fens—an opinion apparently shared by those who designed Bedford's drainage scheme.

Moreover, while Bedford and his partners endeavored to alter the condition of the Great Level, the outcome they envisioned was still limited and conservative in nature. The Lynn Law stipulated only that the drained lands "shall be fit for meadow or pasture, or arable," a target not so very far removed from conditions beforehand. The post-drainage Fens would ideally be drier and more predictable than before, but they would still be flooded in the winter and exploited in the summer much as they always had been. What Bedford sought to do, in contemporary terms, was to turn the Great Level into "summer ground"—land that was dependably dry in the warmer months but still marshy in wintertime. The idea was to control and take better advantage of the annual floods, not to eliminate them altogether. More than this, many believed, could not be achieved; as the anonymous commentator had written of the previous lords of Thorney Abbey, "I doubt whether ever any men of experience and judgment did desire it, knowing it to be impossible to keep [the land] dry winter and summer."[45]

Yet the relatively modest goals of Bedford's drainage project gave rise to confusion and conflict, both in the Great Level and at the royal court. His new drainage works may have resulted in an overall improvement in the Fens, but the benefits were uneven, and the land remained summer ground only. This was a disappointment for those who advocated the much more ambitious goal of turning it into "winter ground"—land permanently suitable for arable or any other purpose, similar in quality to the surrounding uplands. While Bedford deemed his drainage a success and soon sought to claim his reward, others viewed it as an obvious failure, an incomplete effort at best and a fraudulent one at worst. The controversy stemmed from different understandings of what truly "drained" fenland should look like, and as always, the fate of the land itself was shaped by the power to define it. When Bedford and his partners moved to enclose their allotments, the dubious success of their drainage created confusion and unrest throughout the Great Level and gave Charles I the means to oust Bedford from his own project.

CHARLES I AND THE PUSH FOR "WINTER GROUND"

As construction on the new drainage works neared completion, Bedford moved to set up the corporation that was to be responsible for maintaining them, and to take possession of the lands throughout the Great Level allotted to him as the projects' chief undertaker. On 13 March 1635, the king granted Bedford's request for a charter of incorporation to the new "Company of Conservators of the Fens," which gave the shareholders legal standing to manage the drainage in perpetuity.[46] Just more than a year later, at a session of sewers held at Peterborough on 13 June 1636, the commission of sewers for the Great Level declared that Bedford had fulfilled the terms of his contract, and that the level was well and truly drained.[47] The experienced surveyor William Hayward, who had previously made a full survey of the Great Level in 1604–5, was commissioned to carry out another to catalog every plot of land that would now be subject to partition; in his exhaustive report he determined the level's total acreage to be 312,668 acres, 1 rod, 30 polls.[48] Between 27 June and 11 August, the sewer commission formally awarded Bedford his ninety-five thousand acres, and a team of surveyors was dispatched in March 1637 to carve up and allot portions of each several and common, plotting and lockspitting all their work as they went. The royal surveyor general, Sir Charles Harbord, led the surveyors in their work, having been sent to the Fens for that purpose by the king's command.[49]

Yet even as the drainage was declared a success, evidence was already mounting that it was not as effective as many had hoped. A traveler crossing the region in the summer of 1635 described "an equilibrian plantation of earth & water . . . an unhealthful, raw, & muddy land," and lamented that his journey required him to traverse "those shaking quagmires, and rotten fens to Guyhirn."[50] In September 1636, just a few months after the project was supposedly completed, a Dutchman named Henrick van Cranhals wrote to the Privy Council offering to correct a number of obvious shortcomings in the new works.[51] While anecdotal, these scattered observations are an early indication that the drainage was already starting to look like a disappointment.

Trouble erupted throughout the Great Level as soon as the surveyors began to partition the lands. Rioting broke out in several areas during the spring and summer of 1637, as angry fenlanders allegedly "made insurrection & mutinously disturbed the draining of the Fens."[52] The Privy Council informed the lords lieutenant in the fenland counties that "there hath not been that conformity and obedience as is requisite" toward the Lynn Law and other recent acts of sewers, and commanded them to restore order in the region.[53] That was easier said than done, however, as many local officials sympathized with the rioters and did little or nothing to stop them. When several rioters armed with scythes and pitchforks threatened to "let out the guts" of the overseer plotting out the division dikes and driving cattle from the common at Holme Fen in Huntingdonshire, the local magistrate, John Castle, refused to assist the overseer and ordered his own servants to prevent the removal of the fenlanders' cattle. At Ely, Oliver Cromwell—the future lord protector—reportedly told the commoners there that if they would pay him a groat for every cow they kept on their common, he would keep Bedford and his partners tied up in lawsuits for at least five years. Castle was also allegedly involved in the scheme, raising money in Huntingdon "for the recovery of the fens," and he told one of his colleagues that money had already been collected in Norfolk, Suffolk, and Cambridgeshire for that purpose.[54] A few officers even participated in riots themselves; the Privy Council's warrant for the arrest of fourteen rioters at Upwell in Norfolk, for example, included the names of two Upwell constables.[55]

Resistance to the drainage had become a serious problem: it spanned multiple counties, and was showing signs of regional coordination; although it had not yet turned violent, it threatened to do so; and too many local officials either tolerated or encouraged it. Much of the unrest was rooted in the predictable reluctance of the fenlanders to surrender a portion of their common wastes. When several rioters in Norfolk later petitioned the Earl of Bedford for clemency, for example, they explained their actions by pleading that "735 souls" in three townships "subsist merely of the fens, and principally get their fuel and livelihood from thence."[56] In other cases, however, the fenlanders' opposition to the division of their lands was rooted in the belief that Bedford had not carried out his end of the contract as stipulated in the Lynn Law, because the land in question was neither "drained" nor "improved."

In the Cambridgeshire town of Chatteris, for example, located along the newly completed Bedford River, the inhabitants insisted that their lands had seen no improvement, but had flooded badly in each of the past three years. Now, in addition to many acres devoured by the new river itself, Bedford claimed a quarter of their severals and half the town common in recompense for his useless project. The commoners' allegations were supported by the lord of the manor there, who drafted his own similar petition to the commission of sewers.[57] Nor was Chatteris an isolated case. In June 1637, the bishop of Ely petitioned the king on behalf of the inhabitants in parts of Suffolk, Norfolk, Huntingdonshire, and the Isle of Ely, claiming that while Bedford's project might be a good one in principle, their lands had seen no improvement from it yet and were in some cases worse off than before. The bishop asked the king for a commission to investigate and to allow the fenlanders to keep possession of their lands in the meantime. His petition was soon joined by two more, one from the magistrates of Norfolk, Suffolk, and Cambridgeshire, and the other from the inhabitants of several Cambridgeshire communities.[58]

When the Privy Council met in July 1637 to discuss the growing list of petitions alleging that Bedford's drainage was a failure, the king attended the session and was outspoken in his support for the projectors. He declared that "the decrees whereby the fens are adjudged drained shall in no wise be impeached as to the judgment of draining," and ordered that the allotment of lands "shall proceed speedily & effectually without interruption by these complaints or any other." Charles was outraged that some fenland magistrates and constables, those "trusted by His Majesty with the government of the country," had acted instead as "stirrers or inciters of complaints whereby the peace of the country may be disturbed and so great & good a work put in danger." He singled out John Castle, ordering that he be removed from the Huntingdonshire commission of the peace and brought before the council "to answer his misdemeanors."[59] The king's tirade put officials in the Great Level counties in a bind similar to that of Hatfield Level officials like Robert Portington, the Yorkshire justice of the peace. Many sympathized with their neighbors in resisting the division of their lands, having seen little benefit from the drainage they were now required to pay for, but if they did not assist the projectors and suppress the rioters, they risked incurring the king's wrath for their insubordination.

Despite the king's dramatic reaffirmation of royal support, however, Bedford's whole project was on increasingly shaky footing. As his partners grew uneasy about the trouble in taking possession of their allotted lands, they stopped paying in on their shares and thereby cut off the company's cash flow. Bedford soon found it impossible to pay his workmen, who also petitioned the Privy Council for relief.[60] Even more worrying was the fact that the king, like his fenland subjects, was growing frustrated with the apparent inadequacy of the Great Level drainage—the twelve thousand acres allotted to the Crown by the Lynn Law were still flood-prone and had yielded little or no profit to date.[61] To make his displeasure known, the king suddenly withheld his support from Bedford's project at a critical juncture. Several Cambridge colleges and the see of Ely had complained to the Crown regarding unfair allotments from their lands, and the matter had been referred back to the Great Level commission of sewers in July 1637 for resolution. The commissioners met at St. Ives on 12 October and issued an amended version of the Lynn Law to adjust the disputed allotments. The changes they made were minor and were not intended to alter the basic terms of the contract itself, but when they submitted their new decree, which came to be known as the "St. Ives Law," to the king for the required royal assent, Charles declined to give it.[62]

The king may have planned at first only to delay his assent, to put pressure on Bedford to build additional drainage works to improve the condition of the Crown's allotment—a tactic echoing the Crown's insistence that the Dutch Participants in the Hatfield Level drainage should build a new river at their own expense to alleviate flooding in neighboring communities. By early 1638, however, Charles had sensed an opportunity to oust Bedford and his partners from the project altogether so that he could step in and take it over himself, on his own terms. To do so, he would have to show that Bedford had not fulfilled the terms of the Lynn Law contract in spite of the judgments he had already received in his favor from the commission of sewers. The issue hinged on whether or not the Great Level was truly drained, as the commission had declared, and the answer to that question came down to what one meant by "drained."

Bedford seemed perplexed by Charles's unaccustomed disregard. He petitioned the king in November 1637 and again in February 1638 to beg for his royal assent to the St. Ives Law, without which the land allotments could not legally go forward and his investors would be ruined. He pointed out that they had been "encouraged to become adventurers in that vast & hazardous work" based on the king's approbation in 1631, "without which none of the petitioners would have adventured so deeply in so chargeable & dangerous an undertaking." After many years of work and "very great sums of money" expended on it, they had "accomplished the same according to the tenor of their undertaking, as appeareth by four several judgments or decrees of sewers made & given since the perfecting of the said draining." He reminded the king that these judgments had all been "approved by Your Majesty with express order, that the same should not be any way impeached." Without the king's further assent, however, "the same doth now lie dead & fruitless, and the petitioners thereby deprived . . . of the benefit they at last expected by their adventure."[63]

Charles's response placed the whole undertaking in grave doubt. Upon further investigation, it seemed, the Crown's agents had determined that "notwithstanding the judgments & informations that have been given for the draining, yet the same is so imperfectly performed, that as well the country as His Majesty remain much unsatisfied therein." Though the Great Level ought to have been "made fit for culture and tillage," Bedford and his partners had "refused to undertake" to make it so. The king had therefore referred the matter to the lord treasurer, the solicitor general, the attorney general, and the surveyor general, in order to determine "the true state of the same . . . & what course may best be taken by His Majesty for the perfecting, draining & improving of this level which His Majesty . . . is resolved to accomplish."[64] The four men soon certified to Charles that not only was the Great Level drainage "not performed according to the undertaking," but that the Lynn Law itself and the tax upon which it was based were both illegal, because they had not specified each particular plot of land to be assessed. Having denied both the efficacy and the legality of Bedford's project, they advised the king to issue a new sewer commission for the region, and to start anew.[65] Bedford insisted that he and his partners had relied throughout on legal advice they had received from "Your Majesty's then learned counsel & others Your Majesty's ministers," but the king was unmoved.[66] He appointed a new, handpicked commission of sewers to hear the complaints of all of "the freeholders, commoners, & other parties interested," and ordered them to publicize their next session widely so that everyone who wished to speak might be heard.[67]

The new commission met at Huntingdon on 12 April 1638, and the session lasted six days. They heard numerous petitions and complaints from fenland inhabitants, took sworn testimony from surveyors and other witnesses, and also made a view of the surrounding countryside themselves. Despite their predecessors' decrees to the contrary, they determined the Great Level to be "very imperfectly drained." Bedford's new works were deemed to be incomplete and insufficient, "too weak to abide the height and force of the winter floods and waters." The fens remained "subject to be overflown and hurtfully surrounded and are annoyed with water in such ways as are not fit for arable, nor safe for meadow or pasture."[68] But all hope was not lost: in the midst of their deliberations, the commissioners received a letter from the king. Having been informed that the level "hath still great need of draining and may be much improved thereby," Charles offered ". . . to take upon ourself the care to see it speedily and well performed, as a work of the greatest importance that hath been attempted in this kind for our own service and the public good."[69] The commissioners, to no one's surprise, accepted in principle the king's offer to take over the project himself. Nor did they stop there: at another session of sewers held at Wisbech the following May, they further declared that several of the smaller drainage projects in fens surrounding the Great Level were likewise unsuccessful and added these lands into the contract they were preparing to offer the king, bringing an additional eighty-five thousand acres into the new project.[70]

At another session held at Huntingdon in July 1638, the commissioners formally ruled Bedford to be in default of his original agreement for not completing his project within six years, as stipulated in the Lynn Law. The Great Level fens were described as being "imperfectly drained, and were not in many parts thereof any ways meliorated by any of the said Earl of Bedford's works."[71] They also confirmed that the St. Ives Law was invalid, because it had never received royal assent. The commissioners then stripped Bedford of the project and accepted the king's offer to take it up instead, "that it may no longer suffer by the neglect and default of others." The king agreed to drain some four hundred thousand acres of Fens within seven years, and most important, to render them fit for meadow, pasture, or arable *in all seasons*—in other words, to make them into winter ground, as opposed to Bedford's realization of summer ground only. In return for his benevolence, the king would receive 152,000 acres from the Great Level, along with tens of thousands of acres from the smaller surrounding levels. However, insofar as some of Bedford's expensive drainage works would likely prove useful in the king's new endeavor, he and his partners were to receive the generous compensation of forty thousand acres out of the king's allotment, in acknowledgment of their efforts.[72]

The determination that the Great Level was not really drained after all, despite previous judgments to the contrary, and that the king was to take over the failed project, sowed considerable confusion and unrest in the Fens. Bedford's agents were still surveying and enclosing land in the Great Level, but it was no longer clear whether he had rightful claim to ninety-five thousand acres, forty thousand acres, or nothing at all. If Bedford's drainage was inadequate, as the king himself had admitted, then the fenlanders had the right under the Lynn Law to repossess any lands already allotted to the earl and his partners. Acknowledging this, the king "took notice of diverse complaints, that men . . . were excluded from the possession of their lands & commons whose grounds were not drained," and he instructed the commission of sewers "to restore such men to their possession until the lands were adjudged drained." The commissioners sought to manage this tricky process legally, conservatively, and peacefully: "[Y]et so as we admit none . . . but such as make due proof that their grounds are not bettered by the Earl of Bedford's draining, and that they have the special order of the court [of sewers] for it, & that they shall not pull down any hedges, ditches, or fences but make a gap and enter in a peaceable manner." This approach, they believed, was "agreeable to the rules of justice, and hath given a great content-ment to the several counties."[73]

Yet while acknowledging the justice of the inhabitants' grievances, neither the Crown nor the commission was prepared to countenance fenland common-ers taking matters into their own hands, filling in division dikes and driving their livestock back onto their former commons without sanction. The problem was that, under the circumstances, it was very difficult to distinguish between peaceable, upright landowners and commoners who were legally repossessing their lands, and malicious rioters allegedly bent on mayhem. In May 1638, Sir Francis Windebank, the king's secretary of state, issued a warrant for fenland of-ficials to suppress "diverse disordered and mutinous persons in sundry parts of the Great Level" who were vandalizing the new drainage works and harassing the la-borers still digging division ditches.[74] When they came to execute the warrant, however, the commissioners of sewers were concerned that it was not specific enough and might be invoked against even those with permission to reoccupy their lands. Rather than enforce it, they returned it to Windebank and asked him to be more precise, addressing only those "who shall riotously or contemptuously pull down any hedges, ditches, fences, mounds, etc. . . . without special order of the commissioners of sewers."[75]

The difficulties and ambiguities were compounded by the fact that many al-leged rioters insisted they were only acting in accordance with the Lynn Law, which explicitly gave them the right to repossess their allotted lands once the commission of sewers had declared Bedford's drainage a failure, and by the fact that many magistrates, constables, and other local authorities agreed with them. In early June, for example, Sir Miles Sandys wrote to his son describing "a great riot made at Wicken [in Cambridgeshire] by hurling in my Ld. of Bedford's works," and the "treasonable speeches" uttered by some of the rioters. He con-cluded with an alarming postscript: "Whilst I was writing this letter, word was brought by my Ld. of Bedford's workmen, that the country rose up against him, both in Coveney & Littleport, by the example of Wicken men. And I fear if pres-ent order be not taken at the beginning, it will turn to a general rebellion in all the Fen towns."[76]

But the events Sandys described were less rebellious and more complicated than he imagined. At Wicken, several "riotous" inhabitants had peaceably filled in the new drainage ditches dug through their common. The local magistrate, Isaac Barrow, sympathized with the commoners and was reluctant to crack down on them; he refused to bind over anyone involved, and while he persuaded two of them to go to jail voluntarily as a token demonstration of his authority, he promised to intercede in their behalf. When the Privy Council dispatched two messengers to take the rioters into custody, the local constables refused to assist them, claimed that the magistrates, the sheriff, and other local officers were all in London, and warned them not to enter the town. The local curate also tried to dissuade them, warning them that "his parishioners were desperately bent to do mischief," and also speaking at length "against the draining & taking in of the fens." When the messengers finally entered Wicken, they met an assembly armed with pitchforks, poles, and stones. They moved to arrest one of the rioters, John Morcelacke, but he brandished his pitchfork at them and replied that "he would obey no warrant from any lords"—though the curate suggested that "he did not refuse to obey the lords of the councils' warrant, but the lords adventurers." Morcelacke's wife, meanwhile, boldly asserted that "the chief of the parish had a hand in this business though they would not be seen in it." The messengers finally retreated from the town, as the crowd jeered at them.[77]

The Cambridgeshire magistrates later wrote to the Privy Council to justify and explain their alleged acquiescence in the Wicken "riot." They had impanelled a jury to investigate the incident, as was their duty, but the jurors would only indict one man, acquitting all others for lack of evidence. The alleged rioters claimed to have acted "in their own right of commonage," according to the terms of the Lynn Law, once Bedford's project was ruled a failure. If it was truly the king's will that their common should be divided, the commoners vowed that "they would ditch it up again, at their charges."[78] This moderate account was supported by the Earl of Exeter, who wrote from the Great Level to reassure the king that "Your Majesty shall not need to fear a general revolt" in the Fens. The riots there were the result of "the rage the poor people bear to these enclosures," and though local officials "do allow of their works . . . and in that respect will levy no arms against them," they did so "knowing their intents to be no further." Except for their understandable reluctance to prosecute their neighbors for questionable crimes, the officials were "otherwise touching their allegiance dutiful subjects."[79] Though perhaps relieved to hear that the Wicken men did not intend a "general rebellion," as Sandys had feared, the king and his council cannot have been altogether reassured by the fact that fenland officials at all levels had refused to suppress, apprehend, or punish men accused of rioting.

The circumstances surrounding the unrest at Littleport and Coveney, both villages near Ely, were similar to those at Wicken. Some fifty men had gathered at Whelpmore Fen and began "throwing down the undertakers' ditches, but not hurting any man's person or goods." A handful of ringleaders were soon arrested and quiet was restored, though the magistrates reported to the Privy Council that "our warrants that we send in His Majesty's name are resisted by some, neglected by others, and some that are charged in His Majesty's name to aid the constables make light of it and refuse it." One chief instigator, called Edward Powell, claimed when questioned that he was only acting in accordance with the king's wishes, and in defense of himself and his neighbors. "I will not leave my common until I see the king's own signet," he allegedly declared, "I will obey God & the king, but no man else, for we are all but subjects. Why may not a man be inspired? Then, why not to do the poor good about their commons?" The magistrates were alarmed by Powell's "bold & dangerous" speeches, as well as the fact that commoners from multiple communities seemed to be "holding privy intelligence" with one another.[80]

The incidents that took place at Wicken, Littleport, and elsewhere in 1638 were not a harbinger of fenwide rebellion, as Sir Miles Sandys had feared; their cause can be traced at least in part to the confusion created by contradictory official decrees, which suggested that the Great Level was both drained and not drained. Though undeniably "riotous" insofar as they were destructive and not sanctioned by legal authority, the fenlanders' actions were both modest in extent and nonviolent, targeted solely at the division dikes and other drainage works built by the Earl of Bedford as part of a project now formally declared a failure. Magistrates, constables, and commissioners of sewers recognized as much, and were reluctant to treat their tenants and neighbors as criminals merely for taking back what was legally theirs. The perpetrators had good reason to believe they were acting lawfully in reoccupying their commons, at least according to the most recent sewer laws, and their professions of loyalty and obedience to the king were not empty words. Indeed, it is striking to compare their self-justifications with the much more seditious language ascribed to the Axholme commoners a decade earlier. In the Great Level, the Crown was still perceived to be a likely source of relief and protection from the illegal depredations of the Earl of Bedford and his partners. Recognizing this, the commissioners of sewers urged the king to publicize his intent to take over the project himself, as quickly and widely as possible, believing that this would help to dispel confusion and calm the situation.[81]

In the meantime, the Crown began to consider the massive undertaking it had assumed of transforming all of the Great Level and surrounding fens into winter ground. Charles issued a royal writ authorizing his ministers to find, allocate, and disburse funds in pursuance of the drainage project, and miscellaneous state papers illustrate the Privy Council's attention to the matter.[82] Charles's plans were certainly grand in scale. After draining all the land, he apparently intended to build "an eminent town in the midst of the level, at a little village called Manea, and to have called it Charlemont"; he even drew the plans for several of the town's main buildings himself.[83] In September 1639, Charles appointed Sir Cornelius Vermuyden as his agent in the works, with a warrant to hire workmen and purchase supplies using Crown revenues.[84] Delays beset the project, and almost a year later Vermuyden was still vetting his design for the drainage with the commissioners of sewers.[85] Yet even as Vermuyden prepared to begin construction, the Great Level drainage was swept up in the political maelstrom that soon engulfed all of England.

A NEW DRAINAGE PROJECT FOR A NEW COMMONWEALTH

Charles I's relationship with his Parliaments was deeply strained for most of his reign. Throughout the 1620s and 1630s, the Crown's need for increased revenues grew ever more pressing, but Parliament consistently refused to provide additional taxation unless Charles would agree to make some policy concessions and accept certain limitations to his royal prerogative power. They particularly objected to his collection of various prerogative-based taxes and other revenues not specifically approved by Parliament; his support for the conservative, anti-Puritan religious reforms in the Church of England pursued by the unpopular archbishop of Canterbury, William Laud; and his ongoing use of royal prerogative courts such as Star Chamber to enforce his will. Their opposition culminated in the passage of the Petition of Right in 1628, in an effort to rein in what many had begun to see as Charles's growing list of abuses of power. In response to their intransigence, Charles dissolved Parliament in 1629 and determined not to call that body into being anymore if it could be avoided, beginning a period lasting throughout the 1630s that came to be known as the king's Personal Rule.

In April 1640, in the wake of the Bishops' Wars in Scotland, Charles was forced to call a Parliament to fund a military campaign to put down the rebellion in his Scottish kingdom—the first English Parliament to meet in eleven years. This "Short Parliament" was dissolved less than a month later, however, after clashing with the king over religion, taxation, and control of the army.[86] That spring also saw renewed anti-drainage protests and rioting in parts of the Fens, particularly in the Lincolnshire fens outside the Great Level, where various royal courtiers had been unilaterally awarded contracts to drain that land during the Personal Rule. The Great Level was comparatively quiet in 1640, though some rioting did break out in the twelve thousand acres originally allotted to the king by the Lynn Law. The commissioners of sewers there examined several alleged rioters who had "contemptuously intruded upon His Majesty's said grounds," driving their cattle into the enclosures and mowing and carrying away the grass there. The commissioners fined several perpetrators and committed them to jail.[87]

When the ongoing crisis in Scotland escalated with a Scottish invasion of northern England, Charles had no choice but to call what came to be known as the "Long Parliament" in November 1640. As the members wrestled with more prominent complaints and perceived injustices such as ship money, royal prerogative courts, Laudian religious reforms, and evil counselors to the king, they also took up the issue of fen drainage.[88] The outbreak of further anti-drainage riots in 1641, in the Great Level and elsewhere, kept the subject before both houses of Parliament and even created friction between them. The Lords tended to side with the projectors and landowners in the urgent need to quell the rioters, while the Commons insisted that the fenland commoners' grievances deserved a full and fair hearing, with Oliver Cromwell championing the commoners' petitions against the projectors.[89] On 22 November 1641, after a passionate and bitter debate that lasted well into the night, the House of Commons narrowly approved a lengthy catalog of grievances critiquing virtually every aspect of Charles's reign from 1625 onward. Known as the Grand Remonstrance, the long battle over its passage and presentation to the king had irreparably divided the Commons into royalist and parliamentarian factions and thus helped to ignite the Civil War. Of the many popular complaints enumerated in the Grand Remonstrance, one specifically addressed the Fens, alleging that "[l]arge quantities of common and several grounds hath been taken from the subject by color of the Statute of Improvement, and by abuse of the commission of sewers, without their consent, and against it."[90]

Yet even in the House of Commons, not all drainage projects were condemned out of hand. The smaller, courtier-dominated projects in Lincolnshire were deeply unpopular and were viewed as instances of the king's high-handed disregard for the well-being of his subjects in favor of the private interests of his cronies; these obviously attracted little support in Commons. The Earl of Bedford's undertaking, on the other hand, received a much more sympathetic hearing as the house considered bills to strip the project from the king and restore it to the earl.[91] The principal reason for their divergent handling of Bedford's case was almost certainly political. Unlike the various projects in Lincolnshire, Bedford's Great Level drainage was not primarily identified with royal court corruption—it was, after all, sponsored not by the king but by a fenland nobleman. Bedford also had a long history of opposing the abuse of royal prerogative powers, not least in his support for the Petition of Right in 1628. He was viewed as a leader of the parliamentarian faction in both the Short and Long Parliaments, and he was a confidante of John Pym and other prominent figures in the Commons.[92] Parliament's move to restore the Great Level drainage to him can be interpreted as a favor for a valued political ally and a rebuke of the king's unjust and arbitrary behavior in ousting him from it in the first place.

The House of Commons debated three separate bills to restore the Great Level drainage project to the Earl of Bedford (or his son and heir) during the 1640s. The first was introduced on 7 May 1641; it received a first hearing but failed to progress after the earl's sudden death of smallpox two days later. His fellow undertakers—including William Russell, now the 5th Earl of Bedford and already a major shareholder before his father's death—continued to advocate for a new bill, modified accordingly. In February 1642, shortly after the king's withdrawal from London, the House of Commons referred the matter to a newly established Committee for the Fens. In the meantime, Vermuyden had finally begun construction on his new drainage scheme, unsure of which master he served in doing so. Despite the political confusion of the early 1640s and a chronic lack of funds, he managed to widen some of the River Nene and dig a new channel for it to the sea, partially embank Morton's Leam, and begin building a new sluice at Tydd, near Wisbech. By the spring of 1642, he had allegedly drained some 150,000 acres, nearly half the Great Level, at a cost of more than £20,000, though he was desperately short of cash and could not pay his workmen all they were owed.[93] On 31 May 1642, a second bill was introduced in Commons, though the onset of open war with the king a few months later interrupted its progress and put a temporary halt to Vermuyden's works.[94] Finally, the Committee for the Fens took up the matter once again in 1646–47, after fighting in the first Civil War had ended. They held hearings and drafted yet another bill, but the outbreak of the second Civil War delayed matters yet again. The third bill eventually passed, though not until 29 May 1649, just a few months after Charles's final defeat, trial, and public execution.[95]

Throughout this decade-long legislative process, it became clear that the terms of the debate over fen drainage were definitively shifted. No longer were the flooded Fens understood to be a productive and valuable landscape, capable of being exploited so long as the local sewer commissions worked diligently to maintain the necessary conditions. Nor was the limited, transitional improvement of summer ground viewed as an outcome sufficient to warrant the massive effort and expenditure already bestowed on the endeavor. For the councilors of state, members of Parliament, and drainage projectors who debated the matter (if not for all those who actually lived and farmed in the Fens), a successful drainage could mean nothing less than the permanent transformation of the flooded, dysfunctional landscape into dry, arable winter ground. Such a massive undertaking could be achieved only by uniting and bringing to bear the knowledge, skill, and vision of foreign experts; the acquisition of funding from numerous investors both within and outside the fenlands; the central coordination of a permanent, joint-stock corporation; and the robust support of the state, as embodied in the new Commonwealth government.

The principal sign that the bar had been raised for defining what constituted a successful drainage project was in the nature of the three bills submitted to the House of Commons. The 4th Earl of Bedford and his partners had already spent roughly £120,000 to build dozens of miles of new channels and banks, and they had received judgment in their favor from a commission of sewers stating that the land was sufficiently drained. The new earl might have been expected to ask Parliament for a bill simply confirming the commission's prior ruling and awarding him possession of the ninety-five thousand acres he was due. Instead, all three bills acknowledged that the land had yet to be properly drained; the preamble of the third bill described the Great Level as being "of small and uncertain profit, but (if drained) may be improved and made profitable, and of great advantage to the commonwealth."[96] Regardless of what the commissioners of sewers had ruled in 1636, after 1640 the condition of the Great Level as summer ground was no longer viewed as an acceptable outcome or return on investment.

The higher standard for a successful drainage was also made explicit not only in the leading plan proposed for the project but in the various critiques of it as well. When the Commons took up the matter for the second time in February 1642, the members sought advice from what may have seemed the best source of expertise then available: not the new Earl of Bedford and his partners, nor the local commission of sewers, but none other than the Dutch drainage expert, Cornelius Vermuyden. He had been appointed by the king in 1638 to design and oversee the construction of Charles's ambitious effort to transform the Great Level into winter ground, and he had been actively at work on the project for more than a year by 1642 when the Committee for the Fens consulted him. The committee chair, Henry Pelham, ordered that the manuscript plan Vermuyden had prepared for Charles should be printed, so that "all men whom it may concern to take notice of, may thereby inform themselves, and may make their exceptions against it, and likewise may offer any other design." Vermuyden's printed plan thus became the basis for all subsequent discussion of the project.[97]

Vermuyden opened with a brief history of recent efforts to drain the Great Level, from James's abortive attempt in 1621 to "the late Earl of Bedford" who had undertaken to drain the land "so far as to make it summer ground." The latter project was a qualified success, in that "the said Earl did proceed so far in this work, that it was adjudged to be made summer ground, and the recompense thereupon was set out." But that outcome was unsatisfactory, since the partial improvement had brought no real benefit to anyone: "[T]hey find by experience, that the lands can yield little or no profit being subject to inundation still . . . whereby the adventurers not only became frustrate of their expectation, but also the owners in general, who cannot make that use of their lands as they might do, if they [were] made winter grounds and reduced to a certainty." When King Charles took over the project, therefore, he had asked "diverse gentlemen expert in those works to give their advice how these lands might be recovered, in such manner as to make them winter grounds, to the end the work might be performed according to the contract."[98]

This is what Vermuyden's proposal aimed to do; but while the making of summer ground was comparatively easy, securing winter ground was a much greater undertaking that required a correspondingly radical, aggressive approach. The traditional method of scouring and embanking the extant rivers to improve their flow, Vermuyden explained, would not work in this case because of the sheer size of the Great Level and the challenge of transforming it all into arable-quality land. He suggested that for such an ambitious project it would be necessary "to lead most of the rivers about another way." His plan therefore called for the three major river systems of the Great Level—the Welland, Nene, and Great Ouse—to be rerouted into a series of new, artificial channels and penned in by new banks.

Vermuyden proposed to divide the Great Level into three distinct parts: the north level (between the River Glen and Morton's Leam), the middle level (between Morton's Leam and the Bedford River), and the south level (between the Bedford River and the Great Ouse). For the north, Vermuyden wanted to consolidate all of the area's many smaller rivers into the Welland, and then redirect the Welland to flow into the Nene at Morton's Leam. The Nene's greatly increased volume, he believed, would enable it to scour out its own channel and outfall to maintain a healthy flow to the sea. It would also eliminate the need for an additional eighteen miles of artificial banks and might even help to recover Wisbech as a viable river port.

For the south, Vermuyden had to operate within the constraints of the works already built there, specifically the Bedford River. That new channel had helped to take some pressure off the Great Ouse, but it was far too shallow to accommodate the water it was required to carry. Vermuyden therefore proposed digging a second new river, roughly parallel to the first, and rerouting the entire flow of the Great Ouse into it at Earith. The newly built banks on the outer edges of each river would be made higher than the banks on the inner edges, so that in the event of a flood, the excess water would spill not into the surrounding countryside, but into the empty "washes" between the two rivers, effectively creating a single, massive river half a mile wide. This would consume hundreds of acres of land, but the loss would be compensated by much greater security from floods, and in drier seasons these "washes" could still be used for meadow. The remaining tributaries of the Great Ouse below Earith were also to be consolidated and directed to the outfall at King's Lynn. Finally, the middle level would be secured by the much more efficient flow of the new Nene and Bedford Rivers, by a series of new subsidiary drains, and by high banks constructed along its periphery to keep out surging floodwaters.[99]

Vermuyden obviously did not intend to manage and maintain the Fens as he found them, nor did he seek to improve the land incrementally by tweaking its extant drainage system. His proposal called for a massive, aggressive human intervention in the landscape, a redesigned network of artificial rivers and banks, regionally conceived and centrally coordinated, by which the Fens might be made to "imitate nature . . . in the upland countries."[100] Building such a thing would be difficult, time consuming, and very expensive; but it was the only way to progress beyond the modest, partial achievement of the Earl of Bedford, to transform the Fens from semi-drained summer ground into solid, permanent, and profitable winter ground. With a new drainage network in place, the Fens could become a whole new land.

As the Committee for the Fens had hoped, the publication of Vermuyden's design elicited much criticism, especially from some of the overseers of Bedford's previous undertaking, who published pamphlets of their own condemning Vermuyden's plan. But while these authors disagreed with Vermuyden's specific proposals, doubted his overall approach, and deplored his lack of concrete details concerning materials and expenses, they neither challenged the feasibility of transforming the Great Level into winter ground nor questioned the need for it. Edmund Scotten, for example, was an overseer of Bedford's drainage works near Wisbech, and in his 1642 pamphlet he disparaged Vermuyden's unorthodox plan to build expensive new rivers, suggesting instead that the Great Level could be completely drained using the much simpler and cheaper approach of dredging the existing rivers and augmenting their banks. He was passionate in his critique, and in the highly charged religiopolitical atmosphere of the time he did not scruple to liken Vermuyden to "the popish clergy, who keep men as ignorant as they can, that they may the more easily deceive them, and lead them whither they list." Scotten's proposal was conservative, to be sure, linking him to the more transitional approach he had helped to carry out under Bedford. Yet despite his vehement rejection of Vermuyden's radical design, he nevertheless shared the latter's conviction that "the greatest part of the Fens may be made winter grounds."[101]

Vermuyden's most prolix critic, by far, was Andrewes Burrell, a landholder in the Great Level, a commissioner of sewers, a shareholder in the Earl of Bedford's project, and also a former overseer of Bedford's works along the River Nene. Burrell truly despised Vermuyden; he resented what he saw as the latter's contemptuous dismissal of the drainage works he had overseen in the early 1630s and also the fact that he planned to put a receptacle for water on nearly five hundred acres of Burrell's estate at Waldersey.[102] In 1641–42, Burrell published three separate pamphlets criticizing Vermuyden's prior drainage works and his current plan for the Great Level and laying out his own alternative proposal.[103] He believed Vermuyden was a charlatan and had nothing good to say about his expertise, his character, or his design for the Great Level drainage. He predicted that Vermuyden's new rivers would not be able to handle the volume of water required of them; he condemned Vermuyden's placement of banks far removed from the rivers they protected, arguing that it would retard the rivers' flow and cause them to silt up; he believed Vermuyden incapable of building sluices that would survive the first real storm; and he objected to the lack of crucial details in all of what he called Vermuyden's "mystical discourse."[104]

For all his vituperation, however, Burrell never questioned Vermuyden's goal of turning the Fens into winter ground; he simply doubted whether Vermuyden had the skill to accomplish it. And while he took great exception to Vermuyden's criticism of the Earl of Bedford and his partners, "by undervaluing the works which they made," he did not particularly defend Bedford's accomplishment but rather offered a rather dubious reinterpretation of what had actually been achieved.[105] Burrell made no reference to the earl's acknowledged success in making the Great Level into summer ground, claiming instead that he and his partners had "adventured great sums of money to recover them; but *before their works were finished*, they were circumvented and outed of all their intendments."[106] According to Burrell's account, Bedford and his partners had never been satisfied with summer ground, nor had they viewed their undertaking as completed. He blamed Vermuyden and the king for "hindering their proceedings, to their extreme loss and disadvantage," forcing Bedford out before he could finish what he had started.[107] Since winter ground had become the only acceptable outcome

for the Great Level drainage by 1642, Burrell's defense of Bedford's original undertaking had read that ambitious goal back into it.

The tension in Burrell's pamphlets between the need to rehabilitate Bedford's
Great Level project as a putative success while simultaneously stressing the need
for more new drainage works there is also evident in the abundant testimony
heard by the Committee for the Fens. At a series of hearings in 1646–47, as the
committee considered the third bill to restore the drainage project to the new
earl and his fellow shareholders, informants testifying in favor of the bill had to
walk a fine line. The 4th earl had worked for six years and spent vast sums to
drain the Great Level and render it "fit for meadow or pasture, or arable" and had
claimed success in the endeavor. For this to be seen as anything other than a fraud,
real progress and profit had to be demonstrated. On the other hand, the whole
point of the hearings was to debate a bill that would empower the same people to
drain the same land all over again. Advocates therefore had to argue that the
land was both successfully drained and yet still flooded—much good had been
accomplished, but much more work remained to be done. This was the unifying
theme of nearly all testimony offered before the committee in favor of the bill.

The statement from Mr. Glapthorne of Cambridgeshire, heard before the
committee on 25 June 1646, is typical. A landowner and former commissioner
of sewers, Glapthorne insisted that the level was much improved by Bedford's
drainage works. He testified that, "before the undertaking no beasts could tread
but such as waded . . . where now the plough goes & good pasture is"; that land
once worth two pence per acre was now worth ten shillings; that 140 Walloon
families had recently settled there and had employed all the poor people of the
area; that the land now produced "wheat, oats, barley & coleseed, & it hath twenty
times as many cattle as it had before the undertaking"; and most important, that
"the improvement is made by my Lord of Bedford's works of draining." Yet for all
his praise of Bedford's triumph, Glapthorne also testified that the Great Level
was still "not perfectly drained," and warned that conditions would certainly get
"worse & worse if care be not taken & the works do not go on."[108]

Other informants echoed Glapthorne's testimony. Major Underwood, a fen
landowner, reported that while once "the grounds were very bad [and] of little
worth," since the drainage "they are become very good pasture & arable ground"
worth between five and ten shillings per acre. Yet while "the most part is improved to the value before mentioned," he noted that some areas of the level
"are improvable." Captain Roger Pratt testified that he purchased sixty acres
near Welney for £25 when Bedford first began his project in 1631, and after just
one year he was able to rent it for £12 per year. By 1646, however, "the sluice &

sasse being broken," he only received £8 per year, though he believed the estate would be worth £30 annually if the drainage works had only been maintained. And Robert Barton offered the paradoxical statement, regarding his land in Peterborough Great Fen, that "if it were drained it would be of much more value, [but] that it is much drier now than it was before the draining."[109]

The underlying pattern of all the pro-drainage testimony is clear: Bedford's project had done much good in the Great Level, but the works he built were either incomplete or neglected, so that further improvements were not only possible, but likely. The informants echoed the pamphlets of Andrewes Burrell printed in 1641–42, and indeed, Burrell himself appeared before the committee in February 1647 to confirm his views. He testified optimistically that after Bedford's undertaking "the main body of the fens were not drowned in summer time for 7 years together" and claimed to have realized great profits from his own estate after the drainage "by sowing of coleseed upon those lands." At the same time, however, he claimed that Bedford's works "were not finished," implying that if the king had only permitted him to continue, he would have accomplished much more.[110] Winter ground was clearly the only acceptable outcome for such a large and expensive drainage project, and Bedford had always meant to achieve it; but while he had fallen short, through no fault of his own, it remained a feasible and desirable goal.

Not everyone shared the pro-drainage point of view, of course; numerous informants testified against the project at every stage of the debate. They offered myriad reasons for opposing it, but most still rejected the interventionist view of the Fens as a broken and unproductive landscape. They argued that transforming the Great Level into winter ground was neither feasible nor desirable; that the land was already of significant value in its flooded condition, if properly managed; and that the drainage project was premised on ignorance of the Fens' true nature. The bishop of Ely, for example, complained to the Committee for the Fens that the bill in question treated "all low fenny and surrounded grounds" in the Isle of Ely as flooded and worthless, which was a gross mischaracterization. Some fenlands were less valuable than others, yet not only was much of the land productive, it required periodic flooding to make it so: "[T]here are many low grounds which are of the value of 20 [shillings] per annum within the level," he wrote, "and many grounds are . . . better & more fruitful by being sometimes overflown."[111] Other informants likewise questioned the wisdom and value of draining the land, as recorded schematically in the rough minutes of the committee's proceedings: "The work of draining impossible, a spongy bank washed or blown away . . . if drained, those towns can receive no benefit; (1) Quality much of it as good value as the high ground (2) For reed, sedge & osier ground, of as much value as corn land (3) a base moor not mixed the water taken away it produceth nothing"; "All the towns in Huntingdon & [twenty] in Cambridge do disavow the work & do disagree to all manner of draining & that they can receive no benefit by it"; "as necessary for variety of grounds as of trades."[112]

The many objections to the Great Level drainage bill were unavailing. From the very beginning of their deliberations in June 1646, the Committee for the Fens placed the burden of proof entirely on the anti-drainage side, requiring them to demonstrate that "there is not within this level a considerable proportion of land fit to be improved by draining for the advantage, and by the care of the Common Wealth."[113] The committee met only sporadically after July 1646, and some anti-drainage informants complained that they were unable to offer testimony because there were too many seeking to speak against the bill and not enough time allotted to them. Others accused the committee of conspiring to prevent their testimony altogether through a series of unnecessary adjournments.[114] In any case, on 22 November 1647, the committee proceedings recorded that the counsel for the anti-drainage side had failed to appear at a crucial hearing as scheduled, in spite of being given all due notice of it. With no one present to offer any opposition, the committee resolved, "That a great part of the Great Level of the Fens may be drained," and that "this draining will be profitable to the Common Wealth."[115] The outbreak of the second Civil War in February 1648 temporarily interrupted the bill's further progress, and it was not taken up again until after the king's execution. It was not until 29 May 1649 that the House of Commons finally passed "An Act for the Draining of the Great Level of the Fens."[116]

The Drainage Act of 1649 was both the product of a radical, interventionist understanding of the Fens and the means for making that understanding a physical and ecological reality. It declared that the lands of the Great Level "by reason of frequent overflowing of the rivers . . . have been of small and uncertain profit, but (if drained) may be improved and made profitable, and of great advantage to the commonwealth." The language of the act also endorsed the version of recent events put forth by the new Earl of Bedford and his allies: the late Earl, according to the preamble, "did undertake the work" and "made a good progress therein," for which he and his partners "had ninety-five thousand acres . . . decreed and set forth . . . in recompense thereof." Despite his success, however, "by reason of some late interruptions, the works there made have fallen into decay, so that the intended benefit to the commonwealth hath been in a great measure hitherto prevented and delayed." Bedford's drainage project of the

1630s was thus cast as a great achievement, later undermined by his political enemies, including the late king. This account completely overturned that of the king's handpicked commission of sewers at Huntingdon, who had found the land to be "very imperfectly drained" by the earl when they stripped the project from him in 1638. And indeed, so that any lingering doubt on that score might be laid to rest, the new act declared that the Huntingdon sewer law "shall from henceforth be null, void, and of none effect."[117] Any legal record claiming that the late earl's project had been a failure was thus expunged by Parliament.

The act itself was in many ways a reinstatement of the Lynn Law of 1631, with William, 5th Earl of Bedford standing in for his deceased father as the chief undertaker. He was to drain the entire Great Level by October 1656 in return for ninety-five thousand acres in recompense. He was entitled to sell shares in his undertaking, and the shareholders would receive a proportionate amount of land in return for funding the construction work. One crucial difference between the 1649 act and the Lynn Law, however, was that whereas the latter stipulated only that the land should be made "fit for meadow or pasture, or arable," the former explicitly required that "all the said Level . . . *shall be made winter ground,* in such manner as the said rivers or any of them shall not overflow the grounds within the said Level."[118]

The 1649 act also made clear that this drainage project would not be managed locally by a commission of sewers but would instead be a centrally coordinated venture. This is obviously implied by the law itself—previous drainage projects had been set on foot by laws of sewers, not by an act of Parliament—though the act went much further in decreeing "[t]hat no commissioner or commissioners of sewers, by virtue or color of any commission in that behalf, shall at all intermeddle in the said Level, to interrupt, disturb or molest the said William Earl of Bedford, his participants . . . agents and workmen in the carrying on and perfecting of the said work." The traditional duties of the commission of sewers would instead be borne by the earl and his partners, who would have sole authority to order new works, levy sewer taxes, and employ workmen. Moreover, this new arrangement would be permanent: after the work was completed, only those shareholders with at least two hundred acres at stake in the project would be entitled to wield "such and the same power and authority, as commissioners of sewers."[119] After having been lobbied, manipulated, bullied, packed, and ignored in the promotion of prior drainage projects, the commissions of sewers in the Great Level had now been appropriated—swallowed whole by the drainage projectors, as decreed by act of Parliament. Henceforth, the only recognized drainage authorities in the Fens would be the projectors who sought to transform them into dry land.

In the Grand Remonstrance of 1641, the Long Parliament grieved to the king that "[l]arge quantities of common and several grounds hath been taken from the subject by color of the Statute of Improvement, *and by abuse of the commission of sewers, without their consent, and against it.*"[120] In 1649, shortly after the king had been beheaded for ruling England as a tyrant, Parliament not only authorized the Great Level drainage and allotted ninety-five thousand acres to the investors who funded it, but it even forbid any commission of sewers from "intermeddling" in the project. As far as the English state was concerned—whether Crown or Parliament—the Fens were a broken and profitless landscape by the 1640s, a millstone around the neck of the commonwealth. The members of the parliamentary Committee for the Fens were every bit as determined in 1649 to see the Great Level transformed into dry, productive land as the Crown had been a decade earlier, whatever the fenlanders themselves might have to say about it (and they were given little enough opportunity say anything). The consistency of the pro-drainage position throughout the Civil War period is illustrated by the fact that H. C.'s eloquent pro-drainage pamphlet had two printings: the first in 1629, as King Charles and the Earl of Bedford grappled over who would undertake the Great Level project, and the second in 1647, during the parliamentary hearings that restored it to Bedford's son. A project begun under King Charles I and with his blessing was to be completed with the enthusiastic support of the revolutionary Commonwealth regime that had overthrown him. The Civil War had wrought many changes on England's political landscape, but it had not altered the plight or fate of the Fens.

What *had* been permanently altered, even before the outbreak of war, was the dominant political and economic understanding of the Fens, encompassing not just the land but also the people who lived there, the administrative and legal system through which it was managed, and the agricultural means by which it was exploited. The traditional, conservative outlook, in which the annual floods were not only inevitable but beneficial, and experienced local landowners were entrusted to maintain the conditions necessary to take full advantage of the land's natural produce, was decisively superseded by the more radical, interventionist approach. The entire Great Level would be transformed by bold, visionary investors and managed by central authorities. Traditional fenland produce was dismissed as so much worthless trash, to be replaced by more marketable commodities, such as coleseed that would supply valuable oil for England's protoindustrial markets. The lazy and sickly inhabitants would finally be freed from a deleterious and dangerous landscape and set on honest work for the good of the commonwealth.

Yet even as the new drainage project got under way in the early 1650s, the debate about the true nature of the Fens continued, in the Great Level and elsewhere. Though the interventionist view held full sway in circles of power and money, fenland commoners still saw in their watery world a land of beauty and abundance. They sought to wrest control of their fate from the hands of the drainers who would sweep it all away, and the sociopolitical chaos of the Civil War and Interregnum seemed to offer an opportunity for doing so. Against them, however, stood not only the changing regimes of the English state and the corporate interests of a burgeoning market economy, but a dawning religiophilosophical rhetoric of improvement that viewed land drainage as a vital part of a larger project, to perfect the Commonwealth in the eyes of both God and man.

CHAPTER 7

Riot, Civil War, and Popular Politics in the Hatfield Level, 1640–1656

I make no doubt but your honors have all heard of a pretended great riot some years past, committed in the said Isle [of Axholme] . . . but I am persuaded that you have not heard half so much of the proceedings that have been at law by the freeholders there, for the recovery of their ancient right to 7400 acres of commonable grounds. . . . I most humbly beseech you all (as the safest refuge every freeholder in England in this case hath to fly unto) to take care in it, for there is much of the freedom of the laws and liberties of England, in my judgment, either to be preserved or lost in it.

   —*The Declaration of Daniel Noddel*, addressed to the Parliament
        of the Commonwealth of England, 1653

[U]pon full hearing in the Exchequer a decree was made for establishing the possession with the petitioners, which was published upon the place in the presence of divers of the inhabitants who (having now gotten the influence of Lilburne, Wildman, & Noddel) declared that they would not give any obedience thereunto nor to any orders of the Exchequer or Parliament and said they could make as good a Parliament themselves, some said it was a Parliament of clouts, and that if they sent any forces they would raise forces to resist them.

      —Parliamentary committee report, 1653

The manor of Epworth saw considerable unrest throughout the English Civil War and the Interregnum. The chief manor in the Isle of Axholme in northern Lincolnshire, Epworth had been a hotbed of anti-drainage sentiment ever since Sir Cornelius Vermuyden and his fellow Participants had drained the Hatfield Level and taken possession of 7,400 acres of their common waste, in accordance with the terms of the 1636 Court of Exchequer decree. The Epworth commoners had engaged in various lawsuits and occasional bouts of rioting from 1628 onward, and during the chaotic early days of the Civil War in 1642–43, they had even managed to reoccupy some four thousand acres of their common. They set out to recover the remaining 3,400 acres by force in the summer of 1650, igniting the most violent, destructive, and sustained period of rioting ever to take place in the seventeenth-century Fens. Over the next year and a half, the rioters demolished at least eighty houses, a windmill, and a church; destroyed hundreds of acres of crops; and assaulted the tenant farmers who were leasing the land from the Participants.

According to the deposition of Susan Nante, taken in February 1652 before a Parliamentary committee charged with investigating the riots, several dozen Epworth commoners had entered the land leased by her husband and "took their beasts, all armed with clubs, staves, forks, & spades . . . & so continued until they had destroyed all." Six rioters returned later on and "did pull down all the houses," at which time a man called Alexander West reportedly told her, "it was a shame the Parliament should give away their commons, *they were a Parliament of clouts.*"[1] The final report of the investigating committee (quoted above) made much of the remark, emphasizing its seditious implications and claiming that the Epworth rioters had fallen under the influence of John Lilburne and John Wildman, former leaders of the radical Leveller movement.[2] The phrase "Parliament of clouts" was a loaded one. In early modern usage, a "clout" referred generally to any piece of scrap, especially a shred of cloth or a rag; but it was also used figuratively to refer to an imposter or charlatan, someone who was a pale reflection of the person he was supposed to be.[3] In the political context of the early 1650s, this was a powerful and threatening statement for any commoner to make. Only a year after the remnants of the Long Parliament had tried and executed King Charles I for treason in the name of the Commonwealth of England, and while it was still struggling to consolidate its rule, West had allegedly challenged that body's authority and legitimacy—much as Charles himself had done, when he steadfastly refused to enter a plea at his state trial.

Of course, Alexander West may never have said any such thing. The Parliamentary committee investigating the riots was heavily biased in favor of the Participants, who sought to portray the Epworth rioters as dangerous rebels and Levellers in order to convince Parliament to quash them. The veracity of such testimony must be treated with some skepticism when those recording it had an interest in making it sound as seditious as possible, as other historians have noted. Keith Lindley has asserted, "Any challenge to authority or government inherent in the fenmen's actions was incidental" and that more seditious interpretations could be based only on "an uncritical reliance upon unsubstantiated official reports." Clive Holmes has suggested that many of the statements contained in the riot depositions are "tainted by their essentially polemical purpose" and that they should therefore not be taken as an accurate reflection of "either the events they purport to describe, or the ideals which motivated those involved."[4] Yet while the recorded depositions concerning the Epworth riots must be handled with caution, they need not be dismissed as mere fabrications. I argue that they indicate a genuine and intense political consciousness among the Axholme fenlanders. The commoners there expressed their discontent with the political order on numerous occasions from 1628 onward, using both legal and violent means, and whether their expressions were aimed at the Crown or the Commonwealth regime, they were remarkably consistent in their critique.

The Epworth commoners had as much cause as anyone in England to resent the king's abuse of his royal prerogative powers, the clearest manifestation of which was (for them) the Crown-sponsored drainage project imposed on them, the forcible and unjust appropriation of 7,400 acres of their common, and the denial of any means of legal redress. They were convinced that the king had violated the rule of law in enclosing their common and that he had thus behaved tyrannously. Throughout the Civil War, the Epworth commoners staunchly supported the Parliamentary cause, even supplying two companies of infantry for the army. They also repeatedly attacked the drainage works on their lands and destroyed the crops of the Participants' tenants, who had settled in their midst. These actions were taken, so they claimed, in resistance to royalist tyranny. But after Parliament won the war and the monarchy was swept away, the commoners found that little had changed. The new Commonwealth regime was just as enthusiastic about draining the Fens and profiting from the enclosure of improved lands as the Crown had been, and the commoners were once again denied a fair hearing when their lawsuit was stayed over many years. It is not hard to imagine that, after sacrificing so much in the name of the Commonwealth, the Epworth commoners might have expressed their frustration and outrage when the new government seemed to abandon and betray them. They still did not live in a state governed by the rule of law, as they saw it, and the legitimacy of such a state was therefore open to question.

The seditious utterings and riotous events taking place in the Isle of Axholme during the Civil War and Interregnum illustrate the intense, deep-seated, and increasingly radical popular politics found in parts of the seventeenth-century Fens. The culmination of violence there in the early 1650s may be understood as fenland politics by other means, when traditional political institutions had collapsed, both locally and nationally, and legal avenues of opposition such as petitions and lawsuits continued to prove ineffective. The fact that their loyal support for the Parliamentary cause evaporated so rapidly after its victory in 1649 should not be mistaken for political apathy, ignorance, or neutrality in the larger national conflict. Popular politics in Axholme was principally inflected by local interests, to be sure; but it engaged directly with national institutions, pursued a consistent agenda, and articulated it in terms of broader ideological concerns, including the rule of law and the legitimacy of the state. In the case of the Isle of Axholme, at least, the approach was surprisingly effective in allowing the fen-landers to retain de facto possession of land they continued to insist was right-fully theirs.

THE COMING OF THE CIVIL WAR IN THE ISLE OF AXHOLME

The Exchequer decree of 1636 had awarded possession of 7,400 acres of the Epworth common to the Participants in recompense for their drainage project, with the putative consent of the commoners there; the remaining 6,000 acres was left in common. Though the decree was deeply unpopular in the Isle of Axholme, with the vast majority of commoners insisting they had never consented to it, the rest of the 1630s were comparatively calm, and the Participants wasted no time in enclosing and exploiting their improved lands. By the end of 1636, with the help of Dutch investors in the drainage project, they had settled more than two hun-dred families of French and Walloon Protestant refugees in the Hatfield Level, centered at a village called Sandtoft, northwest of Epworth. The foreigners tilled the soil and planted wheat, rye, oats, barley, rapeseed, and flax. Their efforts met with success: by 1639, local inhabitants were concerned that "a general plenty of all sorts of grain . . . where it is much increased by the draining of the Level" had led to a drop in prices, such that "the husbandmen and farmers cannot be able to pay their rents" unless permitted to export some of the excess.[5] At the same time, inhabitants were threatened by overexploitation of the limited common remain-ing to them, with too many householders grazing too many cattle on too little land, whether they had a right to do so or not—a growing problem even before the drainage, much exacerbated by the enclosures.[6]

Tensions simmered between the foreign tenants and the Axholme commoners, as the latter continued to resent the appropriation of their common waste and the high-handed way in which it was accomplished. Within just a few days of the Exchequer decree, the archbishop of York wrote to his counterpart at Canter-bury expressing his concern that the Participants would only "employ French-men, and a few Dutchmen" on their enclosures. Besides the troubling doctrinal implications of fostering a foreign Protestant religious community, outside the established Church of England, he was concerned lest the influx of foreign husbandmen and laborers "take the bread out of the mouths of English subjects, by overbidding them in the rents of the land."[7] Two of the chief Participants, John Gibbon and Robert Long, justified leasing their lands only to the "honest and industrious" foreigners because the native fenlanders "either out of combination or willfulness will not take the same to pay any considerable rent for that which they say have been their commons."[8] Nor was the fenlanders' resentment entirely passive; Cornelius Vermuyden complained to the king in December 1637 that the "many oppositions by the commoners there" had prevented him and his partners from exploiting their lands. He was particularly vexed that "at diverse times in the night, and in times of flood the walls and banks have been cut, whereby they have sustained threescore thousand pounds loss and damage by drowning of rape & corn."[9]

When the Short Parliament met in April 1640, after eleven contentious years without a Parliament, it attracted a wide array of petitions and grievances from throughout England. Drainage projects and the enclosure of common wastes were principal complaints emanating especially from the Lincolnshire Fens. One observer there informed Robert Long, "There is a general league made in all the fen towns to petition and fee counsel against you in the Parliament house," and another warned him, "I believe you shall hear many more complaints in Parliament than for the present you imagine, for such meetings & combinations are among the people as you cannot suppose."[10] The Lincolnshire petitioners complained that the drainage projects had been undertaken in violation of the rule of law, with the aggressive support of the king, and that they had been permitted no legal recourse against them. "[I]t is an invention clearly without precedent invading the ancient well settled laws of this kingdom," they charged, "& so endangering . . . the wonted security & assurance that men have had in their estates."[11]

Drainage projects were also clearly linked in the minds of many with other, more common complaints of royal malfeasance, such as ship money and Archbishop William Laud's conservative governance of the Church of England. A popular rhyme had urged Lincolnshire electors, "Choose no ship [money] sheriff, nor court atheist, / No fen drainer, nor Church Papist."[12] The Lincolnshire MP Sir John Wray, in a speech before the House of Commons in which he also condemned ship money, declared that "we stand not ensured of our terra firma, for the fen-drainers have entered our lands, and not only made waste of them, but also have disseized us of part of our soil and freehold."[13] Fen drainage projects were thus perceived as one of the king's many recent abuses of power, to be remedied at last by Parliament.

Fenland protests did not stop with petitions and speeches; rioting erupted throughout the Lincolnshire Fens in 1640 as the restive inhabitants sought to reclaim their former commons. Investors in various drainage schemes bewailed that the commoners "violently thrust their cattle into our grounds . . . & others of them have cut our banks in several places and have endangered the drowning of a great part of the cole and rape now upon the ground."[14] The Earl of Lindsey even speculated that the convening of a Parliament, after such a long interlude and in the midst of so much popular outrage, had itself helped to incite the riots: "the people who commit these disorders take encouragement by occasion of this Parliament."[15] The Privy Council struggled to restore order in the Fens, writing to the Lincolnshire magistrates that such "insolences and unlawful courses" deserved severe punishment in order to "deter others from committing the like outrages."[16] But the ongoing rebellion in Scotland was a much more pressing problem for the Crown, and in spite of the local unrest, the trained bands in the region were soon dispatched to the north. The burgesses of Boston fretted to the Privy Council that the "multitude of dikers, poor indigent fellows but of strong & able bodies" then at work in the many drainage projects nearby might "grow into riotous disorders so soon as they shall understand that there is no power in the country to subdue them."[17]

The king's preemptive dissolution of the Short Parliament in May 1640, before fen drainage or any other grievance could be addressed, only increased popular political outrage. When Charles was forced to call the Long Parliament in November 1640 to deal with the Scottish invasion of northern England, the fenlanders once again submitted petitions "wherein they did complain of draining of fens and taking their lands from them."[18] The Lincolnshire petitions expressed a comprehensive array of drainage-related grievances, stretching back for fifteen years:

> Our lands & inheritances are taken from us by pretense of laws of sewers wherein the commissioners proceed without inquisitions by jury, but by their own view, who are the commissioners and parties deeply interested; They take the one half of our commons & fen grounds for supposed draining the rest, which they make & leave much worse than they found it. . . . If we proceed at law for trial of our titles, we are sued, pursuivanted, imprisoned, and are ordered by the lords of the Council not to proceed. . . . Our consents are extorted to make us subscribe to their proceedings, and if we gainsay, they imprison us to fit matters for their ends; Our ancient commissions [of sewers] have been put out and foreigners, Adventurers & Participants are put into

commission. . . . Our commons are taken from us under pretense of commissions to improve for the king as lord of the soil, wherein they take more than half, the possession whereof is settled to the king by Exchequer injunctions afore the cause be heard, which when it's done the cause is delayed.[19]

These petitions, and others like them, ensured that when the Grand Remonstrance was drafted in November 1641, giving expression to the collective political outrage of England, it included the fenlanders' grievance that "[l]arge quantities of common and several grounds hath been taken from the subject by color of the Statute of Improvement, and by abuse of the commission of sewers, without their consent, and against it."[20]

The anti-drainage petitions were conservative and provincial in their focus, but they were also very much connected with the broader ideological issues that sparked the Civil War. The principal theme running throughout the fenlanders' grievances is the king's persistent flouting of the rule of law. Their lands had been improved and enclosed without their free consent; their local governing bodies, the commissions of sewers, had been suppressed or co-opted; when they sought a fair hearing of their case, the law courts were barred to them, and they were imprisoned for their trouble; and all of this was done with the active connivance of the Crown. Along with ship money and the abuse of prerogative courts such as Star Chamber, drainage projects were yet another manifestation of the king's intention to rule as he pleased, regardless of the laws of England—they were a symptom of tyranny and of the threat to liberty and property posed by an absolutist king. By 1641, they were the central focus of fenland protests against Charles's abuse of royal prerogative power, as expressed not only in petitions, but also in the riots that erupted in the region on the eve of the Civil War. The rioters targeted the Participants' drainage works and enclosures with their violence and framed their actions as being taken in support of the Parliamentary cause.

Serious and coordinated rioting broke out at Axholme in June 1642, after the king had fled London in the midst of the political crisis but before he had yet raised his battle standard at Nottingham. The rioters threw down the fences around the Participants' enclosures, drove in their cattle to devour the crops growing there, and also destroyed a navigable sasse on the Bickersdike drain. A more sustained series of attacks soon followed in the winter of 1642–43, after the Battle of Edgehill had ignited the Civil War, in which the Axholme commoners seized a sluice along the River Trent, known as Snow Sewer. For several weeks they opened the sluice at high tide to allow the floodwaters into the level and shut it again at low tide to prevent them from leaving. Men armed with muskets guarded the sluice and warned the Participants' tenants to keep their distance or risk being shot. One of the guards was heard to declare that they would keep the sluice open "til they had drowned the said level & had made the tenants thereof to swim away like ducks." The rioters flooded some eight thousand acres of the Hatfield Level, destroying the existing crops of corn and rapeseed and preventing the sowing of new crops the following year, with the Participants claiming damages in excess of £40,000. Having flooded the land, the Axholme commoners then went on the offensive; they forcibly regained possession of roughly four thousand acres of their former common, more than half of what they had lost, and defended it against the Participants' efforts to reclaim it.[21]

Several of the rioters later testified that they had attacked the drainage works not only to drive out the Participants' tenants, but also to defend the area from a royalist military incursion, and that they had acted with the approval of Parliamentary and military authorities. The Bickersdike sasse was destroyed by order of a Captain Kingman, an Axholme native, who in turn claimed that he had been ordered to do so "upon pain of death" by Sir John Meldrum, a Parliamentary army officer.[22] Several rioters said they had opened the sluice at Snow Sewer in order to halt the rumored advance of Sir Ralph Hansby, a royalist officer, with a large company of soldiers. They claimed to have been told to do so by Daniel Noddel, a lieutenant in the Parliamentary army, and with the approval of the Parliamentary committee then governing Lincolnshire "to drown all the level if they pleased." They even pointed out that the Isle's remaining commons had been flooded alongside the tenants' crops, so that everyone had suffered equally (though the flooding of a meadow and the destruction of sown crops were in no way equivalent hardships, as was pointed out by other deponents).[23]

Whether the destructive actions of the Axholme rioters in 1642–43 were really military tactics countenanced by Parliamentary authorities is open to question. Accounts of the riots come mostly from depositions taken years later, in 1652–53. Besides the obvious problem of recounting the details of events after a decade, the testimony may also reflect the political circumstances of the time it was recorded. After Parliament had won the war, the king had been executed, and the Commonwealth regime established, casting the riots as a part of the Parliamentary war effort would have been advantageous for the perpetrators. Other witnesses testifying in behalf of the Participants in their pursuit of damages disputed the commoners' claims. Some said that Sir John Meldrum had never ordered the destruction of the Bickersdike sasse but had actually given order for the drainage works to be left alone. Others said that the committee for Lincolnshire had commanded the men at Snow Sewer to allow the floodwaters to flow back into the Trent, but the latter refused to comply, and the high constable declined to intervene.[24]

Yet even if the Axholme commoners were not following anyone's explicit orders in attacking the Participants' drainage works and flooding their enclosures, their actions should still be viewed as one element of their opposition to royalist tyranny. By 1642, they had been protesting the depredation of their common for fifteen years, and they were well aware of the king's role both in supporting the Participants and in denying them any opportunity to present their case in court. The commoners' general support for the Parliamentary cause cannot be doubted; Daniel Noddel and several other men who opened and guarded the sluice at Snow Sewer also served throughout the war in the Parliamentary army. After the war, the Axholme men reminded the new regime that they had long been "zealous promoters of the Commonwealth's interest . . . at their own charge having raised and maintained many men for your service." They explained their "many illegal and tumultuous actions" of the previous years as having been provoked by "the remembrance of the cruelties exercised upon them by the late king and his pretended drainers, formerly by fines, imprisonments, and bloodshed." While acknowledging that some of their actions were "ignorantly, foolishly, and unlawfully undertaken," they had only sought "to vindicate their own right" against the king's "tyrannical power."[25] They had therefore attacked the drainage works in support of both the Parliamentary cause (as they understood it) and their own—a case of political ideology, military tactics, and local interests being in perfect alignment.

Rioting continued sporadically in the Isle of Axholme throughout the Civil War period.[26] The most serious occurrence took place in 1647, when a group of the Participants, led by John Gibbon, moved to repair their drainage works and take back the four thousand acres that the commoners had seized in 1642–43. Gibbon and his men reentered the disputed lands, impounded the commoners' cattle grazing there, burned the peat turves they had laid out to dry, and assaulted or imprisoned those who resisted them. The commoners responded by attacking the workmen Gibbon had hired to repair the drainage works and destroying whatever they had managed to rebuild. The situation reached a climax when Daniel Noddel marched into Haxey at the head of at least two hundred armed men and confronted Gibbon and his men directly. A deadly conflict was narrowly avoided when the steward of Epworth manor, a Mr. Geery, intervened and persuaded each side to stand down.[27]

Daniel Noddel (1611–72) had emerged as a central figure in the Axholme commoners' efforts to recover their common lands and to resist the incursions of the Participants. He was by this time acting as the commoners' solicitor, a position he would hold for sixteen years, throughout the Interregnum and beyond. Described as a gentleman of Owston in the Isle of Axholme, there is no record of his ever being admitted to a university or the Inns of Court. Yet he clearly had some formal education—he reportedly addressed the pastor at Sandtoft Church in Latin on one occasion—and enough knowledge of the law to represent his fenland clients quite ably. He was also a skilled polemicist and wrote a series of eloquent pamphlets denouncing the Participants and those who supported them. In addition to being a passionate anti-drainage advocate, Noddel was also known for his zeal in promoting the Parliamentary cause during the Civil War. His captain said that he "did always express as much affection by his actions as possibly could be to the service of the Parliament," and another veteran colleague stated that "in all the former wars he did not know any man in Lincolnshire more affected to the Parliament than he was."[28] Noddel's political and military support for the Parliamentary cause and his vehement opposition to the Participants' enclosure of Axholme's common lands were closely linked in his mind, as will be seen.

When the Court of Exchequer investigated the confrontation between Gibbon and Noddel, the latter testified on behalf of himself and his clients that they were only asserting their rightful claim to common lands illegally seized from them. Though he was aware of the 1636 Exchequer decree awarding 7,400 acres of their common to the Participants, he denied its validity on the grounds that it "was not duly obtained for . . . the said cause never received any decree in court upon debate of the title then in question." The 1636 decree had been granted with the alleged consent of only a small minority of the Epworth commoners; it had not resulted from a full hearing of both sides of the case but had in fact forestalled such a hearing; the commoners' consent had been extorted from them through a series of ruinous fines imposed in the Court of Star Chamber; and they had never been given authority to negotiate on behalf of their fellow commoners in any case. The decree was therefore illegitimate. Noddel further testified that he and his men had confronted Gibbon to arrest him for assaulting a commoner, upon a warrant issued by a local magistrate. They had come in force only because they feared Gibbon might react violently, and they had armed themselves in self-defense. Noddel had urged the commoners to proceed peacefully and had told them as their lawyer that "what he might by law do for them he would faithfully perform his duty therein but would not do anything for them but what by law he might safely do." He also pointed out that he wore a sword when he confronted Gibbon because he was entitled to do so as an officer in the Parliamentary army, but he had never intended to commit violence—he was, he insisted, only attempting to enforce the law.[29]

The anti-drainage violence that took place in the Isle of Axholme during the Civil War illustrates both the commoners' capacity for taking action on their own behalf and also their continued insistence on respecting the rule of law. The commoners acknowledged neither the legal claims of the Participants and their tenants to the lands in question, nor the Exchequer decree meant to enforce them. In allowing the Participants to drain and enclose their commons, King Charles had acted arbitrarily, ignoring the terms of the Mowbray indenture that clearly forbid any such thing. The Crown and the Participants had then conspired to ensure that the commoners would never obtain a fair trial in any court, in spite of their many efforts to gain a hearing. In taking back what was rightfully theirs, they had targeted the drainage works and had largely avoided violence or threats against persons, until they were provoked by John Gibbon and his men, at which point they acted in self-defense. The commoners' alleged rioting, in their telling, was really no riot at all, but a legitimate assertion and defense of their customary and legal rights in a context where justice was otherwise unattainable. Not everyone shared this view, of course—the Participants and their tenants certainly did not, nor did the Court of Exchequer—but whatever the merits of their case, the commoners continued to insist that the rule of law and support for the Parliamentary cause were the bases for their actions, even those deemed by others to be riotous.

Beyond their petitions to Parliament and their attacks on the Participants' drainage works, property, workmen, and tenants, the Epworth commoners also renewed their efforts to try their case in court. In October 1645, after another spasm of rioting in the level, the Parliamentary committee governing Lincolnshire sought to broker a settlement in which, among other things, the commoners were given leave to bring suit in the Court of Exchequer to try their title to the 7,400 acres of their former common that were in dispute. In the meantime, each side would retain possession of the lands they then held—some 4,000 acres for the commoners, and 3,400 acres for the Participants. On the basis of this agreement, the commoners commenced their suit in Exchequer in early 1646, with Daniel Noddel acting as their solicitor. The Participants immediately responded by exhibiting a bill of equity, and asking for a stay of the lawsuit until the bill could be acted on.[30] They also presented as evidence the 1636 decree from the same court, granting them possession of the entire 7,400 acres. The court ordered that the Participants' bill be proceeded in, and that the commoners' suit should be stayed indefinitely in the meantime. It further ordered that the Participants should enjoy possession of all the disputed lands while the bill was being heard, but the commoners successfully resisted all attempts to enforce this ruling, including John Gibbon's in 1647.[31]

The legal wrangle dragged on for years, as each side prepared interrogatories and took depositions. The Participants built their case around the typical interventionist argument concerning the improvement of land throughout the Fens: Before the drainage, virtually all of the Hatfield Level (including Epworth common) had been little more than a worthless lake, over which heavily laden boats routinely traveled even in summer. After the Participants had drained it, at great expense to themselves, the same land was made dry and fit for tillage, to the benefit of everyone concerned—their tenants had planted valuable crops, the poor inhabitants had steady employment in the fields, and the commoners' remaining six thousand acres were much improved. The drainage project and subsequent enclosure had been legally commissioned by the king as landlord, upheld in the Court of Exchequer, and consented to by a large number of commoners. Yet since the early 1640s, under cover of the wider unrest, several inhabitants had taken up arms and riotously attacked the drainage works, destroyed the Participants' crops, assaulted their tenants, and ruined their improvement of the land.[32]

The Epworth commoners' response presented not only the traditional, conservative view of the Fens, but also continued their critique that the drainage project had violated the rule of law: The freehold and copyhold commoners had enjoyed customary use rights in Epworth common time out of mind and had exercised them very profitably by grazing livestock, fishing, catching waterfowl, and digging peat turves. Though the common had flooded in most winters, the floods usually subsided by summer and had made the land more productive; far from improving it, the drainage had rendered it less bountiful. The drainage and enclosure were both patently illegal, as they violated the terms of the Mowbray indenture, but the king had conspired with the Participants to ensure that the matter would never be tried in court. The commoners' alleged consent, meanwhile, was illegitimate because it had been extorted in Star Chamber and did not represent the vast majority of commoners. Their attacks on the drainage works during the 1640s were not only a valid assertion of their legal rights but were undertaken in support of the Parliamentary war effort, with the approval of military and civil authorities, and were therefore not riotous. To underscore their legal standing, the commoners had brought their suit in the name of Thomas Vavasour, a gentleman freeholder at Epworth and a direct descendant of one of the commoners named in the original agreement with the Earl of Mowbray.[33]

After five years of investigations and postponements, the Participants' bill of equity finally came to a hearing in the Court of Exchequer in Michaelmas term, 1650. The court ruled against the bill and ordered that any Epworth commoners who had not personally consented to the 1636 decree were free to sue the Participants, giving them a real opportunity at last to prove their legal title to the whole 7,400 acres. This was a significant victory for the Epworth inhabitants, but a still greater victory was yet to come. Nearly one year later, the court ruled in favor of the commoners and awarded them possession of all 7,400 acres in dispute.[34] After waiting so long to obtain a fair trial of their case based on the evidence, the commoners had prevailed.

Unfortunately, the Exchequer verdict of 1651 still did not settle matters, for a series of events both in London and in the Isle of Axholme soon overshadowed the trial. Back at Epworth, just a few months before their lawsuit was finally about to commence, the commoners grew frustrated by the years of delay, and in May 1650 they erupted into the worst series of riots the region had yet seen. In London, meanwhile, Daniel Noddel had sought to strengthen his case by acquiring the assistance of two former leaders of the radical Leveller movement: John Lilburne and John Wildman. Both of these developments cast the fenlanders' cause in a much more menacing light from the point of view of Parliament and the Council of State, and the Participants immediately used them to their advantage in countering the verdict against them.

LEVELLERS IN HATFIELD LEVEL?

The riots that took place in the Isle of Axholme between May 1650 and October 1651 were extraordinary for their duration, violence, and destruction of property.[35] Ironically, the most severe rioting took place in October 1650, just as the Epworth commoners' lawsuit finally proceeded to trial in the Court of Exchequer. But after five years of apparently fruitless delay, and the new Commonwealth regime actively supporting the Earl of Bedford's new drainage project in the Great Level, a favorable outcome probably seemed a remote possibility, and the commoners' long-simmering frustration overcame their patience with the legal process. The rioters were thorough and systematic in their attacks. For a year and a half they broke down fences and turned their cattle and horses into the Participants' enclosures, allowing them to trample and devour the crops growing there. They carted away any valuable commodities they came across, including mown hay and timber. They pulled down at least eighty houses and a windmill belonging to the Participants and their tenants, beating some of the latter and threatening them with further violence if they did not leave. The center of the rioters' attacks was the village of Sandtoft, and particularly the church established there by the French and Walloon Protestants who made up most of the tenants. The church was vandalized on several occasions, as the windows, doors, pulpit, and pews were pulled down and smashed. The rioters allegedly even went so far as to desecrate the church by hanging slaughtered livestock in it and later used the building for a cowshed. When it was over, the Epworth commoners had evicted the Participants and their tenants from their remaining 3,400 acres and had reoccupied all of their former common.

The Council of State was already facing a string of domestic and international crises, with Oliver Cromwell on campaign with the New Model Army in Ireland and Scotland in 1650 and Charles Stuart (the dead king's eldest son) launching his invasion of England from Scotland the following year.[36] In such a turbulent time, and with the army already engaged elsewhere, the councilors could ill afford "riotous and tumultuous gatherings" in the Isle of Axholme and were concerned lest the commoners "begin insurrections and carry on designs, to the interruption of the public peace and danger of the Commonwealth." Noting that "especially in such times as these, a more diligent care ought to be taken to prevent such meetings of the multitude," in July 1651 they ordered the sheriff of Lincolnshire to suppress the riots, authorizing him to call for "such [military] forces of the Commonwealth as shall be nearest you . . . whereby dangerous consequences may be prevented."[37] This order did nothing to quell the rioting, which continued into the fall; if the sheriff ever did call for military assistance, no record survives of it. The rioting finally wound down in November 1651, after the Scots were subdued, Charles Stuart's forces were smashed at the Battle of Worcester, and Cromwell had returned to London.

The riots were chronicled in a series of depositions taken before a Parliamentary committee charged with investigating them in 1652. Dozens of witnesses testified, mostly the victimized tenants, though some of the accused rioters were also heard. The specific events, as given in the depositions, have been thoroughly described elsewhere.[38] Of greater interest here is what those events meant at the time—to the rioters themselves, to the victims who suffered from them, and to the fragile Commonwealth state that had to deal with them. Were the riots just a particularly violent outbreak of anti-enclosure sentiment, limited to the narrow confines of a local land dispute, or did they threaten the security of the state, as the Participants alleged? Did the rioters really utter the seditious and insurrectionary statements ascribed to them, and if so, what did they mean by them? What would have made them turn so suddenly and vehemently against the Commonwealth regime they had fought for so long to bring to power?

Historians have been especially intrigued by the contemporaneous involvement in the Axholme dispute of two former leaders of the radical Leveller movement, John Lilburne and John Wildman (figs. 7.1 and 7.2). Throughout his adult life, Lilburne (c. 1615–57) was something of a professional firebrand—he published dozens of polemical pamphlets and was imprisoned or driven into exile on numerous occasions for his radical writings and speeches. In 1637–38, under Charles I, he was arrested and tried before Star Chamber for printing and circulating unlicensed books and was condemned to being imprisoned, pilloried, and publicly flogged for his "insufferable disobedience and contempt." He was jailed more than once in 1645–48, this time by order of Parliament for criticizing its conduct of the war (among many other offenses), and yet again in 1649 by his former friend and ally Oliver Cromwell, at which time he was tried (and acquitted) for high treason for questioning the legitimacy of the ruling Commonwealth regime.

Though never fully consistent in his politics, Lilburne believed passionately in freedom of conscience, popular sovereignty, and the rule of law. He was deeply suspicious of every regime he lived under, believing that each sought to rob Englishmen of their religious and civil liberties. His writings were most influential in fostering the rise of the Levellers, a political movement supported by radicals in London and soldiers in the New Model Army that advocated a widely expanded suffrage, religious freedom, and the equality of all men before the law.[39] Wildman (1623–93) was also closely associated with the Levellers and became one of the principal articulators of their political agenda; he was a friend and colleague of Lilburne, with whom he was occasionally imprisoned.[40]

The Levellers were most influential during the late 1640s, when political and religious radicals in the New Model Army and in the rising Independent faction of Parliament rejected peace negotiations with the king and pressed instead for his trial and execution. During this period they drafted their most influential manifesto, *An Agreement of the People*, calling for a new English constitution founded on the idea that the legitimacy of the state should be derived from the consent of the governed—Wildman was almost certainly one of its chief authors. Though Cromwell was not altogether unsympathetic to the group's ideology, particularly with regard to freedom of religion, he soon came to distrust their influence within the army and to fear their ultraradical democratic agenda. The Levellers, led by Lilburne and Wildman, accused Cromwell and his allies of plotting to subvert the will of the English people and to rule arbitrarily, prompting Cromwell to suppress the movement.[41]

Figure 7.1. John Lilburne, by George Glover, 1641.
Portrait D10576, © National Portrait Gallery, London.

What exactly was the nature of the old Levellers' interest in the Hatfield Level—were the Axholme riots really an attempted Leveller uprising in the Fens? J. D. Hughes has suggested that the radical political and religious views that predominated in the region would have made the Epworth commoners receptive to Leveller rhetoric, while Lilburne and Wildman may have seen an opportunity to reignite their movement in the Fens after recent setbacks. Keith Lindley has argued, to the contrary, that no political significance should be sought in what was really just a short-lived business arrangement, little more than a "temporary expedient" serving the "mutual advantage" of both parties. Clive Holmes has taken a middle ground, arguing that the involvement of Lilburne and Wildman in the Axholme dispute was neither the start of a Leveller uprising, nor was

NIL ADMIRARI

MAJOR IOHN WILDMAN.

Figure 7.2. Sir John Wildman, by Wenceslaus Hollar and William Richardson, c. 1800. Portrait D28976, © National Portrait Gallery, London.

it "fortuitous or merely tactical," as Lindley asserts. He detects a striking "aware-ness of central politics" on the part of the fenlanders in seeking the two men's aid and suggests that Leveller ideas concerning popular political participation would have resonated with fenland yeomen and husbandmen, men with long experi-ence serving on commissions and juries of sewers, and who resented the state's interference in their local affairs.[42]

For most of the period in question, the influence of Lilburne and Wildman on events at Axholme must have been indirect at best. Despite the Participants' allegations, there is no evidence that either man was present in the region before October 1651, when the worst of the rioting ended, or played any role in events there before that time. Based on witness testimony from both sides, Noddel prob-ably first approached the two in London in the fall of 1650, long after the rioting

began, to ask for their help in preparing the commoners' lawsuit once it was finally allowed to proceed. It is quite possible that he was already acquainted with
Lilburne from their contemporary service in the Parliamentary army. One witness,
George Wood, testified that Noddel "wanted one to balance Mr. Gibbon," and
Wood "wished him to take Lilburne & he thought he would match him." Another witness, John Thorpe, said Noddel had sought the Leveller's assistance
"because Lilburne was a powerful man & he having friends would give a sooner
end to the business"—though by that time Lilburne's political influence was very
much diminished.[43] Still others confirmed that Lilburne was not in the Isle of
Axholme during most of the rioting. He and Wildman most likely remained in
London, preparing the commoners' case, as Noddel had engaged them to do.

In October 1651, both Lilburne and Wildman were at Axholme and had become personally involved in the case. They each signed a deed pledging to continue representing the commoners in return for one thousand acres apiece out
of the Epworth common.[44] The precise terms of the deed are uncertain; some
claimed that the two men agreed to defend the commoners from prosecution for
all their past actions, but others suggested they had deliberately avoided any such
obligation, insisting "that they would not intermeddle with former riots." Thorpe
later said of Wildman that "he was so far from animating the riots that he told
them they would undo themselves."[45] In any case, Lilburne soon found himself
at the center of a tense confrontation between the Epworth commoners and the
Participants' tenants.

On 19 October 1651, a Sunday, Lilburne and Noddel led an armed contingent of Epworth commoners to the settlers' church at Sandtoft, just as they were
gathering for worship. After posting two men with swords at the door to intimidate the worshippers, Lilburne declared himself to be a freeholder of the manor
of Epworth (based on his agreement to accept one thousand acres there) and told
the settlers, "this is our common, you shall come here no more unless you be
stronger than we." He then expelled the minister and preached in the church
himself, after which the commoners smashed the windows, pulpit, and pews.
Lilburne subsequently had the minister's house repaired (it had been heavily
damaged in previous rioting), placed his servants in residence there, and kept his
cows in the church.[46]

Some witnesses testified that Lilburne next tried to incite riots at some of the
calmer neighboring manors in Hatfield Level. He allegedly went to Crowle and
informed the tenants there that unless they agreed to pay their rent to the fenland commoners instead of the Participants, their crops would be destroyed,
their cattle impounded, and themselves forcibly evicted. Noddel also supposedly

declared that after settling matters in Lincolnshire, Lilburne would proceed to the Hatfield and Thorne manors in Yorkshire to spark an uprising there as well, "& then they should give the Attorney General work enough." When it was pointed out that the manors there did not have nearly as strong a case in law as Epworth had, he answered, "no matter for that, we will make something of it." Other witnesses testified, however, that Lilburne's actions after the episode at Sandtoft church were moderate and peaceful: the commoners of Crowle had consulted with him only after the rioting had ended, at which time he merely advised them to impound the tenants' cattle and get them to seek a writ of replevin, which would enable the commoners to present their case in court.[47]

Lilburne and Wildman ceased to have any further involvement in events at Axholme just a few months later; Noddel carried on by himself. There is little to suggest that either man envisioned the riots as the prelude to a Leveller uprising in the Fens. Their involvement was mostly limited to preparing the commoners' legal case, and is probably best understood as part of their efforts to rebuild their lives and careers after a political reversal—Lilburne by developing his law practice, Wildman through speculation in real estate. The most suggestive bit of evidence—that the most severe rioting in the Fens took place in October 1650, when Noddel first made contact with the two men in London—is circumstantial at best. While they may have sympathized with the commoners' rage, they were understandably reluctant to involve themselves with riots that occurred mostly in their absence. In his pamphlet of November 1651, Lilburne excused the rioters' actions during the Civil War, explaining that "the commoners being in arms for the Parliament . . . did take advantage of the time; and as they had been put out of possession by force, and could not through the tyranny of those times have any legal remedy; so by force they put themselves into possession again." But he disavowed the most recent riots as the work of "the poorer sort," who "did foolishly throw down many poor houses . . . (but that folly of the multitude none of the most discrete commoners and tenants of the Isle do justify)."[48]

Yet while the Axholme riots of 1650–51 were not a Leveller-inspired insurrection, they were nevertheless intensely political—and even seditious. The Epworth commoners were not focused solely on local affairs but were aware of and engaged with the broader political context. Their knowledge of Lilburne, their belief that he was "a powerful man" and "one to balance Mr. Gibbon," indicates a consciousness of national politics (albeit an out-of-date one). Nor were the fenlanders' pleas for assistance misplaced, given that the Levellers had taken some notice of their plight before the movement's collapse, as in a 1648 petition attributed to Lilburne condemning "all late Enclosures of Fens, and other

Commons."[49] It may even be speculated that the coincidence of the former Levellers' initial involvement and the timing of the most severe rioting in October 1650 may have reflected the commoners' being emboldened by knowledge of their new allies, even without the latter's explicit encouragement.

But the fenlanders did not require Leveller encouragement to express their displeasure with the Commonwealth regime and their disdain for its authority, as several witnesses testified. Noddel allegedly thumbed his nose at the letter of the law by instructing the rioters to throw down the enclosures while working only in pairs, so they would not technically be guilty of rioting (a "riot" legally requiring the participation of at least three people), and then all but daring the sheriff to intervene. Many local officials refused to suppress the riots, and some even participated in them: Richard Mawe, a constable of Belton, helped to pull down several tenants' houses and looted their property, while the high constable Peter Bernard allegedly assaulted one of the tenants. Several tenants said they were physically barred from seeking redress from any local magistrates except those who supported the commoners. One justice of the peace, Michael Monckton, abetted the rioters by refusing to bind them over, fining those who were convicted only twelve pence apiece, and informing the tenants that "there was no law" for them.[50]

The rioters were also accused of making inflammatory and seditious statements, scorning the authority of both Parliament and the law courts. Alexander West's remark that "it was a shame the Parliament should give away their commons, they were a Parliament of clouts," has already been discussed. Daniel Noddel allegedly said that he would have the commoners' case printed and "nail it up upon the Parliament doors & make an outcry and if they will not hear us we will pull them out by the ears," and that "when Lilburne came to London . . . there would be a new Parliament would call the old ones to account." Richard Ile said ominously that "if we lose our commons whilst this government is, the government may alter & then we shall have it again." John Barrow was heard to say by two witnesses that "he fought for the Parliament, but if they did take away their commons, he would bring a 100 men to fight against them." Several other rioters exclaimed that "they would not obey any order or decree of the Exchequer nor any order made by the Parliament for they could make as good a Parliament themselves." Robert Forster claimed that "if we cannot get our commons by no other law, we will get it by club law," and that if soldiers should come to Axholme to quiet them, "it will be a bloody day."[51] Several Epworth rioters allegedly threatened to march to London with 1,500 men and "pull the Parliament out by the ears and sit as a Parliament themselves in the Isle of Axholme."[52] Such examples could easily be multiplied.

Historians must handle statements such as these with caution, for their reliability is open to question. The only sources for them are the witness depositions given during the Parliamentary committee hearings a year or more after the events in question. The witnesses were mostly the Participants and their tenants—the victims of the riots—and they had every incentive to make events in Axholme seem as incendiary as possible in appealing for Parliamentary support. Casting the rioters as dangerous rebels who posed a grave threat to the peace and security of the Commonwealth served two purposes for the Participants: it shifted the discussion away from the weakness of their legal claim to any of Epworth common, and it distracted the Commonwealth regime from the Participants' (now embarrassing) history of supporting the royalist cause in the Civil War. The investigating committee itself, moreover, was strongly biased in favor of the Participants, not least because of Parliament's anxiety over the former Leveller leaders' involvement in the affair. Its members were particularly receptive to testimony alleging sedition among the rioters and may have selectively recorded the testimony they heard in order to justify their eventual report condemning them. The many seditious statements ascribed to the rioters may therefore represent nothing more than the Participants' cynical attempt to curry favor by playing on the insecurities of the Commonwealth regime.

But I would argue that this interpretation is short-sighted and unnecessary. The Axholme commoners had allegedly expressed similarly seditious sentiments during the anti-drainage riots of the late 1620s, aimed in that case at King Charles I. They were well aware of England's unsettled political context before the start of the Civil War and proved adept at taking advantage of it for their own ends once hostilities began. They were active supporters of Parliament's military efforts, and their wartime service may well have strengthened their political awareness by bringing them into contact with the radical ideologies of the Levellers, among others. Their fervent support for the Parliamentary cause was born in part from a conviction that the monarchy under Charles I had ceased to govern according to the rule of law and had become tyrannous and illegitimate. In fighting and sacrificing so much to bring the Commonwealth regime to power, the fenlanders believed they were helping to restore just and lawful governance. Yet by the middle of 1650, with the king dead and the monarchy abolished, they had little to show for their loyal service. Their efforts to obtain a fair trial were still stalled in the Court of Exchequer, after years of delay. They had recovered de facto possession of their common, but only because they had managed to seize it for themselves, not because they enjoyed any political backing.

And at the same time, Parliament was throwing its full support behind the Earl of Bedford's new scheme to drain the Great Level.

In the political context of 1650–51, it is not hard to believe that the Axholme commoners might have perceived the Commonwealth regime they helped put in power to be indifferent, or even hostile toward their interests. The rule of law remained elusive in the Fens, from the commoners' point of view, and they may have started to wonder whether the new regime was really much of an improvement over the monarchy. The seditious words attributed to them by their enemies ring true in their frustration, outrage, and sense of betrayal. Their implied critique of the Commonwealth state's legitimacy is consistent with their prior critique of Charles's legitimacy as expressed in their petitions to the Short and Long Parliaments, and it was provoked by the same concerns. All politics is local, ultimately, and the Commonwealth regime had failed to redress the fenlanders' long-standing grievances and continued to deny them a full and fair hearing. The Axholme commoners' abrupt desertion of the cause they had supported enthusiastically for a decade should not be attributed to political ignorance, indifference, or myopic provincialism. It was based on a consistent and principled stance in favor of the rule of law, their desire to recover their land and their rights through legal means, and their disillusionment with a new regime that seemed no better than the one it replaced.[53]

THE PRICE OF POPULAR POLITICS

John Lilburne's political fortunes took another turn for the worse at the end of 1651—a Parliamentary committee was investigating him for his published attack on Sir Arthur Haselrig's administration of the sequestered estates of accused royalists. The committee found him guilty of libel, fined him £7,000, and banished him from England, after which he spent the next two years in exile in the Low Countries. He took no further part in events at Axholme. For his part, John Wildman tried to disavow any involvement in the Axholme dispute and continued to speculate in real estate—more profitably, in general, than he had at Epworth. Yet while the two men were only briefly and tangentially involved in fenland affairs, the Participants saw an opportunity to regain the upper hand after the Exchequer verdict recently handed down against them. They chose their moment carefully: on 10 January 1652, the day after Lilburne's libel conviction, the Participants submitted a petition to Parliament asking for relief from the damages they had suffered.

The petition rehearsed the Participants' version of events, including their successful drainage of more than 60,000 acres at great expense, their settlement of hundreds of Protestant refugees on the 24,500 acres allotted to them, and the settlers' industrious cultivation of the newly improved land. They pointed out that, before the recent unrest, they had paid £1,228 annually in fee farm rent, first to the Crown and then to the Commonwealth state. In 1642, however, the Epworth commoners "did rise in tumults . . . too great to be suppressed by the ordinary courts of justice" and laid waste to four thousand acres of their allotted lands. Having escaped punishment during the Civil War "for their former inso-lencies and rebellions against the law and government of this state," the com-moners came under the influence of Lilburne and Wildman and were embold-ened to seize by force the remaining 3,400 acres of their former common. The sheriff in Lincolnshire had attempted to intervene, but they continued to destroy the enclosures before his eyes. The rioters were also heard to use "high reproach-ful, and seditious language against the Parliament and present government," say-ing that "they would obey neither Barons [of Exchequer] nor Parliament, and that they could make as good a Parliament themselves, and that if the Parliament sent forces against them, they would raise forces and resist them." The Partici-pants claimed to have suffered some £40,000 in damages, "besides the damage to the Common-wealth in general by destruction of so many habitations and ploughs for tillage and husbandry." They asked the members of Parliament to consider three matters:

1. The great disobedience to the Parliament and courts of justice, and contempt of the present government.
2. The great damage to the Common-wealth in general, and in particular to your petitioners and their tenants, to the value of at least £40,000.
3. The consequences of such unparalleled and rebellious riots, and to direct such a way for the petitioners' reparation, and damages, and future preservation of this level, as to your grave wisdom shall be thought meet.[54]

The Participants did all they could to connect the Axholme commoners with the now-disgraced Levellers and to portray them as dangerous rebels. They invoked Lilburne's name four times in their short petition, Wildman's twice, and presented it to Parliament just one day after Lilburne's conviction for libel. They emphasized the commoners' disobedience and disregard for law and order at every opportunity and complained not only of the riotous attacks on their pri-vate property, but of the grave damage done to the Commonwealth, from the loss of £1,228 in annual rent to the destruction of growing crops and prosperous farmsteads. Their petition had the desired effect; six days after they received it, the members of Parliament referred it for investigation to the very same com-mittee that had just fined and banished Lilburne. In the meantime, before the committee had even started their investigation, Parliament specifically excluded the Axholme rioters from the Act of General Pardon and Oblivion.[55]

The chair of the investigating committee, Sir William Say, was in no way a disinterested arbiter: he had owned land in the Hatfield Level as recently as March 1650, which he had purchased from John Gibbon, and still owned land near Peterborough connected with the Great Level drainage.[56] Daniel Noddel later referred to him as "a lawyer, a drainer, and Mr. Gibbon's usual chapman for land in those parts," and noted that he took down the witnesses' testimony in his own hand, "for he would not suffer the clerk chosen by the committee to pen them, such was his zeal to promote the Participants' complaints."[57] The depositions Say recorded would be used to compile the committee's final report, and they remain the principal source of historical evidence concerning the Axholme riots. Over thirty folio pages of recorded testimony, the victims of the riots described the violence with which their crops were devoured, their cattle impounded, their homes destroyed, their bodies beaten, and their lives threatened, and they accused the rioters of making seditious statements about the Commonwealth regime.

Noddel mounted a vigorous defense of the Axholme commoners, presenting his own witnesses whose testimony countered that of the Participants' tenants. Several testified that Epworth common had been productive and valuable land before the drainage project and that the Participants had done much more harm than good, because the land was no longer so fertile without the annual floods. Some freehold commoners, unable to keep as many livestock on what was left of their common, had been forced to sell their lands. Other witnesses tried to deflect accusations of sedition and treason away from themselves and onto the Participants. They testified that while Noddel and others had served Parliament long and faithfully during the Civil War, the Participants had raised a troop of horse for the royalist cavalry. One witness claimed to have heard some of the tenants say "what was the Parliament, they were traitors & deserved their throats were cut." Another alleged that John Gibbon had pledged to supply horses, arms, and ammunition to Charles Stuart as he advanced his royalist army toward Worcester in 1651, declaring that "there be many Parliament rogues in the Isle," but "Prince Charles will come & then all in the Isle shall be put to fire & sword."[58]

Despite Noddel's efforts, the investigating committee's final report endorsed the Participants' version of events and all but condemned the Epworth commoners as rebels and traitors to the Commonwealth. The report, compiled by Say as chairman, established that the Participants had expended more than £175,000 to drain some sixty thousand acres in Hatfield Level, improving their productivity and value at least fivefold. They had received 24,500 acres in recompense for their worthy endeavor but had been long disturbed in their possession of 7,400 acres of it, formerly part of Epworth common. The report catalogued the long history of rioting in and around Epworth, going back to 1642 and the start of the Civil War and culminating in the rioting of 1650–51; these latest riots, it was alleged, were instigated by the notorious Levellers, Lilburne and Wildman. Say recounted all of the most egregious acts of violence committed by the rioters, as well as the seditious statements they were alleged to have made; he not only quoted from the tenants' depositions, he included marginal annotations cross-referencing each incident with the witnesses who testified to it. On the other hand, the report all but ignored the fifteen folio pages of testimony taken from the Axholme commoners. Say did acknowledge their claim to have been acting in support of the Parliamentary cause during the Civil War and mentioned the several witnesses who said that Noddel was "well-affected to the Parliament." But regarding all of the most serious allegations, including "the language spoken against the Parliament," the report flatly stated that "no defense is made."[59]

Noddel was furious with the flagrant partiality of the investigating committee, and having failed to obtain a fair hearing from them, he turned to the printing press to tell the commoners' version of events. Over the next year and a half he published three dense, strident pamphlets addressed to the members of Parliament, denouncing the Participants' drainage project, their supposed claims to Epworth common, and especially Say's report.[60] In the wake of such a deeply biased official investigation, Noddel wanted to ensure that the commoners' side of the story was made public, and he emphasized the rule of law as the basis of their legal case and justification for their conduct. "I make no doubt but your honors have all heard of a pretended great riot some years past, committed in the said Isle [of Axholme]," he wrote, "but I am persuaded that you have not heard half so much of the proceedings that have been at law by the freeholders there, for the recovery of their ancient right to 7400 acres of commonable grounds." The successive pamphlets repeat, but also build on one another, as Noddel grew more and more outraged.

Noddel laid out the commoners' legal case against the Participants in detail, from a full translation and transcription of the original Mowbray indenture, to the several points of law and equity raised in the depositions given by the Epworth commoners to the Parliamentary investigating committee. The Participants' claim to any part of Epworth common was based entirely on "the late king's illegal patent" for the drainage project and on the commoners' alleged consent to the 1636 Exchequer decree. Neither document had ever been valid, Noddel argued—the former because it clearly violated the terms of the Mowbray indenture, the latter because the consent had been coerced. The commoners had tried for years to seek redress for their grievances in court, but the Crown and the Participants conspired to thwart them at every turn, "in the times of tyranny, by means of the [Privy] Council table and Star Chamber, not suffering the commoners to enjoy the benefit of law."[61]

When the Civil War began in 1642, the frustrated commoners took action both "in the just defense of their undoubted right of possession, when they could have no proceedings at law," and also in support of the Parliamentary cause. Noddel contrasted their loyal backing of Parliament with the royalist tendencies of the Participants. He marveled at "the impudence of [John Gibbon], how under the notion of riots he condemns . . . things done for the Parliament's service" and suggested that "no doubt but if it were in his power, [he] would make all the Parliament's battles riots."[62] He was more apologetic concerning the riots of 1650–51, which could not be justified as a military tactic during wartime. But having been forced from their common by a tyrannous king and his cronies, denied a trial at law for nearly twenty years, and still unable to obtain justice even under the new Commonwealth regime despite their patience and loyal service, it could only be expected that "some of the ruder sort will grow disorderly and endanger both themselves and those that deal for them."

Noddel condemned the Parliamentary investigating committee for being thoroughly biased and unfair, noting Sir William Say's prior relationship with John Gibbon and complaining, "how ready they are in their report to mention the least unbeseeming word that any particular inhabitant has spoken, but altogether omit to make out the inhabitants' defense." Say's report, he asserted, had "in ten lines locked up and imprisoned the truth of above twenty witnesses" for the inhabitants, yet "stretched out near two hundred lines . . . for the Participants' advantage," based solely on the testimony of "tenants and farmers to the Participants" who were testifying in the hope of receiving damages.[63] Say's report had unjustly condemned the commoners as lawbreakers and rebels, yet it was they "who all along have desired trials at law, but could not have them," and had long endured "the villainous, tyrannical, and forcible entries, & intrusions, arrests, and imprisonments, brought upon them by the said Participants, contrary to law." The Participants were responsible for all the disturbances in the Fens: "[L]et but self-interest be laid aside, and conscience must needs conclude the Participants to be the rioters. And . . . whosoever they were that had their hands in these horrid things, are well they have hitherto escaped the rope, for committing such unparalleled actions against the law."[64]

Noddel's passion and eloquence were all for naught. Sir William Say had submitted the investigating committee's report directly to Cromwell and the Council of State in June 1653. It did not go to Parliament for consideration as intended because Cromwell had forcibly dissolved that body two months earlier, ending the Commonwealth government and replacing it with the Protectorate. This left Cromwell himself (now as lord protector), the Council of State, and the army as the only governing authorities in England. The council reviewed the report and the depositions it was based on, reaffirmed it as a fair and accurate account of events, and made it the basis for all further action. In doing so, they ignored the Court of Exchequer's verdict in favor of the Epworth commoners, the testimony of the commoners' witnesses, and their long record of loyal service to the Parliamentary cause.

The council condemned the commoners' "tumultuary proceedings and evil practices . . . to the great affront of justice, damage to the Commonwealth in general, and the prejudice of the interest of the state," all of which had been "promoted by the countenance and encouragement" of Lilburne, Wildman, and Noddel. Possession of the entire 7,400 acres was to be restored immediately to the Participants. A commission of oyer and terminer was granted to the assize judges so that they might investigate the matter, punish the rioters, and award damages to the Participants and their tenants. Finally, given the notably "turbulent and seditious spirits" of the rioters in this case, the councilors were moved to "apply the military power in aid and assistance of the civil government and execution of justice." They instructed local officials to call for troops quartered in the region to help restore possession to the Participants, execute the orders of the law courts, and prevent any further outbreaks of rioting in Hatfield Level.[65]

This was the second time that the Council of State had authorized the use of military force to restore order in the Isle of Axholme—the first came two years earlier, in the midst of the rioting itself, in July 1651. Troops were needed, the council concluded, to remedy the apparent collapse of law and order in the region, reestablish the power of civil authorities there, and safeguard the interests of both property holders and the state. As before, however, the order to send in the army neither overawed the fenlanders nor restored the peace. The Epworth men held their ground, keeping possession of all 7,400 acres of their disputed common, and there is no evidence that troops ever intervened. When the Participants tried to execute the council's order themselves and seize the disputed land again in April 1654, the inhabitants destroyed their fences, beat their workmen, and drove them away.[66]

This set the pattern for violence in the Isle of Axholme for the rest of the Interregnum period and beyond. From the fall of 1655 onward, the Participants made repeated attempts to recover their allotted lands in Epworth Common, but the commoners there stood firm and continued to occupy them. The Participants' forces were led by one of their own, Nathaniel Reading, who seized hundreds of livestock from the commoners and worked to rebuild the Participants' fences, but each time the commoners were able to recover their herds by force and tear down whatever Reading had managed to rebuild.[67] Though matters never again reached the level of sustained violence and destruction that had prevailed in 1650–51, as many as a hundred commoners occasionally took up arms to rescue their distrained cattle and sheep, assault local officers attempting to restore the Participants' enclosures, or sack and burn Sandtoft church once again. In their appeals to the Council of State, the Participants continued to portray the commoners as rebels and traitors, stressing their violence toward local officials, their "discomposure of the public peace," and their "scandalous & rebellious speeches" regarding Cromwell in particular.[68]

Witnesses testified to the commoners' ongoing tendency to question the Protectorate government's legitimacy. One witness allegedly heard a rioter say that they were "resolved to make opposition against all orders and decrees of His Highness's [i.e., the lord protector's] courts of justice," and that "they will defend their common with their swords which they say they may as lawfully do as the Protector may the government he hath taken upon him."[69] Another claimed that several commoners "did declare that if His Highness were there they would make no more matter of him than of an ordinary person," and that "if His Highness would make choice of one hundred men, one hundred of the Isle aforesaid would fight with them for their possessions." When the witness replied that they would surely not dare to rebel against the lord protector, the commoners retorted, "they had rebelled against a better man and would not care to rebel against him."[70] As always, such comments must be read with caution, since the only surviving evidence of them comes from hostile sources, with an interest in depicting sedition everywhere in Axholme to win the support of the state against their adversaries. And yet the commoners' alleged expressions are remarkably similar to those they had been accused of making for nearly thirty years, against whatever government was in power at the time, and for the same essential reason. If the state— whether monarchy, Commonwealth, or Protectorate—was prepared to assist their enemies in usurping their common and to brand them as criminals for defending their rights as Englishmen, then that state was illegitimate and unworthy of their loyalty.

Given the political instability of the time, the Protectorate government could not allow its authority to be flouted in the Isle of Axholme. In April 1656, the Council of State referred the matter to General Edward Whalley, the army officer responsible for keeping the peace in Lincolnshire during the year and a half of direct military governance in England known as the "Rule of the Major-Generals." The council ordered Whalley "to prevent all such further riots and disorders, and . . . to punish such scandalous disobedience." He was specifically urged "not to suffer the said inhabitants to keep by them any arms or other instruments, for the further aiding of such disorders; as also to give order to some of your regiment to be effectually aiding and assisting" the sheriff and other civil authorities in the county.[71] Once again, however, this failed to quiet the Isle of Axholme or to restore the Participants' enclosures. Although Whalley reported to Parliament in December 1656, "Our forces have been troubled to suppress the tumults," the Epworth commoners held on to the whole of their common, and the dispute continued to be waged through lawsuits, petitions, and occasional violence on both sides.[72]

The tumultuous events in the Isle of Axholme during the Civil War and Interregnum are important for many reasons. The riots there were the most sustained and violent of any fenland anti-drainage or anti-enclosure riots in the seventeenth century, as well as the most successful in achieving their aims. The Epworth commoners reoccupied the whole of their common and managed to keep de facto possession of it throughout the rest of the seventeenth century, in spite of the Participants' ongoing efforts to drive them from it and the state's condemnation of their actions. One perplexing issue is the failure of the New Model Army to restore order: the Council of State authorized the use of military force at Axholme on at least three occasions during the 1650s, but there is little evidence troops ever intervened. The Protectorate regime may have been unstable, but it was not militarily weak; given the army's proven ability and the proximity of garrisons in the region, it seems implausible that professional soldiers would not have made a decisive impact if they had been deployed aggressively.

It is possible that, in spite of his public pronouncements, Cromwell really had little sympathy for the Participants and never intended to use the army to aid their cause. But this would be inconsistent with Cromwell's actions in the Great Level, where he was perfectly willing to make troops available to secure the interests of the Earl of Bedford and his partners, as will be seen in the following chapter. Another, more likely possibility is that, while Cromwell may have wanted to aid the Participants and threatened the use of military force toward that end, he was reluctant to commit troops to a pitched battle in the Isle of Axholme. The local political context there was vastly different from the Great Level, where the inhabitants tended to be royalists and the drainage projectors had mostly supported the Parliamentary cause. At Axholme, many of the rioters the army would be suppressing were themselves wartime veterans of that same army, who had sacrificed much to put in power the state that Cromwell now struggled to govern. Ruthlessly crushing them and giving their lands to former royalists might have lent too much validity to their critique of the Protectorate state's legitimacy, vindicating and perhaps propagating the sedition that Cromwell needed to subdue. This theory is speculative, to be sure, but it would help to explain why Cromwell was willing to threaten the use of military force at Axholme but never actually committed any troops to settle the conflict.

Perhaps the most interesting aspect of the affair is the way each side appealed to the Interregnum state for support. Like all drainage projectors of the era, the Participants based their appeal on their costly efforts to improve the land and benefit the commonwealth. They described Hatfield Level as a flooded and valueless waste before the drainage, peopled by "beggars and idle persons," and claimed to have vastly improved the quality, productivity, and value of the land. They wished to be seen as projectors in the positive sense of the term—enterprising entrepreneurs whose vision, hard work, perseverance, and willingness to assume the risk had ultimately benefited everyone involved. They also cast themselves as the proponents of order, progress, and social hierarchy, as against the rebellious and leveling tendencies of the obstinate and backward fenlanders who violently opposed them. This latter rhetoric became a central part of the Participants' case especially after 1650, when they needed to shift attention away from their support for the royalist side during the Civil War.

The Axholme commoners, on the other hand, articulated their case primarily in terms of the rule of law. Their legal claim to their common waste ought to have been unassailable, given the terms of the Mowbray indenture, but they were denied a fair hearing and trial through the machinations of the Participants and the tyranny of a greedy king. For many years, they also argued that the Participants had not improved their land, but diminished it, and thus had no basis for laying claim to any part of it. They cast the Participants as projectors in the negative sense—charlatans, frauds, and crooks, who had happened on a perfectly good and productive common that rightfully belonged to others, did their best to ruin it through their needless schemes, and then helped themselves to the best portions of it, all without the consent of those who lived and depended on it. And while the commoners made a conservative argument, they were not static in their own defense; by the early 1650s, they had even started to come to terms with the drainage of their common as permanent, irremediable, and inevitable. In a drafted petition to Parliament, the commoners pledged that if their cause prevailed, they would not tear down the drainage works but would keep the land drained and enclosed so long as sufficient common lands remained for their needs. They also reassured the MPs, who were at that time debating the Great Level project, that their own situation was entirely unrelated to that one, and a ruling in their favor would have no impact on Bedford's undertaking.[73]

This unusual and unexpected stance may or may not have been a majority view in the Isle, but it appears in unexpected places beyond a single drafted petition. John Lilburne acknowledged in his pamphlet that with "the River Idle being now destroyed, and the old drains also stopped up," restoring the old common waste would be impossible. Realizing that "the new drains . . . must be maintained to keep the commons dry . . . the commoners are resolved to contribute their share to maintain the drains . . . [and] to keep up the improvement, if any be, to the best advantage of the Commonwealth."[74] Even Daniel Noddel, in his 1654 pamphlet, conceded that though the drainage had been unnecessary and counterproductive, it could no longer be reversed. The commoners, therefore, "desire to do the drainage no hurt" but rather stood ready to "improve the grounds both for the good of the Commonwealth and themselves." He also reassured his readers in Parliament that "the manor of Epworth differeth much in the title from other drainings, which no doubt in some places are lawful and laudable." The Epworth men's case, in other words, would have no bearing on other drainage efforts such as Bedford's Great Level project, which enjoyed strong parliamentary support.[75]

The shift in the Epworth commoners' position concerning the drainage, from outright rejection to grudging acquiescence, represents a remarkable change in fenland popular politics. It suggests a recognition that, at least among the highest levels of political power in England, the only good fens were drained fens—just as the only acceptable outcome in the Great Level was the making of permanent "winter ground." The commoners were no longer arguing from a purely conservative, traditional understanding of the Fens, condemning all drainage projects as a mutilation of the land and a violation of their customary use rights. They were challenging the legality of *this* project as a particular case: the drainage and enclosure of Epworth common was unlawful and unjust not because of any inherent problem with drainage projects per se, but rather because of the unique circumstances of the manor and the illegal way in which the project was driven forward. Since that drainage could no longer be undone, it was now a fact of life to be accommodated rather than resisted. They therefore asserted their legal right to the land, even in its drained condition, and sought to reassure Parliament that they would be good stewards of their putatively improved Fens. Their new approach illustrates that fenland popular politics was not invariably provincial, conservative, or rooted in custom; the fenlanders were capable of discerning and engaging with central political priorities and shifting their tactics to adjust to changing circumstances.

What did not shift over three decades of petitions, pamphlets, lawsuits, riots, and seditious utterings was the Epworth commoners' assertion of the importance of the rule of law, the king's arbitrary and high-handed neglect of it, and the Commonwealth's mandate to restore it. This was at the heart of their fervent support for the Parliamentary cause during the Civil War and also their seditious critique of the Commonwealth regime they fought to bring to power but which, they felt, had betrayed them. In making the commoners' case to a newly sovereign Parliament, Daniel Noddel sought to remind the MPs of their weighty obligation. He implored them to respect the law and to restore it as the cornerstone of just governance: "The Lord direct your Honors to hold forth the law to them; for there is nothing in this case, that is not determinable in law. So shall the hearts of many thousand men, women, and children, in the Isle of Axholme, have occasion to bless God for his deliverance, when they see that through your means the law of the land is become their protection in their estates, against usurpers and wrongdoers."[76] The violent events at Epworth common were never just a series of provincial anti-enclosure riots, divorced from the broader political context of the time. The commoners' case, as Noddel presented it, was not their own affair alone, but touched on the rights and liberties of all freeborn Englishmen. The state was legitimate insofar as it protected those rights and liberties, and it became tyrannous when it ignored them. Parliament had a duty to ensure the Epworth commoners their day in court, Noddel insisted, "for there is much of the freedom of the laws and liberties of England, in my judgment, either to be preserved or lost in it."[77] In founding his argument squarely on the rule of law, and putting it to Parliament in the wake of the Civil War and the king's execution, Daniel Noddel was speaking to something far greater than just the draining of the Fens.

The Second Great Level Drainage, 1649–1656

The undertakers have always vilified the Fens, and have misinformed
many Parliament men, that all the Fens is a mere quagmire, and that
it is a level hurtfully surrounded, and of little or no value: but those
which live in the Fens, and are neighbors to it, know the contrary. . . .
What is cole-seed and rape, they are but Dutch commodities, and but
trash and trumpery. . . . The undertakers talk of great matters that will
accrue to the Commonwealth by rape; I am sure they have committed
a rape upon the republic, in ravishing the good people of this nation
(by their tyranny and oppression) out of their properties and liberties.
                    —[John Maynard?], *The Anti-Projector: Or the History*
                                            of the Fen Project, c. 1653

[V]ery great therefore is the improvement of draining of lands, and
our negligence very great, that they have been waste so long, and as
yet so continue in diverse places: for the improving of a kingdom is
better than the conquering a new one.
                              —*Samuel Hartlib His Legacie*, 1651

In 1658, the mathematician and surveyor Jonas Moore published *A Mapp of ye
Great Levell of ye Fenns*.[1] Moore was the principal surveyor employed by the
group of investors who had succeeded at last in draining the area in question,
commonly referred to as the Adventurers, and they had commissioned him "to
survey the Great Level of the fens and to make an exact map. . . . And to serve
the company in such other things belonging to a surveyor within the said level."[2]
For nearly eight years Moore had plotted out the various drainage works and
division dikes the company needed to construct, measured out their allotted
lands, and produced maps of the region for the use of both the company and the
state. *A Mapp of ye Great Levell of ye Fenns* was his cartographic masterpiece,
meant to showcase not only the company's triumph in taming the fenland floods,
but his own talent and skill as a surveyor.[3]

The map is an enormous work, befitting the size and grandeur of the drainage project it depicted in such minute detail. Printed over sixteen plates, when fully assembled it measures nearly two meters by one and a half; its scale, roughly two inches to a mile, was not improved upon for a map of the Fens before the end of the nineteenth century.[4] It was, and remains, an impressive visual spectacle, at once practical, beautiful, and overwhelming (fig. 8.1). The map's dominant features are certainly the two great artificial drainage channels that were the cornerstone of the project—the Old and New Bedford Rivers toward the center, both newly constructed by the company; and Morton's Leam to the west, now rebuilt and much expanded. On the hand-colored copy in the British Library's

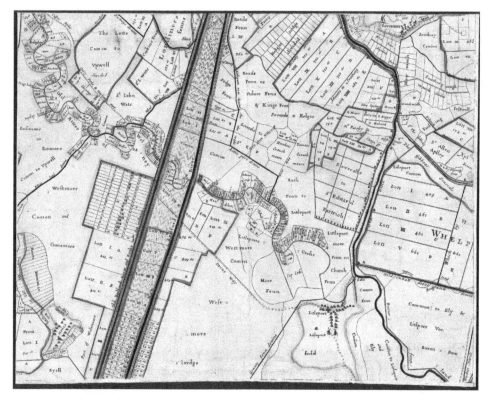

Figure 8.1. Detail from Jonas Moore, A *Mapp of ye Great Levell of ye Fenns . . .* (London, 1658; 3rd ed., 1706). The image is one of sixteen plates that together constitute the entire wall map. It depicts a section of the Old and New Bedford Rivers, between the villages of Littleport and Manea in Cambridgeshire and Welney in Norfolk, as well as part of the River Great Ouse.
© The British Library Board, Maps 184.L.1, plate 7.

collection, with bodies of water highlighted in blue, these channels stand out in part because of how little blue there is on the rest of the map; next to the company's new drainage works, the few remaining pools and meres seem insignificant. The channels' prominence lies also in their unnatural shape: sharp, razor-straight, two arrows pointed directly toward the North Sea, leading the floodwaters there by the most direct route possible. The contrast with the lazy, meandering rivers that had previously defined the landscape of the region could not be greater.[5]

Besides the new drainage works, the map's most dominant feature is the intricate, weblike geometric pattern into which the surrounding land has been carved—this is where Moore's skill as a surveyor was really brought to bear. The landscape has been divided and subdivided, with the ninety-five thousand acres allotted to the company's shareholders outlined in red on the British Library copy. The allotments, scattered over dozens of plots of every shape and size, are all neatly labeled with their acreage and the lot number to which they belong. Many of the surrounding lands are also labeled, either as severals belonging to a specific landowner or commons attached to a particular village. The challenge of depicting so many plots of land, many quite small and usually of irregular shape, was immense; and yet the measurements are impressively accurate. The complex geometry of the new Fens is visually striking, and the map's message is impossible to miss: efficiency, rationality, and mathematical order have been imposed on a formerly unruly landscape by the minds and hands of men, transforming chaos and waste into a fixed, measured, controllable, and productive resource.[6]

Moore's map makes clear where the credit lay for such a grand achievement. In the bottom left panel, next to the map's title and Moore's assertion of authorship, is the company's coat of arms. The crest, a lion passant atop three wavy blue lines, is flanked by two figures dressed as laborers carrying a trenching spade and a scythe, and surmounted by a bountiful cornucopia; the motto inscribed below is "Arridet aridum," or "The dry [land] pleases."[7] And around the entire periphery of the original 1658 map are the arms of all the most prominent officers and shareholders in the company, who were responsible for bringing the project to fruition—just as Moore was the author of the map, so these men were the authors of the drainage project itself.[8] This feature was omitted from subsequent editions of the map, printed after the Restoration of the English monarchy in 1660, almost certainly for political reasons. For besides the arms of the Earl of Bedford, several of those depicted belonged to important members of the Commonwealth and Protectorate regimes, or to senior officers in the Parliamentary army. These include Oliver St. John, lord chief justice of the Court of Common Pleas (as well as cousin and close friend of Oliver

Cromwell); John Thurloe, secretary to the Council of State; Major Generals Edward Whalley and William Goffe; and the regicides Robert Tichborne, William Say, and Valentine Walton, among others. Although not formally a state undertaking, the arms surrounding the original map communicate unmistakably the Protectorate regime's support for and endorsement of the project, as well as the powerful political connections the company enjoyed and relied on throughout the Interregnum.

The Adventurers' triumph in the Great Level of the Fens was by far the largest drainage project in early modern England, and perhaps the largest in all of Europe. To achieve it, the shareholders first had to identify and secure the expertise needed for the job. Though they eventually hired Sir Cornelius Vermuyden to direct the works, they were not happy about it. They did not trust him and could not control him, but because his valuable skills and knowledge were so rare in England, they had little choice but to depend on his services. Equally important was the multifaceted support of the Commonwealth and Protectorate regimes, without which the project could never have been attempted, let alone accomplished. The state's involvement was pivotal in many ways, perhaps most importantly in forestalling and overcoming the opposition of the fenland inhabitants. It also underscored just how political drainage projects really were, as part of a larger movement toward centralized, unitary government and a nascent English empire. During the Interregnum, the dominant intellectual discourse of the era had come to view agricultural improvement generally, and land drainage in particular, as vital parts of a program for utopian reform. The "improver" literature, especially works produced by Samuel Hartlib and his wide network of correspondents, promoted drainage projects as rational, productive, and patriotic ventures, certain to enrich the commonwealth, foster trade, support settlement, and advance civilization both within England and throughout its colonies.

DIRECTING THE DIRECTOR: CORNELIUS VERMUYDEN AND THE COMPANY OF ADVENTURERS

The leading figure in the Great Level drainage during the Interregnum, and the eponymous cornerstone of the company of investors responsible for it, was William Russell (1616–1700), 5th Earl (later 1st Duke) of Bedford (fig. 8.2). Bedford had come into his title, lands, and wealth after the unexpected death of his father, the 4th Earl, of smallpox in 1641. He had been elected to the House of Commons in both the Short and Long Parliaments, and followed his father in siding with the Parliamentary cause on the eve of the Civil War, maintaining that position after his translation to the House of Lords. Though still in his twenties, Bedford was initially given positions of real responsibility in the Parliamentary war effort, including the lord lieutenancy of the counties of Devon and Somerset, and he fought with the Parliamentary army at the Battle of Edgehill in 1642. He favored a peaceful resolution to the conflict almost from the beginning, however, and once it became clear that his colleagues in Parliament would not seek a negotiated settlement, Bedford abandoned the cause in 1643; he joined the royalists and was pardoned by King Charles. He fought briefly with the king's army, but Charles's other advisers did not trust him, and he was given only trivial responsibilities. Bedford soon grew disillusioned with the royalists and attempted to return to Parliament's service, only to find that his former colleagues there no longer trusted him either. They would give him no command, refused to allow him to resume his seat in the House of Lords, and even imprisoned him briefly, though he was released in the summer of 1644.

From that point onward, distrusted by leaders on both sides of the conflict and alienated from the ever-more radical course pursued by the Parliamentarians, Bedford played no further military or political role in the Civil War or Interregnum. He retired instead to his estate at Woburn, where he spent his time managing his extensive landholdings. His political exile allowed him ample time to reinvigorate the Great Level drainage project, which had fallen into abeyance after the death of the 4th Earl and the start of the war. From 1642 onward, Bedford had sought an act of Parliament granting him permission to secure investors and resume construction work on his father's project. Though he was no longer a central figure in the counsels of the Commonwealth regime, he was not without powerful friends in the government. After years of trying, he finally got his wish with the passage of "An Act for the Draining of the Great Level of the Fens" in May 1649, more commonly known as the "Pretended Act" since the Restoration in 1660, because it had never received the king's assent. He soon recruited all the "Adventurers" he needed to get things started, and he set to work.

After resolving "to return a thankful acknowledgement" to Oliver Cromwell and his son-in-law Henry Ireton for facilitating passage of the 1649 Drainage Act, the company's first item of business was to hire a director to design the new drainage network for the Great Level and oversee its construction.[9] This would be the largest and most challenging drainage project yet undertaken in England: five times larger than the Hatfield Level project and much more ambitious than the 4th Earl's attempt in the 1630s, insofar as it explicitly required turning the land into "winter ground." In 1649, there was only one obvious candidate for the job—Sir Cornelius Vermuyden. He had a number of advantages to recommend him: he was the architect of the (partially) successful Hatfield Level drainage; he

Figure 8.2. William Russell, 5th Earl (later 1st Duke) of Bedford, by George Glover and Peter Stent, c. 1643–67.
Portrait D19960, © National Portrait Gallery, London.

had been studying the problem of how to drain the Great Level since the early 1620s and had actually directed Charles I's project there before the Civil War interrupted it; he already had a plan for the work, published in 1642; and though his reputation was decidedly mixed, he had more firsthand experience in building large-scale drainage works than anyone else in England.[10]

On the other hand, Vermuyden also had his detractors, who were quick to point out that none of his drainage projects to date had been wholly successful and that those investing in them had lost more money than they had made. He was accused of running up costs to line his own pockets, and his preferred methods were denounced as being too similar to those practiced in the Low

Countries, not well suited for draining England's wetlands. His stubborn, contentious personality and resistance to sharing authority had also won him few friends. The company's shareholders knew all of this, but they invited Vermuyden "to propound his demands for perfecting the work" nonetheless.[11]

Negotiations between Vermuyden and the company were difficult, and the parties failed to reach an agreement over the next several months, losing the entire 1649 working season.[12] The most contentious issues had to do with authority, accountability, and trust. The shareholders recognized Vermuyden's experience and skill, but they were leery of his quarrelsome nature and reputation for producing less than satisfactory results. They did not trust him to direct the project without their active oversight and insisted that all of his decisions and expenditures be subject to review by a committee of control, nominated from among themselves. They demanded that Vermuyden provide detailed plans for all of his proposed works, including their precise locations, routes, and dimensions—information not included in his printed plan—before any agreement was reached. And while they were prepared to give Vermuyden four thousand acres of improved land for his services—one full share of the company's allotment, out of twenty—they insisted that he not sell any of it until after the work was completed and that he must keep at least half of it for seven years afterward. They feared that if he liquidated his share before the work was finished, he would have little personal incentive to secure a permanent drainage—his former partners in Hatfield Level had accused him of neglecting that project once he had sold his interest in it.

Vermuyden, for his part, resisted all efforts to curtail his absolute authority in designing and directing the works, and he bitterly resented the implication that he could not be trusted. He rejected the company's demand that he be subject to a committee of control, fearing that they would only urge him to cut costs at every turn and then blame him when the cut-rate works proved insufficient. He grudgingly consented not to sell half of his allotment for five years after the works were finished, though the demand clearly rankled him: "[M]y present resolutions are not to part with any part of it, neither now nor hereafter, but under favor I desire to be master of my own." As for revealing all the details of his plans, he thought the requirement "needless" and ranted that "my experience is so well known, that under favor it is frivolous and I cannot but much wonder that you put me to replication having hitherto prosecuted your business so long . . . with care, travail, and charge, and withal considering that in my desired recompense I become a joint Adventurer. . . . And so upon that small advantage I recover your lost estates and gain to the Common Wealth a great & vast country."[13]

Each party broke off negotiations more than once. The company "thought it not fit to depend upon Sir Cornelius Vermuyden any longer but make choice of some other to go on with this summer's work," while Vermuyden "declared to the company that he disclaimed to meddle or have anything to do with the draining of the fens."[14] The company consulted with multiple candidates, both Dutch and English, and at one point even offered the directorship to Sir Edward Partherich, one of the shareholders, who had submitted his own drainage plan for consideration. Nevertheless, Vermuyden and the company always resumed negotiations within a few weeks or months after each break. Despite his resentment of their mistrust and threatened interference, Vermuyden wanted to direct the project; and though the shareholders were obviously uncomfortable entrusting him with the work, they never identified a suitable replacement. Vermuyden's unrivaled experience, his general knowledge of the principles of land drainage in England and elsewhere, and his prior involvement in the Great Level combined to make him the only truly viable candidate. When he approached the shareholders on 1 January 1650 offering to demonstrate that "Sir Edward Partherich his design is destructive to the works of draining in diverse particulars," they quickly seized on his overture, and a few weeks later Vermuyden was named "sole director of the work of draining the Great Level," a post he held from that point forward.[15]

On paper (or vellum), Vermuyden had capitulated on nearly every point of contention. The agreement authorized him to build whatever works he deemed necessary, provided always that he "be subject to and observe such orders and directions" as would be provided by a nine-member committee of control. All power to appoint officers and overseers and to make contracts with workmen remained with the company. Vermuyden was to receive four thousand acres of improved land, plus a stipend of £1,000, and he was to be exempt from paying any sewer taxes on his allotted lands until total expenditures on the project surpassed £90,000; but he was barred from selling more than two thousand acres until seven years after the land was drained.[16] Yet while the company prevailed in most of the battles, Vermuyden may be said to have won the war. The protracted negotiations had underscored just how reliant the company was on his rare expertise, and after his appointment he knew the balance of power was in his hands. For the next three years he carried out his duties in the Fens more or less as he pleased, ignoring the repeated demands of the company that he submit to its control.

Having finally hired a director, construction could begin at last in 1650. The company decided to tackle the massive project in two phases, first building the new drainage works to the northwest of the Bedford River (the areas that Vermuyden had deemed the North and Middle Levels) followed by those to the southeast (now called the South Level) (fig. 8.3). The works proposed for the North and Middle Levels were extensive, but they mostly built on existing drains, and Vermuyden's efforts before the onset of the Civil War had been concentrated in those areas, so he did not have to start from scratch. The first phase required rebuilding and reinforcing the banks of the Welland River near Crowland; rebuilding and reinforcing the banks of both the Nene River and Morton's Leam, leaving a wide "wash" area between them, from Whittlesey near Peterborough to Guyhirn, and thence to Wisbech; improving and lengthening Bevill's Leam; dredging the channel of the Nene from Wisbech to the sea and building a new sluice to restrict the incoming tides; raising a high, sturdy bank along the northwestern side of the Bedford River to protect the Middle Level; and digging several new drains to draw water from the Middle Level into the Bedford River. These new drains included one that was ten miles long near Chatteris, known variously as Vermuyden's Drain or Forty Foot Drain, and another between Vermuyden's Drain and Popham's Eau, known as Thurloe's Drain or Sixteen Foot Drain. Popham's Eau was also substantially rebuilt, and was linked with the Great Ouse by a new drain called Marshland Cut.

While the first phase largely involved the modification and reinforcement of existing drains, the second phase, in the South Level, required fewer works but was more novel and ambitious. It required building an entirely new river, one hundred feet wide and more than twenty miles long, half a mile southeast of the Bedford River and roughly parallel to it. This channel, to be known as the New Bedford River, was the most important and impressive work in the entire project. It would handle the huge volume of water from the redirected River Great Ouse that the original (Old) Bedford River had never been equal to carrying. Vermuyden also planned to build a high bank along the southeastern side of this river, to mirror that on the northwestern side of its counterpart. When floods threatened, the high outer banks of each river would keep the excess water out of the Middle and South Levels, spilling it instead into the extensive "wash" lands between them and temporarily turning them into a single, massive river, half a mile wide.[17] Other new works included erecting large and complicated sluices at either end of the New Bedford River—one upriver at Earith to shunt the waters of the Great Ouse into the new channel, and the other downriver at Denver, near Downham Market, to prevent tides from entering. The Great Ouse below Denver was substantially widened, in order to accommodate all of the water now vented directly into it at that point, by a new cut called Downham Eau. Finally, several smaller rivers flowing into the Great Ouse from Norfolk were to have new and reinforced banks, including the Grant/Cam, Brandon, Stoke, and Mildenhall Rivers (fig. 8.4).[18]

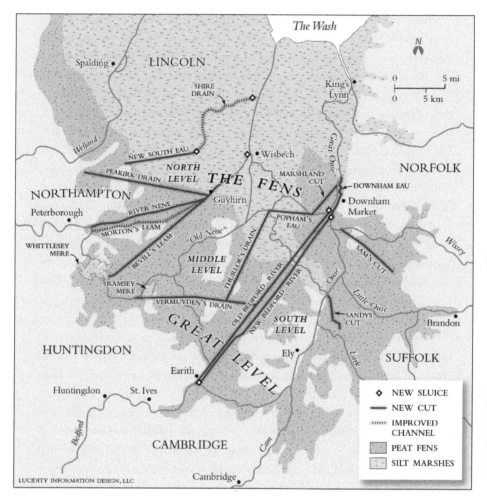

Figure 8.3. Map of the second Great Level drainage project, undertaken 1649–56.

For better or worse, Vermuyden's drainage scheme employed his preferred methods throughout: redirecting the rivers into new, straighter courses; placing his riverbanks at some distance from the rivers themselves, leaving a "wash" area in between; and using sluices to control the rivers' flow. His approach was controversial—it was expensive to build so many new banks and rivers, while his reserved washes ate up a sizable quantity of otherwise valuable land and diminished the rivers' ability to scour out silt by slowing their flow. Vermuyden, however, believed that the benefits of his strategy outweighed the costs. The new drainage channels would take full advantage of the land's limited gradient and

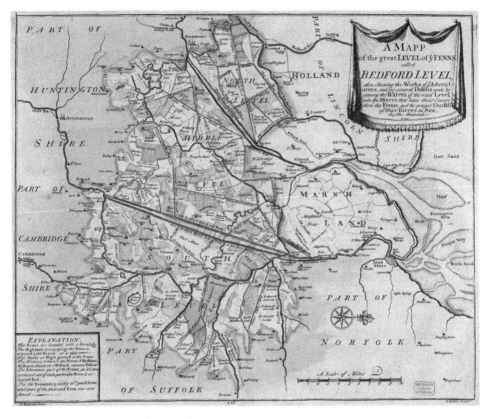

Figure 8.4. A map of the completed Great Level drainage project; north is to the right-hand side of the image. This map was printed in Thomas Badeslade's *The History of the Ancient and Present State of the Navigation of the Port of King's-Lyn . . .* (London, 1725), as a post-drainage companion to figure 1.1. It is a much-modified version of the lost Hayward survey map of the pre-drainage Fens (c. 1605).
Reproduced by kind permission of the Syndics of Cambridge University Library, Cam-a-725-1.

shorten the distance to the sea, while the washes would help to contain surging floodwaters in wetter seasons. The new sluices were intended to keep out the tidal silt during high tide and to release the pent-up river waters all at once during low tide to scour out the riverbeds.

As construction work proceeded, the company was plagued throughout by financial and administrative problems. Building new rivers, banks, and sluices was labor intensive, and therefore costly. Constructing such massive engineering works without the use of modern, heavy machinery meant that all of the backbreaking work had to be done by hand, shifting one shovel- or wheelbarrow-full of muddy, peaty earth at a time. While unskilled manual labor was cheap on an individual basis, the thousands of men required for such a vast undertaking constituted a huge and ongoing expense, and the largest budgetary item in the project, by far, was the cost of labor. The money was supposed to come from the company's shareholders, each of whom owned at least a partial share of one of the twenty lots of soon-to-be improved fenland, four thousand acres each, allotted to the company. The shares were to be taxed as needed to cover the costs of building and maintaining the new drainage works, with each shareholder's liability proportional to how much land he owned.

But the shareholders could not actually take possession of their allotted lands before the Great Level was drained, and the flooded land would not have generated much revenue in any case—the investment lay entirely in its future potential. Shareholders were thus expected to pay hefty taxes on land of little or no present value to them, and many were soon in arrears, creating a shortage of ready cash and hampering the company's ability to hire, pay, and retain laborers. The list of delinquents was long and included some of the most prominent shareholders, including Lord St. John (£106 5s.), the Earl of Arundel (£675), and John and Edward Russell, brothers of the Earl of Bedford (£528 15s.).[19] The only way to force them to pay up was to threaten to sell their allotments to more willing investors. But with seemingly endless expenses and no immediate hope of profits, buyers were always hard to come by and the threat, while often invoked, was usually an empty one.

Equally serious were the company's ongoing difficulties in overseeing the construction work. Most of the shareholders were based in London, and nearly all of their regular meetings throughout the 1650s took place there. The work was overseen on-site by Vermuyden, a handful of surveyors and other hired officers, and one or two shareholders who agreed to stay in the Fens to keep an eye on things. The officers too often proved to be untrustworthy, incompetent, or both, so that the company was chronically ill-informed about vital matters—what works were being built, by whom and under whose authority, how much they would cost, and how much money was available to pay for them. Monitoring expenditures proved at least as hard as collecting taxes in arrears, and the company's proceedings are filled with complaints about overseers making contracts without approval, paying exorbitant rates for materials and labor (and perhaps receiving a kickback), pilfering supplies, paying the workers in clipped coin or failing to pay them at all, and so forth. During the first two years, the only man the company really trusted in the Fens was Anthony Hamond, a shareholder who acted as their principal agent and informant. But though he was a competent overseer and a diligent correspondent, Hamond could not be everywhere in the Great Level at once.

Among the company's most frustrating problems was its inability to control Vermuyden. Despite his having agreed to allow a committee of control to monitor and approve his decisions, the committee members found it all but impossible to make him obey their orders when his opinion differed from theirs. This was most evident with respect to when construction should begin on the second, southeastern phase of the project. Vermuyden recognized that this would be the most challenging part of the whole endeavor, and he was eager to get started on it as soon as possible, taking advantage of an unusually dry summer for the work. He was less concerned about the works planned for the northwest—most were already well in hand, and of those not yet under way he had come to believe many were unnecessary. The company, however, insisted on the full completion of the northwestern works as soon as possible, so that they might request a formal adjudication of the North and Middle Levels and receive a proportional allotment of the lands due to them. This would provide a partial return on their huge investment to date, mollifying the current shareholders, attracting new ones, and encouraging delinquents to pay the taxes they still owed, before taking on the even more challenging and expensive works to the southeast.

From September 1650 through all of 1652, the shareholders in London repeatedly ordered Vermuyden to leave off working on the southeast side of the Bedford River until he had finished the work to the northwest. They were incensed that "Sir Corn. Vermuyden should take upon him to dispose of their money," buying materials and hiring laborers to build the New Bedford River without their approval.[20] Hardly a month went by in which they did not send him another stern warning to obey their directives, but in vain—the very ubiquity and stridency of their demands gives ample indication that Vermuyden felt free to work when and where he pleased. The shareholders asked the Earl of Bedford to take a more active hand in managing the company's affairs in March 1651, hoping this would cow Vermuyden, but the earl's regular attendance at meetings in London made little difference in the Fens.[21] The following July they dispatched John Thurloe, secretary to the Council of State and a prominent shareholder, into the Great Level to bring the director to heel, but their letter rebuking Vermuyden yet again just one month later suggests that even Thurloe's intervention had made no impact. The shareholders' collective political and economic clout still did not give them sufficient power to gainsay Vermuyden's expertise, and he was confident in believing they could not dismiss him and hope to complete the project.

The shareholders also wrote in frustration to the other officers and overseers in the Fens, commanding them to rein in the director that they themselves could not control. They leaned particularly hard on Anthony Hamond, threatening in August 1652, "if yet our orders in this behalf are still laid by and no observance had thereupon, we shall know upon whose account to put all the inconveniences and evils that will ensue thereupon."[22] But Hamond and the other officers on-site were reluctant to override Vermuyden's decisions, and they tacitly deferred to his greater experience and authority. At a rare company meeting held in the Fens at Ely, for example, just a few days after the shareholders had rebuked him, Hamond argued alongside Vermuyden that some of the works planned for the northwest were not needed.[23] Jonas Moore, the company's principal surveyor, told the shareholders a few months later that he also agreed with Vermuyden.[24] If Hamond and Moore would not challenge the director's judgment, the nine-member committee of control was in no position to do so. In their review of several disputed works in August 1652, the members declined to overrule Vermuyden on any of them, adopting a wait-and-see approach instead.[25] By the end of the year the company had stopped even referring matters to the formal committee of control. It consulted instead a new, ad hoc committee including Hamond, Moore, Edward Partherich, and others with actual experience overseeing the project on-site.[26] This was a much less distinguished, but far more knowledgeable group, and one that had worked harmoniously with Vermuyden for the most part.

The company's difficulty in controlling Vermuyden was rooted in the fact that, even three years into the project, the English overseers were still remarkably ignorant of the actual work required to dig a new drainage channel or build a secure bank. At a meeting in London in September 1652, with Vermuyden in attendance, the shareholders received a letter from William Drury, one of the men overseeing work on the Old and New Bedford Rivers. The old river had to be dredged, but Drury did not know how or where to do it, and with Vermuyden in London he was forced to stop work and ask for guidance. The instructions Vermuyden sent back were recorded in the company's proceedings: "You are to sludge the same and to do it by long poles and a piece of plank at the end, and to road it, *the workmen will understand what I mean*, do it exactly in the places where the mud is driven in about Vermuyden's Eau . . . go from Welsh's dam unto half a mile near Welney, but do not touch nor stir the ground below Welney, otherwise it will be dangerous."[27] That Drury, an experienced overseer, had to send to London for instructions in carrying out such a basic operation reveals much about the empirical knowledge and skill the company might expect in its officers. Even more revealing is the fact that Vermuyden did not expect Drury to understand his instructions and implied that he should just pass them along to the workmen, who would understand what he meant. From 1649 onward, there was no one in the company at any level with the experience, knowledge, skill, and authority—the expertise—to challenge Vermuyden's directorship.

This does not mean, of course, that Vermuyden's judgment was necessarily correct. Historical opinion of his Great Level drainage scheme has been largely negative, with various engineers and historians criticizing all of his choices, from the Restoration onward.[28] But whatever Vermuyden's merits as a drainage engineer, during his tenure the company could neither compel his obedience to its directions, nor evaluate his plan except by judging the results on completion. The company did not trust him, yet it could not help but rely on him. In this, its experience was similar to other cases of strained expert mediation, such as the copper mining venture undertaken at Keswick during the reign of Elizabeth I. The English investors in the Company of Mines Royal found it all but impossible to oversee and trust the German miners, assayers, and mine managers whose rare knowledge and skill they necessarily depended on.[29]

In the end, Vermuyden's directorship was a qualified success. Construction proceeded quickly—at the height of the work in the summer of 1652, at least ten thousand men were employed in the project, with a weekly payroll of £8,000. The works they built were simply immense: the project required moving millions of cubic yards of earth by manual labor and building several sluices that were among the largest such works ever constructed in the early modern period.[30] Yet despite its size and complexity, the bulk of the project was completed more than three years ahead of the October 1656 deadline imposed in the 1649 Drainage Act. In the spring of 1651, the company appealed to a parliamentary commission specially created for the purpose and asked that the North and Middle Levels, an area of some 170,000 acres, be declared drained. The commissioners met for several days at Wisbech and Peterborough, heard witnesses and considered a number of petitions from fenland inhabitants, and finally handed down a judgment in the company's favor on 26 March.[31] The shareholders were awarded a proportional share of the ninety-five thousand acres due to them, giving a much-needed boost to their finances and their morale. The same process was repeated just two years later for the South Level; after a series of meetings at Ely, the commissioners declared that this area was also well drained, and awarded the company the remainder of its ninety-five thousand acres on 26 March 1653.[32]

The project's early completion and successful adjudication are only part of the story, however. The shareholders still had much to be unhappy about, including the astronomical expense. In the plan he presented in December 1649, just before the company appointed him director, Vermuyden had estimated that the cost of labor and materials for building the works on both sides of the level would be just under £120,000.[33] The actual cost of the work is very hard to determine, but it was certainly more than double Vermuyden's estimate—the company claimed in 1653 to have spent £300,000 altogether.[34] Nor was this the final figure: even after the adjudication, much work still had yet to be completed, including the many miles of division dikes that would separate each shareholder's allotted lands from the surrounding commons and severals, and from his colleagues' neighboring plots. Ongoing maintenance costs were also a real burden. In February 1655, nearly two years after "completion" of the project, the company estimated that £23,000 was needed to repair it, and it levied a tax of five shillings per acre to cover the expense, one of the highest figures it ever assessed.[35] Subsequent maintenance costs continued to range between £5,000 and £10,000 annually throughout the 1650s. The company was chronically short of funds for years after the adjudication, and though the project was a technical success, many of the original investors were ruined by it.

Yet despite the ballooning expense and the fragility of the works, the Great Level had been drained—the vision first articulated by Humphrey Bradley to Lord Burghley in the 1580s was realized at last. The company steadily diked out and enclosed its allotments, planting them with corn, onions, peas, rapeseed, coleseed, hemp, and madder. The drained lands were not of uniform quality, but most were improved in terms of productivity and value, and they were let out for good rents. The landscape of the Fens had finally been transformed through the construction of massive new drainage works that were well beyond the reach, if not the imagination, of any previous projectors or commissioners of sewers. Jonas Moore's wall map of 1658 illustrated the degree to which the company had imprinted its stamp on the fenland, in terms of both geography and landholding patterns. And while the Great Level drainage had been privately funded by London-based investors and a few large landowners in the Fens, it could never have come to pass without the active and ongoing support of the Interregnum state.

THE STATE AND THE GREAT LEVEL DRAINAGE

State support for the drainage project was multifaceted and began well in advance of the project itself, with parliamentary consideration and passage of the Drainage Act of 1649. The many hearings on the merits of the act, convened by the specially created "Committee for the Fens" during the late 1640s, facilitated the project's success not least by accepting dozens of petitions from fenland inhabitants and allowing both sides of the debate a chance to be heard. Some anti-drainage partisans, to be sure, insisted they were not given sufficient opportunity to testify. Yet the hearings gave at least the appearance of impartiality and genuine public interest to the committee's legislative deliberations and provided a solid political and legal basis for passing the act as a measure intended for the good of the commonwealth and nation.

The law's preamble, indeed, expounded on all of the explicitly *national* benefits that would arise from the project: transforming the Great Level into winter ground would render it

> fit to bear coleseed and rapeseed in great abundance, which is of singular use to make soap and oils within this nation, to the advancement of the trade of clothing and spinning of wool, and much of it will be improved into good pasture for feeding and breeding of cattle, and of tillage to be sown with corn and grain, and for hemp and flax in great quantity, for making all sorts of linen cloth and cordage for shipping within this nation; which will increase manufactures, commerce, and trading at home and abroad, will relieve the poor by setting them on work, and will many other ways redound to the great advantage and strengthening of the nation.[36]

The language indicates the degree to which the Commonwealth state was a projecting state; it echoes the many projectors of earlier decades who sought to convince the Crown that draining the Fens would benefit not just the fenlanders, but all of England. It invokes the good of the nation repeatedly—in the economic and social benefits borne of increasing domestic manufactures, the stimulation of trade and commerce, and the supply of vital wartime materiel during a prolonged period of international conflict. The act not only promoted the Great Level drainage project; it showed how a wealth of other projects would also be promoted thereby. After provoking so much political opposition over the previous half century, much of it aimed at the Crown, projecting had become the national policy of the new Commonwealth regime, literally the law of the land.

The 1649 act marked one of the first times a drainage project was granted legal standing by parliamentary statute rather than a local law of sewers, and it spelled out the legal terms under which the project should proceed. It invoked prior laws of sewers to justify the claims of the Earl of Bedford and his fellow shareholders (such as the Lynn Law of 1631), and declared others to be null and void whenever they contradicted those claims (such as the Huntingdon Law of 1638 that had awarded the project to the king). It described with great specificity the geographical boundaries of the Great Level to prevent any confusion or challenge about whose lands were, and were not, to be included in the venture.

Bedford and his partners were authorized to make new drainage works or repair old ones as needed and to raise money by selling shares and levying taxes among themselves on the ninety-five thousand acres to be allotted them in recompense. They were also given sole authority to govern and manage the Great Level drainage as they saw fit. The existing commission of sewers was barred from "intermeddl[ing] in the said level, to interrupt, disturb, or molest the said William Earl of Bedford . . . in the carrying on and perfecting of the said work," while the company's shareholders were authorized to act as commissioners of sewers henceforth.[37]

In addition to giving the project full legal standing, the 1649 act also provided an administrative apparatus to ensure the drainage was a success and to safeguard the interests of all parties as conflicts inevitably arose. It created an independent parliamentary "commission of adjudication" to determine whether the land was drained; to provide a forum for hearing the myriad complaints, petitions, and disputes the project generated; and to settle "all such points, matters and things which in their judgments are or shall be necessary . . . according to the true intent and meaning of this act."[38] The nominated commissioners were all high-ranking members of the government, including the speaker of the Parliament; the lord keeper, lord chancellor, or lords commissioners of the Great Seal; all of the justices of the Upper Bench (formerly the Court of King's Bench) and Court of Common Pleas, as well as the barons of the Court of Exchequer; and some fifty-two members of Parliament and the Council of State, including Oliver Cromwell, Richard Cromwell, and Henry Ireton. The commission thus formally oversaw the project on behalf of the Commonwealth state, and though in practice its meetings usually included only between six and ten commissioners, both of the lords commissioners of the Great Seal were regular attendees, ensuring that the Council of State was well represented.

The commission of adjudication was supposed to guarantee that, while Bedford and his partners had the power to build and maintain the new Great Level drainage network as they saw fit, they would not act as judges in their own case. The commissioners met regularly throughout the project, mostly in London, though they briefly adjourned to the Fens in 1651 and 1653 to observe and judge in person the status of the drainage and to award the company its allotted lands. A copy of their proceedings survives for the period 1650–56, from shortly after the inception of the 1649 Drainage Act until three years after the final adjudication of the second and last phase of the project—the window of time in which the law permitted alterations to the company's allotments.[39] They considered dozens of petitions and worked to resolve numerous disputes. Most petitions were from fenland inhabitants and dealt with complaints regarding remuneration for damage done by the company's works to private landholdings; adjustment of land allotments considered unfair; requests for highways or bridges to be built to restore access to common lands cut off by new drainage ditches; complaints that drainage works were ineffective or decayed; and similar matters. Some petitions came from laborers claiming they had not been paid promptly for their work, while others came from the company itself, seeking redress after rioters had vandalized its works or damaged its allotted lands.

The commissioners were thorough in their work, and by and large impartial, though they were not speedy. They settled a few of the simplest petitions outright, usually by ordering the company to make restitution, but in most cases they delegated each matter to a small group of referees to help broker a resolution. This process could take months, or even years, and in many cases no settlement was ever recorded in their proceedings. But where a final determination is recorded, the commissioners often found in favor of fenland petitioners and against the company. They were most likely to do so in cases dealing with unfair allotments, claims of damage to private lands, and requests for a new bridge or highway; those who denied the efficacy of the drainage itself were far less successful, though such petitions were received and referred for mediation like the rest.

Though never as troubled as the Isle of Axholme, the Great Level saw significant anti-drainage sentiment at times—the dozens of complaints and petitions addressed to the commission of adjudication are a good measure of the fenlanders' discontent. Yet the existence of a truly independent, state-sponsored body to review their complaints may have helped to mollify the inhabitants by providing an impartial forum in which to plead their case. The exalted political and legal stature of the commissioners gave the body real gravitas and clout, and their independence appears to have been taken seriously by all concerned. Two of the nominated commissioners who qualified by virtue of their offices of state, Lord Chief Justice Oliver St. John and MP John Trenchard, were also shareholders in the company; but neither was ever recorded in attendance at a session, perhaps because of the obvious conflict of interest. And given the commission's tendency to rule against the company in routine matters, fenland petitioners could have some faith that they were getting a real opportunity to obtain justice—as long as they did not oppose the drainage itself. The commission thus represented not only the public interests of the Commonwealth state but the benevolence and impartiality of that state in balancing private interests against one another. This, in turn, helped to lend a sense of fairness and legitimacy to the entire project.

Beyond the Drainage Act of 1649, which provided the legal authority and key administrative apparatus for the project, much of the state's support came in the form of cheap, conscripted labor. Wages for unskilled manual labor were by far the largest portion of the company's expenditures, and its spotty record of paying its workmen, in good coin and on time, made it harder to attract native fenlanders to take employment in the works. The Council of State thus provided a much-needed boost to the company's fortunes when it agreed to allow the company to put prisoners of war to work in the Fens. During the Anglo-Scottish War (1650–52), Cromwell's New Model Army had captured several thousand Scottish soldiers, especially after the Battles of Dunbar and Worcester. Feeding, housing, and clothing so many idle men was a burden the struggling Commonwealth hardly needed, yet they could not be allowed to return to Scotland during the war. The government addressed the problem by selling many Scots into indentured servitude in the West Indies and American colonies to help improve the land and work the plantations there.[40] The Great Level drainage was very similar to these imperial projects, with a pressing need for strong, unskilled laborers to perform backbreaking work at minimal cost. Whether the government or the company first proposed the idea is uncertain, but with state officials such as John Thurloe and Oliver St. John among the company's shareholders, a mutually beneficial arrangement was soon concluded. In October 1651, the company took delivery of hundreds of Scottish prisoners, relieving the state of the expense of caring for them and putting them to work in the Fens instead.

The Scots performed much of the manual labor in building the Great Level drainage works until at least September 1652 (the Anglo-Scottish War ended in May 1652). They were involved in the most intensive period of work, leading up to the second and final adjudication in March 1653. They labored not only in building the New Bedford River and other drainage works but also in preparing the already improved lands of the northwestern level for tillage. The company was responsible for feeding, housing, and clothing the men, and further committed to giving them "a fitting reward from us for their labor," so that "they might have had just cause to bless God for the Parliament's mercy to them." At the same time, the shareholders always sought "to make the best contract they can touching the said prisoners and with least prejudice to the company." Even after seeing to the men's basic needs and paying them a small wage, the Scots must have been a considerable bargain for labor, given the company's eagerness to take on as many as possible.[41]

The Scots' departure in fall 1652 did not end the state's provision of conscripted labor for the project. During the first Anglo-Dutch War (1652–54), English naval victories led to the capture of several hundred Dutch seamen, and the state immediately moved to shift as many of them as possible into the company's service. Five hundred Dutch prisoners were put to work in the Fens in the summer of 1653, on terms similar to those provided to the Scots. The Dutchmen proved to be a more refractory labor force, however; many refused to work or ran away, often with the encouragement of fenland inhabitants, who hid escapees from the authorities.[42] But by the time the Dutch arrived in the Great Level, most of the new drainage works were already either completed or well under way, so their impact was comparatively small in any case.

The most challenging (and expensive) aspect of employing prisoners of war in the Fens was ensuring they would not escape—this was particularly true for the Scots, who could flee homeward without having to obtain sea passage. The company took pains to prevent their escaping; it committed to providing the Council of State with a monthly census of the conscripted labor force, hired special overseers to guard them, and even dressed them in distinctive clothing to facilitate recognition and recapture of any runaways. The state did its part to secure the prisoners as well, allowing the use of regular army troops to transport them (at the company's expense) and passing an ordinance decreeing "death without mercy" for escapees. Several did escape in spite of this, but while some may have made it back to Scotland, those recaptured were executed as a deterrent to the rest.[43] The vast majority of the prisoners stayed put, however, and the company was soon forced to consider that several Scottish women had come to join their husbands in the Fens and were now "big with child," so that local parishes worried about the possible impact on their poor rates. At least a few of the Scots appear to have settled permanently in the Great Level after their release.[44]

Transporting prisoners of war was not the only occasion for which the state provided military assistance to the company, and the regular use of troops was yet another important aspect of state support for the project. Despite the best efforts of the commission of adjudication, the undertaking remained extremely unpopular in parts of the Great Level, and attacks on drainage works and the laborers building them were a frequent problem, particularly after the adjudications in 1651 and 1653, when the company moved to enclose its allotted lands. Rioting erupted on numerous occasions during the early 1650s, and the company moved aggressively to identify and prosecute the offenders, even seeking a parliamentary ordinance declaring it a felony to vandalize its drainage works.[45] The Commonwealth was eager to assist in suppressing riots and punishing those who incited them, given the pressing need to maintain order while the nation's political and military situation was still so unsettled, and especially with hundreds of prisoners of war at work in the region. In September 1651, as Charles Stuart led a royalist invasion force into England from Scotland and advanced all the way to Worcester, the Council of State worried that he might take his army into the Isle of Ely and "stir & join with such disaffected persons who are now there at their draining work and bring trouble thereby to those parts."[46]

The mutual interest of the company and the Commonwealth in keeping order in the more restive parts of the Great Level led the state to make troops available, at the company's expense, to provide additional security for the project. By the end of 1651, the company maintained companies of soldiers at Ely, Lynn, Haddenham, and Ramsey, among other places.[47] They instructed Anthony Hamond to have soldiers ready to prevent any "clamors or mutiny" among workmen who had not yet received their promised wages. They received permission from Cromwell himself to give the commissioners of adjudication a military escort when they came to judge the second phase of the drainage in 1653.[48] Most important, however, was the company's ability to call on troops to prevent and suppress anti-drainage rioting. In January 1653, for example, after rioting took place near Swaffham Bulbeck in Cambridgeshire, the company told Hamond to get "a squadron of horse to quarter thereabouts" from the chief officer at Ely, "as formerly you have done." It urged him "to requite the soldiers with such satisfaction and gratuity as you shall think fit . . . conceiving this to be a principal means to suppress riots for the future."[49]

THE RESPONSE TO RESISTANCE

The state's active and multifaceted support for the drainage project, and the company's dependence on it, are most in evidence in the spring and summer of 1653, when anti-drainage resistance reached its peak in the Great Level. The trouble began in March, when the commission of adjudication convened at Ely to judge the second phase of the work. Anti-drainage sentiment was already running high in the region, as suggested by the company's provision of a military escort for the commissioners. Their proceedings for the three days they met at Ely, 24–26 March, fill twenty folio pages; they received a total of fifty-four petitions, more than twice as many as when they had judged the first phase two years earlier. The nature of the petitions was also strikingly different. In 1651, nearly all the petitions addressed minor grievances—requests for allotments to be more fairly apportioned or demands for compensation for lands lost to the drainage works. They neither challenged the company's contention that its project had improved the land, nor questioned its legitimacy.

Some of the 1653 petitions were of a similar nature, but others were far more critical of the whole endeavor. Many petitioners complained that the drainage had not improved their lands, or that it had actually made conditions worse, and thus the company had no right to expect any compensation in the form of allotments and enclosures. Towns and villages throughout Norfolk, Cambridgeshire, and the Isle of Ely collectively argued that the company's works had "not secured or meliorated" their commons, or that they had "received much damage and no good" from them, and they asked for the adjudication to be postponed until they might be heard at greater length. The commissioners, in keeping with their standard practice, referred most of the petitions for further consideration, but this did not prevent them from ruling that the entire Great Level was drained or from awarding to the company the remainder of its ninety-five thousand allotted acres.[50]

The commission of adjudication was never a "rubber stamp" for the company, ruling against it more often than not in more incidental disputes, and the company's proceedings show how genuinely concerned the shareholders were to demonstrate to the commissioners that the drainage was a success. The 1653 judgment in the company's favor was probably a reflection of the massive, impressive-looking new rivers, banks, and sasses that had already been constructed at great expense, which most likely improved the surrounding lands when taken as a whole. Yet the sheer number of petitions presented at Ely indicates that the drainage was hardly an unequivocal triumph, and that some areas probably benefited more than others. The company's own proceedings after March 1653 also make it clear that, whatever the region's status at the time of adjudication, a great deal of work still remained to be done to secure it permanently from flooding. After the adjudication, when the prescribed means of legal redress had offered them no relief, the aggrieved fenlanders resorted to more violent forms of protest.

Rioting erupted in Cambridgeshire and Norfolk as soon as the company moved to enclose its allotted lands and continued throughout 1653. The violence was aimed at the company's drainage works and at the laborers who were still building them, and it was concentrated in the communities that had petitioned in vain for redress.[51] The most serious violence took place at Swaffham Bulbeck and Bottisham in Cambridgeshire, where in April some 150 rioters threw down the nearby works and beat the workmen they found there. The riots continued during the summer, culminating in August when at least eighty men, armed with muskets, swords, and pikes and "giving out very high and insolent speeches," attacked some soldiers guarding the works, severely wounding one of them.[52]

The fenlanders also tried to incite resistance among the Dutch prisoners of war who had been dispatched to work in the level that summer. The Earl of Bedford complained to John Thurloe in July that the prisoners "are encouraged by the country people . . . to run away, hiding them in the corn; and many of them are run away." Those who did not escape stubbornly refused to work, "being possessed by the country, that they being prisoners of war, they are to be maintained by them that keep them; so that barely to keep them alive, they spend more than they earn."[53] The company probably bore some responsibility for the situation, having advised Anthony Hamond in June to quarter the Dutch prisoners in some of the most restive parts of the level as a punitive measure.[54] This would undoubtedly have fostered more resentment among the fenlanders and allowed two hostile groups to unite in common cause.

The company appealed to local authorities for assistance in quelling the riots but received little help from them. It complained that its efforts to subdue the rioters too often "proved fruitless and ineffectual by reason of the contrary interest and disaffection of the civil magistrates in those parts."[55] The company tried to prosecute rioters to the fullest but had a great deal of trouble identifying offenders and locating witnesses who would testify against them, and even the few they managed to prosecute were punished with trivial fines. Anthony Hamond informed John Thurloe that the Cambridge magistrates were "wanting will and affection to do us any good," and predicted that a commission of oyer and terminer would only backfire: "I doubt the jury will serve us, as they did in the case of John Lilburne, and there we are worse than we were before."[56] He even suggested that the only way to quell the riots for good would be to remove the perpetrators from the Fens by pressing them into the navy and sending them to fight the Dutch—a neat symmetry with the captured Dutch seamen who were supposed to be building drainage works in the Fens: "I have been thinking that nothing would fright & quiet them more, than if there were 100 of these desperate fellows pressed for the sea-service, they being all water men and having little to do at home, do make these night excursions and show their valor against my Lord General's men, which were much better employed against the Dutch."[57]

With riots breaking out across the southeastern Great Level, and expecting no aid from fenland officials in suppressing them, the company turned to the Council of State for assistance. Oliver Cromwell had forcibly dissolved the Rump Parliament on 20 April 1653, the same day as the first major riot at Swaffham Bulbeck. Without a sitting Parliament, the stability and constitutional legitimacy of the Interregnum state were both in considerable doubt, and the council could ill afford to have "several persons disaffected to the peace of this Commonwealth,

[who] upon occasion of the present change of government, do assemble together in a riotous and tumultuous manner in the Great Level of the Fens."[58] Still, the regime's initial response to the riots was measured. In April, Cromwell dispatched "one of my troops with a captain who may by all means persuade the people to quietness, by letting them know they must not riotously do anything for that must not be suffered." But he also reassured them that "if there be any wrong done by the Adventurers, upon complaint such course will be taken as appertains to justice, and right will be done."[59] More troops were dispatched in May to deal with riots in Norfolk, though again Cromwell pledged that anyone with a legitimate grievance to present "may be heard therein, and justice done unto them." The Council of State was prepared to ensure a fair hearing for petitioners "upon complaint made in a regular way," but violence and disorder must and would be suppressed, the councilors "not holding it fit that the people should right themselves in that way."[60]

As rioting continued through the summer, the council's patience wore thin, and it moved more aggressively to assist the company to restore order. Several companies of foot soldiers and cavalry were active in the Great Level for the rest of 1653, at the company's expense, suppressing riots and guarding the works against further attack. Troops were deployed wherever rioting occurred, and after putting down the riots, they were often quartered in the most troublesome communities, partly to prevent a recurrence but also as a punitive measure. Though their wages were paid by the company, the soldiers were still under the command of their regular officers, who took their orders in turn from the Council of State. All of the company's requests for military assistance had to go through the council, with influential shareholders such as John Thurloe acting as emissaries. The army's activities in the Great Level were thus a cooperative peacekeeping effort between the company and the state.[61]

Yet even as it put military force at the company's disposal, the Council of State also moved quickly to augment the administrative apparatus for dealing with the fenlanders' many petitions and complaints. The commission of adjudication was supposed to serve this function, but by 1653 its limitations had become apparent. Except during an adjudication, the body always met in London, "which is a great distance from the level, and thereby the petitioners put to great expense, and travel, in seeking their relief." This was probably the reason so many petitions were presented to them all at once, on the eve of the adjudication in March 1653; it was only the second time in four years that the commission had convened in the Fens. Moreover, the body was empowered by law to hear appeals and adjust land allotments for only three years after the final adjudication. The process of mediation was so slow, and the demand for it so overwhelming, that "it will not be possible, through the multitude of the petitions that are daily exhibited to them, to hear and determine the several cases and complaints."[62]

The council therefore created a second body, the "committee of petitions," to deal with the backlog. The new committee was composed of fenland officials living in and near the Great Level. They met regularly for one year, exclusively in the Fens, and were empowered "to hear and determine in a short and easy way, particular complaints and grievances . . . [presented] in a peaceable manner" that would otherwise have gone to the commissioners in London, as long as the latter group had not already taken up the matter in question.[63] They could refer complicated cases to the superior jurisdiction of the commission if they wished, but they were under no obligation to do so, and whenever they saw fit to make a determination, their decision was binding. The committee's membership included those sympathetic to the drainage, along with a few who opposed it.[64] They were active throughout their one-year tenure, and like the commission of adjudication, they endeavored to provide a fair hearing in all cases; the company had to defend its interests vigorously in that forum, and sometimes came out on the losing side. By the end of 1653, the commissioners in London were referring cases to the fenland committee that they felt could best be handled locally.[65]

The Council of State thus adopted a two-pronged approach for dealing with the anti-drainage crisis of 1653; if the threat of military force was the stick, the new committee of petitions was the carrot. The council had pledged that all fenlanders with a legitimate grievance could appeal for redress "if complaint be made in a regular way." When the councilors realized that the existing apparatus for doing so was inadequate, they augmented it so that "all pretenses may be taken away" for further violence or unrest. They then issued a proclamation announcing the new committee and declaring that henceforth the state would "straightly forbid . . . all such unlawful assemblies, and riotous actings." From that point onward, rioters would "be looked upon as disturbers of the public peace, and be proceeded against accordingly."[66] The company had 350 copies of the council's proclamation printed and distributed throughout the Great Level, hoping that this would calm the situation.[67] When sporadic rioting continued in 1654, despite the proclamation, the Council of State issued an ordinance making it a felony to maliciously damage the company's drainage works. This time, the company printed and distributed eight hundred copies of the ordinance.[68]

The Interregnum state's support for the Great Level drainage was neither blind nor absolute, and it could never be taken for granted by the company. Yet while doing what it could to ensure that the project went forward according to law and that disaffected parties received a fair hearing, the state's principal goal was always to have the drainage completed as thoroughly and rapidly as possible. The company could never have achieved what it did without the legal sanction, administrative apparatus, conscripted labor force, and military assistance provided by Parliament and the Council of State. The shareholders embraced and publicized the state's ongoing involvement, reinforcing the notion that their interests and those of the state remained in complete alignment with respect to draining the Great Level, in spite of all the political uncertainty in London. This tactic served them well in responding not only to riots, but also to a series of polemical pamphlets condemning their entire project.

SIR JOHN MAYNARD AND THE POLITICS OF DRAINAGE PROJECTS

Sir John Maynard (1592–1658), lawyer and politician, had long been a thorn in the sides of both the Adventurers and the Cromwellian regime. Once a prominent courtier under Charles I and a protégé of the powerful Duke of Buckingham, Maynard nevertheless sided with Parliament during the Civil War. He became a leading figure in the moderate Presbyterian faction of the Parliamentary cause, which favored a negotiated settlement with the king, and was a harsh critic of the radical Independent faction and the leaders of the New Model Army, who pressed for the king's defeat, trial, and execution. As the radicals gained dominance, Maynard acquired powerful enemies; he was impeached in 1647 and ultimately purged from Parliament, though no specific charge was ever made against him. He spent several months in the Tower of London, where he met and formed a friendship with John Lilburne. Though differing in their politics, Lilburne and Maynard agreed about the dangers of burgeoning state power, and both took a dim view of state-sanctioned drainage projects. Maynard had testified before the parliamentary Committee for the Fens in opposition to the new Drainage Act in 1646 and was briefly appointed as a committee member himself before his ouster from Parliament the following year.[69] It was even suggested that the Independents had him impeached, "only for opposing the Fen project wherein L[ieutenant] G[eneral] O[liver] C[romwell] is so deeply engaged."[70] He remained an outspoken critic of drainage schemes, writing a series of petitions and pamphlets against the Great Level project and others between 1650 and 1653.[71] The Adventurers thus had their own legally trained authorial gadfly, just as the Hatfield Level Participants had been forced to contend with Daniel Noddel.

A gifted polemicist, Maynard's condemnation of fen drainage projects was passionate, comprehensive, and intensely political. Some of his critique was quite conventional, rooted in a conservative understanding of the traditional fenland ecology and economy. Drainage projects, he wrote, were little more than corrupt boondoggles, driven by greed and ignorance. The projectors "have always vilified the Fens, and have misinformed many Parliament men, that all the Fens is a mere quagmire . . . of little or no value," he wrote, "but those which live in the Fens, and are neighbors to it, know the contrary." In truth, the undrained landscape was both productive and valuable, helped rather than harmed by the annual floods. The abundant grazing lands of the Fens he described as "the rich ore of the Common-wealth," while the much-vaunted crops usually planted on drained lands, such as coleseed, rapeseed, and hemp, he characterized as "Dutch commodities, and but trash and trumpery."[72] Such arguments had, of course, been made many times over the past half-century.

Far more innovative and inflammatory was Maynard's legal and political critique, in which he condemned the Great Level drainage as an attack on local governance, a violation of the common law, and a tyrannous threat to the liberty and property of English landowners. Maynard recognized the degree to which the success of the company's venture depended on the support of Parliament and the Council of State, and especially leading figures in the Independent faction— Cromwell, St. John, and Thurloe among them—who had purged him from his seat in Parliament. Maynard took advantage of the state's deep involvement in the project to formulate a critique of it based on the radical republican rhetoric of the Civil War period. By casting the drainage as a corrupt, tyrannous, and even treasonous scheme, he aimed to undermine the company and embarrass his political enemies in one stroke. His attacks on fen drainage thus cannot be separated from the Interregnum political context in which he wrote and published them.

Maynard mounted a full-throated defense of the rights of local landowners and commissioners of sewers to manage their own drainage affairs without unwanted interference from London. He argued that drainage projects could only be legitimate if undertaken in accordance with one of two long-extant statutes: 23 Hen. VIII c. 5 (1532), which gave local commissions of sewers the power to drain lands within their jurisdiction as they saw fit; and 43 Eliz. c. 11 (1601), which permitted outside projectors to drain lands only with the approval of the landowners and a majority of the commoners there. Either scenario was "legal and good for the Common-wealth, because it is with their free consents." The Great Level drainage, however, had taken neither of these paths. It was sanctioned instead by a new law—the Drainage Act of 1649—which he claimed was

passed at the urging of "parties interested . . . when many of the uninterested
Parliament-men were absent." There had been little pretense of consulting the
fenland inhabitants or securing their approval before its passage, and instead of
working through the commission of sewers, the act actually barred local commissioners from interfering. If the drainage projectors could defraud the people of
the Fens this way, with the authority of a corrupt Parliament behind them, there
would be no limit to their potential mischief, even to the point of "draining" and
enclosing the entire nation: "[T]hey may make *England* the level, *England* to be
surrounded, and take one third this year for draining, and another third the next
for melioration, and the third take all by their prerogative."[73]

The 1649 act, and the drainage scheme it authorized, were pernicious in part
because they tended to "destroy propriety, and take away land from the owners
without and against their consents."[74] But they were also dangerous because they
obliterated existing laws and fatally undermined local institutions in favor of
centralized control from London, all for the profit of outside investors. For
centuries, each fenland locality had maintained its own commission and jury of
sewers, composed of local landowners, yeomen, and husbandmen, who knew the
land intimately and maintained it with their own best interests at heart. For the
past fifty years, however, these institutions had been systematically "obstructed
by powerful undertakers, courtiers, Lord-keepers, Attorneys-General, and projectors, and dissolutions, and intervals of Parliament," to the detriment of fenland
drainage generally. The 1649 act had finally administered the coup de grâce, su-
perseding the commission of sewers and subverting local governance altogether.
Local officials were replaced by a parliamentary committee meeting in London,
"to make the people dance attendance a hundred miles from their homes . . . to
attend them for four or five hours, to spend their moneys, and have no grievance
redressed, sometimes no committee appearing." All of this had overturned centuries of established sewer law, and "destroyed a chief branch of the common law, in
depriving the people of their juries of the neighborhood."[75]

The Great Level drainage was thus every bit as threatening to the liberty and
property of Englishmen as were the tyrannous policies of Charles I: "The judges
for ship money were accused for treason," Maynard wrote, "by reason it was
destructive to propriety, yet that was not three [shillings] in the pound; but the
fen-project cuts our estates asunder at a blow." Such injustices were nothing less
than "the height of arbitrary government, and tyranny itself, which is a speedy
destruction to a nation, and the pest of the Common-wealth."[76] Little wonder,
then, that so many drainage projects had been initiated under Charles, shortly
after he had cast aside the Petition of Right and embarked on the Personal Rule:

"then the projectors swarmed again, for they knew the king was irreconcilable to Parliaments, and all things were carried with a high hand. . . . This was a fit season to monopolize."[77] The Grand Remonstrance had plainly said as much, calling the drainage schemes "an injustice, oppression, violence, project, and grievance," one of many royal provocations that had sparked the Civil War.[78]

Yet having now struggled for so long and sacrificed so much to overthrow a corrupt and tyrannous king, having finally abolished the monarchy and established a republican Commonwealth government in its place, the fenlanders found that they were no better off. Drainage projectors continued to rob and persecute them with impunity, "confident to pass a law, and to enslave us, who have conquered them." This was all "contrary to the principles of a republic, who acknowledge the supreme power resides in the people, and the supreme authority is derived from them, and the soul and heart of the Common-wealth is *liberty* and *property*."[79] For the new regime to permit such a thing was a betrayal of "300,000 people concerned in it, which have furnished the Parliament with many thousand men, horses, and vast sums of money, besides the prayers of a numerous, godly, precious people." The drainage projectors falsely claimed to be acting for the good of the Commonwealth, but what good could possibly come from such a clear violation of justice and right? "The undertakers talk of great matters that will accrue to the Commonwealth by rape; I am sure they have committed a rape upon the republic, in ravishing the good people of this nation (by their tyranny and oppression) out of their properties and liberties."[80]

The company recognized Maynard's politically charged pamphlets as a real threat to order in the Great Level, and to its credibility everywhere. The shareholders overseeing matters in the Fens informed their colleagues in London in August 1653 that "the people [are] beginning to be very high and riotous . . . and do all talk of Sir John Maynard's petition being read and that the whole matter is to be reexamined and they hope revoked and made null." The volatile political circumstances of the time made them especially nervous, and they worried that the fenlanders might "take the advantage of any trouble or disturbance in the state as they did in Yorkshire and Lincolnshire"—a clear reference to the recent rioting in the Hatfield Level.[81] Anthony Hamond reported that the fenlanders "do combine against us, do raise money to bear the charge, and are framing very large and clamorous petitions against us," and he too was anxious that they might "take the advantage of any trouble or misfortune that may happen to the state, for I conceive we are in the same danger of [Maynard] that the state is of Lilburne."[82] The company responded to Maynard's public condemnation of its project by refuting his polemics with its own pamphlets, answering his accusations point for point. Whereas Maynard had worked to politicize the drainage project, the shareholders turned the tables by politicizing his strident opposition to it. They rejected his depiction of the project as illegal and tyrannous, and countered by portraying him as a dangerous Leveller, a firebrand who questioned the authority of Parliament to govern, and whose loyalty to the Commonwealth state was therefore doubtful.[83]

Even in rebutting Maynard's most conventional critique, that the drainage was unnecessary and destroyed common wastes, the shareholders made sure to highlight the project's valuable contributions to the Commonwealth, especially in a time of war. Their efforts had secured more than three hundred thousand acres for tillage and pasture, by means of which the Fens would yield "many hundred thousand pounds" in annual revenue, with "thousands of poor set on work." They scorned Maynard's contemptuous dismissal of crops such as coleseed, rapeseed, and hemp as mere "trash and trumpery," correctly pointing out that the oil from coleseed and rapeseed was useful for England's cloth dyeing industry and was thus "no contemptible commodity to this nation," while flax and hemp would supply the navy with vital domestic sources of linen cloth and cordage. "[T]o such as know the concernments of the Common-wealth," they wrote, such crops were "if not of absolute necessity, yet certainly of great and high advantage to this nation."[84]

To counter Maynard's accusations that the drainage was an illegal and tyrannous attack on the liberty and property of English landowners, sanctioned by a corrupt Parliament, the company pointed to the 1649 Drainage Act and the circumstances of its passage. The parliamentary Committee for the Fens had held more than forty separate hearings on the bill over three years, as circumstances permitted during the Civil War. The hearings had been widely publicized throughout the Great Level, and anyone who wished to be heard was encouraged to attend. Dozens of witnesses had testified both for and against the bill, including Maynard himself. The bill was finally passed by the whole Parliament, and while some MPs were indeed shareholders in the company, they had been barred from voting by order of the House expressly to avoid the appearance of partiality. Furthermore, an independent commission for adjudication had been created to handle any and all complaints against the company, and those few commissioners who also happened to be shareholders had declined to attend its sessions for the same reason. The act had thus been passed according to due process, after every effort had been made to give the country a full hearing, and with provisions made for settling grievances fairly, through the impartial judgments of disinterested men.

Nor did Parliament lack the authority to enact such a law, especially when previous acts had all failed in their purpose. The old system of sewer law, as laid down in 23 Hen. VIII, had not kept the land drained and productive; the commissioners of sewers had lacked both the revenue and the consensus among themselves to accomplish what the company had now achieved. The General Drainage Act of 43 Eliz. was entirely unworkable for an area the size of the Great Level, in which even Maynard admitted there lived "many thousand persons," making it "almost, if not altogether impossible to know who were the major part, or to have the consent of the major part."[85] This, indeed, was the whole point of the 1649 act—it remedied the shortcomings of former laws by substituting Parliament's approbation for the expressed agreement of the fenlanders. That Parliament could compel individual landowners and commoners to part with a portion of their property for the good of the Commonwealth could not be doubted—this was the legal basis for all taxation. Parliament spoke with the voice of the whole nation, and whatever a majority of its members decreed was binding to all, even if they had not personally consented to it.

The company also defended the appropriation and centralization of control over the Great Level drainage and the marginalization of the local commission of sewers. It was only right and proper that the leading shareholders should wield the powers of the sewer commissions, since it was they who had paid for the successful drainage, and it was their allotted ninety-five thousand acres that would be taxed to maintain it ever after. This differed but little from the old system, in which prominent local landowners had served as sewer commission-ers and assessed taxes on their own and their neighbors' lands to pay for neces-sary drainage works; the new Drainage Act simply streamlined and corporatized the process. Since the company would now bear the sole financial burden for keeping the land dry, its shareholders should have the sole right to determine how best to do that. Anyone else who wished to have a say was welcome to pur-chase the two hundred acres from the company's allotments required to obtain a voting interest.

There was thus no corruption of the Parliament by private interests, no sub-version of the common law, no tyrannous landgrab. The works of the Great Level drainage were "such as are not elsewhere to be seen in England, scarce in the Christian world, worthy the care and countenance of a Parliament." The company had built them at its own cost, "upon the authority and credit of the said Act of Parliament," a law that was "just, honorable, and necessary, and [made] sufficient provision thereby for just complaints." Anyone who doubted this was, in effect, questioning the very basis and legitimacy of the Commonwealth state: "And men which broach such principles, as *that Acts of Parliament* (wherein all men's consents are included) *destroy propriety*, would be carefully heeded in time, lest the consequence may tend to bring in question, not only many things well settled by Acts of Parliament for public good in most (if not all) former Parliaments, but interrupt the great affairs of this Common-wealth, if such an unsound, destructive principle (*that a commoner's right cannot be bound by an Act of Parliament*) should be taken up by tumultuous giddy people."[86] Sir John Maynard, then, ought not to be viewed as a defender of the liberty and property of Englishmen, but as a dangerous and seditious malcontent, casting doubt and opprobrium on a worthy endeavor and challenging the authority of a sovereign Parliament to authorize it in the name of the Commonwealth.

FENS, COMMONWEALTH, AND EMPIRE: THE DOCTRINE OF IMPROVEMENT

The Adventurers might well be expected to insist that their project was legal, just, and good for the Commonwealth. But in doing so, they tapped into a rising intellectual current in England that saw the improvement and enclosure of waste lands as the greatest possible benefit, a sound investment for landowners and a sure means to increase the nation's wealth and population. Such views were part of a wider enthusiasm for utopian reforms of all kinds that was especially prevalent during the turbulent years of the Interregnum. Among the most zealous, imaginative, and prolific reform advocates were Samuel Hartlib and his extensive network of correspondents. Born in Elbing to a prominent Baltic merchant and his English wife, Hartlib (c. 1600–1662) emigrated to England in 1628 and remained there until his death. He was widely known among the educated social elite, in England and elsewhere, as an "intelligencer"—one who collected, connected, and circulated valuable information. He was inspired by the philosophy of Francis Bacon and believed that valuable practical knowledge ought to be shared as widely as possible. Throughout his life, Hartlib exchanged letters with hundreds of correspondents all across Europe and served as the central node of a prolific communication and publishing network known today as the "Hartlib Circle." The group had strong interests in agricultural improvement and the rational management and exploitation of natural resources, among many other things.

The Hartlib Circle was most active and influential during the Interregnum, when so many among England's political and intellectual elite believed that a new age of the world was at hand and that the opportunity to transform the nation into a true, godly Commonwealth was at last within reach. Hartlib was on good terms with key figures in the Interregnum state, including John Thurloe and Oliver Cromwell himself, as well as many members of Parliament; the House of Commons voted to award him an annual pension of £100 in 1649 for his work promoting "the advancement of arts and learning." Hartlib saw it as a sacred duty to disseminate useful information and to promote reformist projects of all kinds. Throughout the 1650s, he printed and circulated his friends' reform proposals and lobbied on their behalf through his connections in government. The height of his influence in England was also the most intensive period of work in the Great Level drainage. Given the strong interest of the Hartlib Circle in agricultural innovations and improvement schemes of all kinds, it is no wonder that some of their most important treatises fixed on fen drainage and the improvement of waste lands generally as a means to enhance the nation's prosperity and virtue.[87]

One such text was a lengthy letter written by Robert Child, published by Hartlib in 1651 as part of a compendium on agricultural improvement titled *Samuel Hartlib His Legacie*. The letter, which discusses various agricultural improvement techniques practiced in the Low Countries with the hope of introducing them to England, runs to more than one hundred printed pages. It covers a wide array of topics and criticizes throughout the ignorance and backwardness of English farmers compared with their Dutch and Flemish counterparts. One of Child's chief concerns was England's "deficiency, concerning waste lands," an issue he considered vital given England's rising population, growing need for food, and high proportion of marginal farmland that was not being exploited to its full potential. But a rare bright spot, for Child, was England's recent interest and activity in draining its Fens. Though he mentions *"Axtel-holme Isle* in Yorkshire" as one of many areas still sadly neglected, he contrasts these with "that great *fen* of *Lincolnshire, Cambridge,* [and] *Huntingdon* consisting as I am informed of 380,000 acres, which is now almost recovered." Such sweeping improvements as these were the best and godliest way for England to prosper, because they increased the size and productivity of the nation through peaceful means, rather than conquest. "[V]ery great," Child writes, "is the improvement of draining of lands, and our negligence very great, that they have been waste so long, and as yet so continue in diverse places: for the improving of a kingdom is better than the conquering a new one."[88]

Walter Blith (c. 1605–54) was another of Hartlib's correspondents who saw great potential in fen drainage. The son of a small farmer in Warwickshire, captain in the Parliamentary army, and by 1649 the owner of an estate in Lincolnshire, Blith was one of the most important and influential writers on agricultural improvement during the seventeenth century. He published a substantial treatise discussing several means of improving farmlands called *The English Improver, or a New Survey of Husbandry* in 1649. Three years later, he had accumulated enough additional material to publish a much-expanded third edition, appropriately titled *The English Improver Improved or the Svrvey of Hvsbandry Svrveyed.*[89]

Blith was initially ambivalent toward fen drainage projects. He lamented in his first text that, for want of proper drainage, an "abundance of the best land of the kingdom is hereby lost, and much more corrupted with coldness and bogginess." But prior efforts, particularly those of certain Dutch projectors, had yielded little real improvement, and he had harsh words for those "men out-landish" who had "held forth wonders, but ever upon the charge and expense of others. And have produced little but to themselves."[90] By 1652, however, Blith had apparently spent some time around the Great Level drainage works and had altered his opinion. His revised and expanded text omitted most of his earlier criticisms and included instead a lengthy new chapter offering, "a brief and plain discovery of the most feasible way of fen-draining, or regaining drowned lands."[91] The book's new frontispiece also indicated the author's newly optimistic attitude toward land drainage and its connection to a prosperous Commonwealth. Among the myriad laborers and tools depicted to illustrate good husbandry, Blith included a man wielding a trenching spade to dig a new drainage channel, a surveyor measuring out newly allotted lands, and several other implements used in ditch digging. All of this was depicted beneath the arms of the Commonwealth of England and a banner reading "Vive La Re Publick" (fig. 8.5).

Blith was no respecter of blind tradition when it came to agricultural practice. He believed that English farming could be vastly improved through any number of ingenious innovations, and he criticized English farmers for holding the entire Commonwealth back through their "sloth, prejudice and ill husbandry." He had an especially negative view of the Fens and their inhabitants, and argued that age-old fenland practices such as intercommoning and unrestrained grazing of livestock had damaged the land and prevented more enterprising farmers from improving it. Both practices he viewed as ignorant, ill-considered, and irrational. Fenlanders, he scoffs, were all but unanimous in claiming that their traditional practices had already "brought the Fens into as good posture as all the undertakers' works," and that prior drainage efforts "had no whit at all advantaged them." But their conservative outlook, low standards, general laziness, and defeatist attitude had obviously discouraged innovation, serving only to "weary the minds, and weaken the hands of others that would endeavor it."[92]

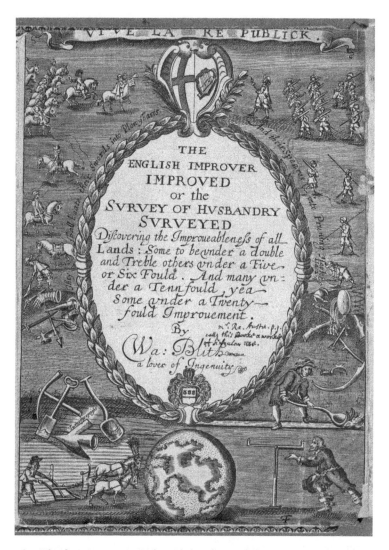

Figure 8.5. The frontispiece to Walter Blith, *The English Improver Improved . . .*
(London, 1652). Blith's keen interest in contemporary fen drainage projects is
indicated by the laborer prominently displayed at the lower right, digging a new
drainage channel with a trenching spade, and the surveyor at work just below him,
who may be plotting out a new drain or dividing newly drained lands for enclosure.
The caption along the top reads "VIVE LA RE PUBLICK"; the coat of arms at the top is
that of the government of the Commonwealth of England.
RB 600165, The Huntington Library, San Marino, California.

Alongside the ignorant and idle fenlanders, Blith also criticizes self-interested and small-minded drainage projectors, who would employ half-measures seeking to improve the best lands and abandon the rest to the floodwaters. Such a projector, he writes, is "an enemy to the common good," driven by "a corrupt selfish spirit." An imperfect, piecemeal drainage was better than none at all, perhaps, but to reach their full potential the Fens would have to be drained completely and all at once. "[W]e never accomplish *the end*," Blith insists, "until we have brought it to its best perfection, that is not only to recover it from drowning . . . from which the rude, ignorant fen-man desires no appeal . . . but the perfection is in the reducing it to soundness and perfectness of mold and earth, whether sand, clay, gravel, or mixed. . . . These lands thus perfectly drained, will return to be the richest of all your lands, and the better drained the better lands." With a sound understanding of the land, a thorough knowledge of the principles of drainage, a willingness to innovate, and the drive to work hard and persevere, the Fens could not only be improved, they could be made as fertile and productive as lands anywhere else in England. They could be restored not just "to their natural fruitfulness," but brought "to a more supernatural advance than they were ever known to be." Real improvement was possible in the Fens, Blith was convinced, and his new chapter was intended to "dismystery" the process for those courageous drainage pioneers who wished to pursue and perfect "those profitable works, the Common-wealth's glory."[93]

Yet another Hartlib Circle correspondent interested in the Fens was Cressy Dymock, a Lincolnshire farmer, inventor, and projector who promoted various agricultural improvement schemes.[94] Dymock was less interested in fen drainage per se, than in how the land might be better organized and exploited once the drainage was complete. In 1651 he wrote to Hartlib outlining "a plain discovery of that prudential contrivance for the more advantageous setting out of lands," which Hartlib published two years later in another compendium titled *A Discoverie for Division of Setting Out of Land, as to the Best Form*. The title page indicated that the plan was intended "for direction and more advantage and profit of the adventurers and planters in the Fens and other waste and undisposed places in England and Ireland." It had been "offered in vain to some of the company of drainers of the great fen," who ignored it, at which point Dymock sent the plan to Hartlib to publicize instead, "That I may in some sort be blameless to all posterity, though those lands be not well divided or sub-divided."[95]

Dymock's grand objective was to optimize the efficient use and exploitation of every acre of land in England. His treatise concentrated on the drained Fens because they represented a unique tabula rasa—a newly created landscape ready

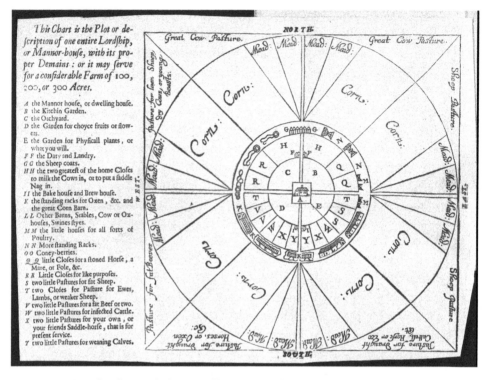

Figure 8.6. Plan for an ideal, rationally organized farmstead in the newly drained Fens, printed in Cressy Dymock, *A Discoverie for Division or Setting Out of Land, as to the Best Form* (London, 1653).
RB 88473, The Huntington Library, San Marino, California.

to be surveyed, divided, enclosed, tilled, and planted with novel crops. The Fens were thus an ideal forum for reorganization, experimentation, and reformation. The centerpiece of Dymock's proposal was a plan for a maximally efficient farmstead on a square lot, with a farmhouse in the center surrounded by concentric rings of fields, each devoted to a specific purpose (fig. 8.6). The circle nearest the house was to be planted with gardens and orchards; next came a ring of closes for keeping lambs and calves, saddle horses, pregnant or injured animals, and others in need of the farmer's attention; then came a ring of outbuildings including poultry coops, rabbit hutches, a bake house, a brew house, a dairy, and a laundry. In the largest, outermost ring Dymock placed cornfields in each quadrant of the circle, separated by meadows at the cardinal points. The corners of the square, where the circular fields did not extend, were designated as pastures for various livestock.

Dymock's treatise criticized the most common layout for newly enclosed fenland farmsteads, which favored long, rectangular lots with the main dwelling sited at one end, next to a drainage canal for easy water access. He believed that his circle-in-square plan with the farmhouse in the center offered numerous improvements and efficiencies. The arrangement of land into concentric circles based on usage allowed for the tidy organization and separation of arable, pasture, and meadow, making the best use of all three. The central farmhouse dramatically cut the maximum distance a farmer would have to walk to reach the farthest extent of his lands and allowed him the greatest facility in surveying his estate to identify problems as they arose and to encourage further improvements. Dymock also put arable fields closer to the main buildings, minimizing the distance (and therefore time and cost) for hauling harvested crops to storage barns, while locating pastures in the far corners because livestock were self-ambulatory and could transport themselves for free.[96]

Another benefit of Dymock's square-lot arrangement was that it could be easily scaled up or down to meet the different needs of lords, gentry, yeomen, and husbandmen, as shown in a second diagram depicting several square farms clustered together in a neighborhood (fig. 8.7). Dymock's larger plan, as his captions explain, showed a large fenland lot of two thousand acres abutting a main drainage channel, divided into sixteen "great farms" of one hundred acres each, and another sixteen "lesser farms" of twenty-five acres each. Every farmstead featured a centrally located farmhouse of appropriate size for the acreage it commanded, and its fields could thus be laid out in the most efficient circle-in-square pattern, as Dymock advised. The four central "great farms" might also be combined into a single estate of four hundred acres, "a grand farm, or the lord's demains [demesne]," from which the owner of the entire property (equal to half of one full share in the Adventurers' allotted lands) might easily survey all of his surrounding tenants.

Dymock's idealized vision of the improved and reorganized Fens is thus a neat geometrical pattern of fields, drainage ditches, and navigation canals, weaving together a network of orderly and hierarchically arranged farmsteads, intended to achieve optimal efficiency through rational, mathematical organization. Dymock believed that a farm arranged according to his plan, "merely in the contrivance, without or besides any other improvement, shall make 100 acres, to all intents and purposes as useful and profitable, as 150 acres can be." If comprehensively adopted, Dymock's geometrical plan promised almost miraculous returns: "the most perfect, right, and ample use of every foot of ground enclosed entire." Besides being more profitable, moreover, his rationally organized estate

This Chart contains 2000 Acres, confisting of or divided into 16 great Farms, conteining 100 Acres apiece, and 16 leffer Farms, confifting of 25 Acres apiece : And that fo as each thoufand Acres may be confidered apart, as being divided in the middle by the great Bank or high way, with the two great Drains on each fide of the fame.

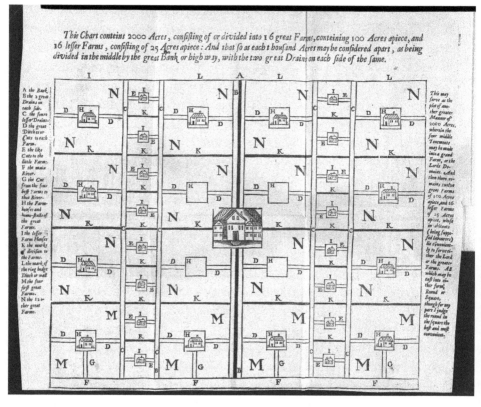

Figure 8.7. Plan for a series of neighboring farmsteads in the newly drained Fens, rationally organized and hierarchically arranged, printed in Cressy Dymock, *A Discoverie for Division or Setting Out of Land, as to the Best Form* (London, 1653). RB 88473, The Huntington Library, San Marino, California.

would also be more pleasurable for its inhabitants: "[Y]ou may at all times with ease view and take account of your business, and yet be as neat and sweet as in a *burgemasters* house in *Holland.*" The circular layout in particular was not only more efficient but had an undeniable aesthetic appeal for Dymock, even with respect to the manor house itself, "which also I would build round, which form I suppose to be of most beauty, use, and least cost to him that will give his mind to consider it rightly."[97] For Dymock, the rational, mathematical organization of one's farm was literally its own reward.

The Hartlib Circle's interest in draining the Fens was part of their broader program for promoting agricultural improvement throughout England, but it was also linked to their promotion of English overseas trade and imperial settlement in Ireland and the New World. Hartlib and Robert Child were closely involved in the production of Gerard Boate's *Irelands Natvrall History*, the publication of which was exactly contemporary with the Circle's greatest attention to fen drainage. Like the Fens but on a larger scale, they viewed Ireland as a great tabula rasa on which their theories for "manuring and improving" the land could be tested, not only making Ireland more productive, rational, and civilized, but also laying the groundwork for further improvements in England. Though their prejudiced notions of Ireland's poverty and backwardness led them to omit or misinterpret much of the data they collected there, their explorations reinforced their belief that Ireland could be transformed from a rude, stagnant waste into a prosperous beacon of rational and orderly management.[98] Beyond Ireland, Hartlib also supported a project to create a godly Protestant settlement in North America. The colony was to be made self-supporting and even profitable through the implementation of innovative agricultural practices, which would in turn provide an example for farmers back in England.[99] Both imperial projects were founded on the Hartlib Circle's view of the natural world as a disorderly, irrational place and their belief in the capacity for human reason and hard work to reform and improve it, ideas that also governed their understanding of the Fens.

The Hartlib Circle's view of fen drainage as part of a larger program of English agricultural improvement and imperial settlement was mirrored in the involvement of some fen drainage investors in other imperial and mercantile ventures. Edward Gorges and his son, Richard, are prime examples: both were among the foremost shareholders in the Bedford Level drainage, and Richard even served as the company's surveyor general for thirty years after the drainage was completed. Beyond the Fens, they were also large landowners in Ireland and members of the Irish peerage, and before purchasing his shares in the Great Level, Edward Gorges had invested heavily in North American plantations.[100] Gaulter Frost is another example: secretary to the Council of State (a post also held by John Thurloe) and formerly the commissary for provisions in Ireland, at his death in 1652 he held investments in the Great Level drainage as well as the East India and Guinea Companies.[101] Admiral William Goodsonn, governor of the Jamaica colony after its capture from the Spanish in 1655 and thus an important figure in managing England's nascent imperial holdings, also owned a five hundred–acre share in the Great Level drainage.[102] And Samuel Fortrey, a prominent investor and office holder in the Great Level project, was also the author of *Englands Interest and Improvement*, an influential treatise promoting the enclosure of common lands, the development of manufacturing, and the expansion of England's overseas trade.[103]

Besides recouping a healthy return on their financial investment, the support-ers of fen drainage projects and England's other imperial adventures seem to have shared some larger goals in common: transforming the natural environment to make it more productive; overcoming the resistance of backward, ignorant na-tives to turn them into rational, governable, and civilized English subjects; and doing all of this in the name of promoting English trade and economic growth. Whether undertaken at home or abroad, such projects were all linked to the same utopian, reformist, and imperialist worldview, propagated by Samuel Hartlib's net-work of would-be improvers and projectors. Fen drainage can thus be understood ideologically as an early modern instance of "internal colonialism."[104]

THE MAPPING OF MASTERY

In his insightful analysis of the many English plans to conquer, colonize, and "civilize" early modern Ireland, John Patrick Montaño has argued that surveying and cartography were vital tools of imperialism. English landowners had long relied on surveyors, whose mathematical and cartographic skills enabled them to manage their estates more rationally and more profitably. But surveying in Ireland went much farther, taking an entire landscape perceived by the English as chaotic, irrational, unsettled, and wild, and imposing on it a rigid order and geometry that worked to "civilize" it visually, even as Protestant English settlers strove to civilize it in practice through cultivation. English surveys in Ireland were thus "another component in the triumph of civility over savagery, this time in the guise of reason over chaos and disorder." The maps the surveyors produced were meant to be reassuring to the English viewer, "discerning order from the topographic confusion" and "imposing a civilized regularity on the land." En-glish settlements, the imposition of English landholding patterns and divisions such as shires, the conversion of pasture into arable land for growing grain, and the use of maps to allot and manage estates were all linked components of a broader program of civilization and imperial state building. The new English surveys and maps of Ireland represented "the cultural construct of English civil-ity and order, of a cultivated landscape, that was inscribed in the version pro-duced by the surveyor."[105]

This same use of cartography and surveying as tools of imperialism, state building, and civilization in order to force rational order on an unruly landscape and the backward people who lived in it applied to the seventeenth-century English Fens as well. Jonas Moore's impressively detailed wall map of the post-drainage Great Level, dominated by the enormous artificial drainage works and complex geometry of land allotments, was meant to illustrate the company's imposition of order and reason on a wild, unpredictable, and unprofitable waste. Cressy Dymock's plans for an idealized fenland farmstead did the same at the micro-level, using reason and geometry to maximize not only efficiency and productivity, but beauty, harmony, and virtue as well. Dymock's geometrical plans resemble another from the same period, an estate map of Munster in Ireland, and the similarity was not a coincidence. Promoters of agricultural improvement saw direct parallels between Ireland and the Fens, with Irish improvement little more than fen drainage writ large.[106] In 1652, Robert Child wrote again to Samuel Hartlib from Ireland's County Antrim, where he had recently settled, asking for news of "what other things in husbandry are coming forth," since "husbandry beginneth to flourish here very much, & men are desirous to see books of that kind." He particularly mentioned that "if Blith do set forth more concerning draining, it will be very acceptable, for in these parts, the English are very busy in draining bogs which at length prove the best land."[107]

The same themes or order, reason, geometry, and improvement prevail in William Dugdale's maps of the Great Level, included in his historical account of English land drainage. Dugdale, already a well-known and highly regarded antiquarian by the late 1650s, was commissioned by the Adventurers in 1657 to tell the story of their triumph in the Fens. The company paid him to compose a sympathetic history of the enterprise and to situate it within a long historical context of drainage projects, taken from throughout Europe but concentrating mostly on England. Dugdale researched the book thoroughly and wrote the first history of its kind, publishing it in 1662 (after first removing any untimely references to the Protectorate regime and adding a dedication to the restored King Charles II). He fulfilled his commission admirably, producing a scholarly and detailed history that presented land drainage as a mark of civilization and virtue—the first drainage projector in his account was God, who had separated the earth from the water at the Creation. "[T]he most civilized nations," Dugdale wrote, were those who had "by so much art and industry endeavored to make the best improvement of their wastes, commons, and all sorts of barren land," and he was happy to report that England had at last succeeded in joining their ranks.[108]

Dugdale included maps of all the major drainage projects discussed in his text, and his account of the Great Level was bookended by two maps, before and after the drainage. The first, "A Mapp of the Great Levell, Representing It as It Lay Drowned," is dominated by a massive pool of shading that represents the lands subject to regular flooding (fig. 8.8). Though most of the shaded area was not underwater for more than a few months each year, the map's visual impact

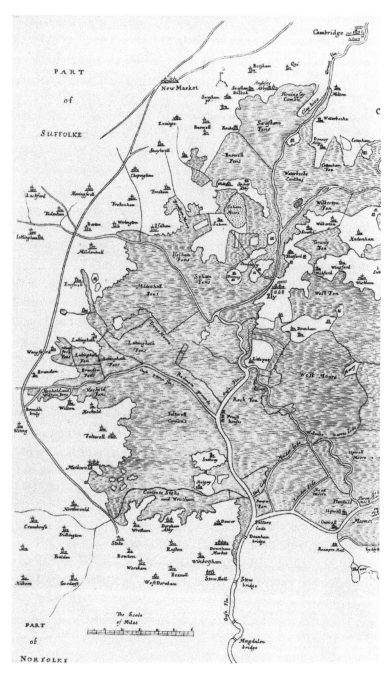

Figure 8.8. "A Mapp of the Great Levell, Representing It as It Lay Drowned," printed in William Dugdale, *The History of Imbanking and Drayning of Divers Fenns and Marshes . . .* (London, 1662); north is toward the bottom of this map. Houghton f EC65 D8787 662h, Houghton Library, Harvard University.

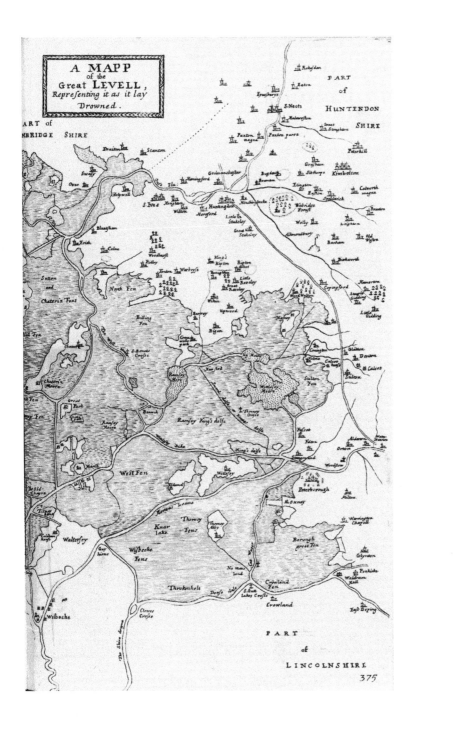

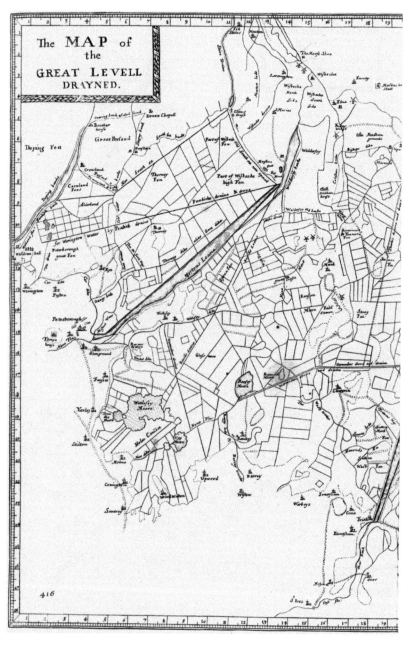

Figure 8.9. "The Map of the Great Levell Drayned," printed in William Dugdale, *The History of Imbanking and Drayning of Divers Fenns and Marshes . . .* (London, 1662); north is toward the top of this map. It is based on the much larger and more detailed map by Jonas Moore (see fig. 8.1, which corresponds to the right-center part of this map).

Houghton f EC65 D8787 662h, Houghton Library, Harvard University.

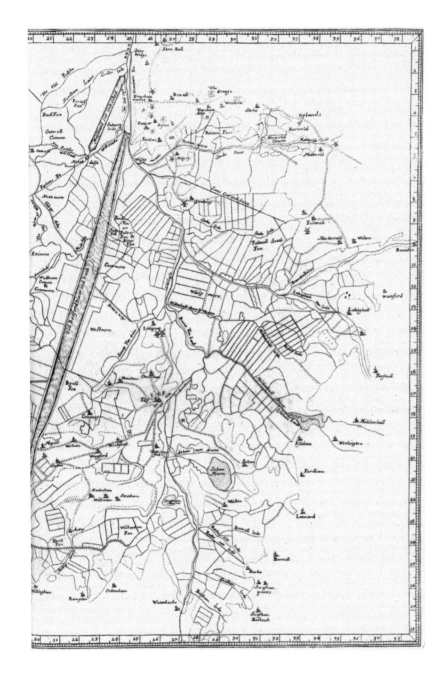

suggests it was all one massive quagmire. Several villages are depicted, but almost
none are in the shaded area; they cling instead to the periphery and the small
patches of white that represent the few "isles" within the flood zone. The map is
also marked by the region's many rivers, which meander along their own twisted,
uncertain paths to the sea. There is almost nothing about the land depicted in this
map that looks man-made, or even inhabitable; it is a region over which a capricious
nature has full sway, uncontrolled and apparently uncontrollable.[109]

Dugdale's companion map, "The Map of the Great Levell Drayned," provides a stark contrast; it is clearly based on Moore's massive wall map, the first
edition of which was published four years earlier. Here the shading is all but
gone, confined mostly to the "wash" areas deliberately left between the wide
banks of the company's new rivers, which cut through the landscape like scars, or
sharp arrows pointing toward the North Sea (fig. 8.9). The old, meandering rivers
have faded into obscurity and are difficult to discern. The most noticeable feature of this map is the intricate spider web of surveying boundaries that mark all
of the shareholders' allotted lands. The map's small size made it impossible to
reproduce all of Moore's careful and exact detail, but it still conveys the same
sense of geometric order. This is a man-made, and man-dominated, landscape: if
the first map made the Fens appear to be a giant swamp, this one resembles a city
street plan; if the first map represents the disordered tyranny of unruly nature, this
map conveys the triumph of man's reason, persistence, labor, and civilization. It
also illustrates both authorship and ownership—in transforming the landscape,
the Adventurers demonstrated their mastery of it, and their delineated allotments
have both redefined and colonized the region.[110]

The Great Level of the Fens was drained at last, transformed from a disorderly
and barren waste into a rational and cultivated landscape. It was, indeed, no
longer the "Great Level," but had become the Bedford Level, named for the 4th
and 5th Earls, who were responsible for draining it. The vast expanses of shared
common wastes were gone, carved into innumerable enclosures by an artificial
network of drainage channels and division dikes. The fenland inhabitants would
no longer be suffered to lead idle, unproductive, disreputable lives, but would
become honest, hardworking, civilized farmers. This was accomplished by
means of technical expertise brought over from the Low Countries, which swept
away the traditional methods and patterns of fen drainage and superseded the
conservative commissions of sewers that oversaw them. The region's drainage
would now be governed and maintained, at great annual expense, by a centralized, corporate body based not in the Fens but in London. The company, in
turn, continued to be supported by the full legal and administrative apparatus of

the early modern English state, both before and after the collapse of the Protectorate and the Restoration of King Charles II. It was, to paraphrase Robert Child, both "the improving of a kingdom" and "the conquering a new one."

And yet this apparent triumph over unruly nature was never as deep or as durable as the surveyors portrayed or the shareholders required. The land was soon enclosed and tilled; the riots ended as fenlanders grudgingly accepted the new order of things; but the Fens could not be so easily reduced to order and civility, and they remained disruptive and unpredictable. Problems soon emerged with the new drainage network that could not easily be solved with any amount of regular maintenance and taxation. By century's end, much of what Vermuyden and the Adventurers had achieved at such great expense lay once again at the mercy of the floodwaters.

The Once and Future Fens

Unintended Consequences in an Artificial Landscape

[O]f late the *Fens* nigh *Cambridge* have been *adjudicated* drained, and
so are probably to continue. . . . [T]he country thereabout is now
subject to a *new drowning,* even to a *deluge* and *inundation* of *plenty,*
all commodities being grown so cheap therein.
  —Thomas Fuller, *The History of the University of Cambridge,* 1655

[B]y the skill of these Adventurers, and, at a prodigious expense, they
have cut new channels, and even whole rivers . . . to carry off the great
flux of waters . . . [but] notwithstanding all that hands could do, or art
contrive, yet sometimes the waters do still prevail, the banks break,
and whole levels are overflow'd together.
  —Daniel Defoe, *A Tour thro' the Whole Island of Great Britain,* 1725

At Lakenheath Fen, the [Royal Society for the Protection of Birds] has
converted an area of arable farmland into a large wetland, consisting
mainly of reedbeds and grazing marshes. The new reedbeds have
attracted hundreds of pairs of reed warblers and sedge warblers, as
well as bearded tits and march harriers.
  —"About Lakenheath Fen," RSPB website, 2016

In May 1657, shortly after the Adventurers in the Great Level project had commissioned him to produce a learned history of land drainage in England as a
monument to the company's triumph in the Fens, William Dugdale embarked
on a personal tour of the region he was to write about. His itinerary took him
throughout the Great Level and the Hatfield Level, as well as the fenlands of
northern Lincolnshire that lay between them, and he kept a detailed diary of all
that he saw during his twelve-day journey. In addition to marveling at the impressive new drainage works built by his patrons, Dugdale was struck by the evident fertility he witnessed throughout the former wastes. Near the head of the

Old and New Bedford Rivers at Audrey Fen, he noted "a fair plantation of onions, peas, and hemp," as well as "extraordinary numbers of cattle feeding," while
at the other end near Denver sluice, he saw "several rich plantations of flax,
hemp, coleseed, etc.," and a "plantation of fruit tress of all sorts, and garden stuff;
and likewise of woad." He traveled through "exceeding rich grounds, both corn
and meadow" on his way to Peterborough, and at Thorney, the estate of the Earl
of Bedford, he described "all sorts of corn and grass, now growing thereon,
the greatest plenty imaginable."[1]

Dugdale's account of the rich farmland he saw wherever he traveled in the
Fens may have been somewhat colored, like his published history, by the hagio-
graphic agenda of his commission, for which the Adventurers ultimately paid
him a well-earned £150. Yet his observations can be confirmed to some degree
from other contemporary sources. The Cambridge antiquarian Thomas Fuller
was warm in his praise of the drainage in 1655: "[T]he best *argument* to prove that
a thing may be done, is actually to do it," he wrote. "The *undertakers* in our present age, have happily lost their first name, in a far better, of *performers*; and of
late the *Fens* nigh *Cambridge* have been *adjudicated* drained, and so are probably to continue." The most serious complaint now to be heard in the Fens, he
wittily suggested, was that the countryside was "drowning" in new crops, "a *deluge* and *inundation* of *plenty*."[2] In October 1669, the justices of the peace in
the Isle of Ely petitioned the company to establish a third market town for the
region at March (in addition to Ely and Wisbech). They wrote that "since the
draining of the Great Level of the Fens the said Isle is become much more
populous than before, and the trade of the said country in coleseed, all sorts of
grain, hemp, flax, cattle, butter, cheese, and other commodities is greatly increased."[3] And in 1685, an anonymous verse lauded the drainage as a truly epic
achievement:

> I sing floods muzzled, and the ocean tam'd,
> Luxurious rivers govern'd, and reclaim'd,
> Waters with banks confin'd, as in a gaol,
> Til kinder sluices let them go on bail;
> Streams curb'd with dams like bridles, taught t'obey,
> And run as straight, as if they saw their way.
>
> I sing of heaps of water turn'd to land,
> Like an *elixir* by the chymists hand
> Of dropsies cur'd, where not one limb was sound,
> The liver rotted, all the vitals drown'd.

No late discovered isle, nor old plantation
New christened, but a kind of new Creation.[4]

Things certainly seemed to be going exceptionally well for the shareholders in the company. The drainage was a success, and their allotted lands had been parceled out, enclosed, and "sown already with wheat, rye, barley, beans, oats, coleseed, rapeseed, hemp and flax."[5] The company's proceedings indicate that it had reached an agreement with the Greenland Company merchants in 1655 to export coleseed from England, rather than importing it as had been done in the past, and in 1661, it permitted James Smith to erect a kiln on his fenland estate to process his abundant madder crop.[6] The Protectorate regime also continued to support the company in its venture. The Council of State issued *An Ordinance for the Preservation of the Works of the Great Level of the Fenns* in May 1654, decreeing among other things that anyone caught damaging the drainage works could be fined double damages, jailed, or even prosecuted as a felon. The council's ordinance was endorsed by Parliament when it was next convened in 1657.[7]

When the Protectorate collapsed after the death of Oliver Cromwell and the monarchy was restored under King Charles II in 1660, the company remained in the new regime's good graces even though many of its Interregnum shareholders were not so fortunate. The new king lent the support of the Crown to ensure the company's continuance, and Parliament passed a new General Drainage Act in 1663. This law confirmed the terms of the 1649 Drainage Act (known thereafter as the "Pretended Act") and later ordinances, which had never received royal sanction and had thus become legally dubious in light of the king's return. The new act also formally established the shareholders as the Company of Conservators of the Great Level of the Fens (more commonly known as the Bedford Level Corporation) and set forth the terms according to which the company would govern itself and manage the drainage henceforth, the basic structure of which lasted into the twentieth century.[8]

The financial benefits of the drainage were realized primarily by the company's shareholders and the other large landowners in the region, who were in the best position to take advantage of them.[9] Tens of thousands of acres, once common wastes, were enclosed and tilled. Cereal crops were among the principal commodities produced initially, but as grain prices steadily fell during the latter half of the seventeenth century, fenland farmers planted ever more nonedible industrial crops such as coleseed, flax, hemp, madder, and woad, as well as market gardens. The new bounty boosted land values and rents, yielding a healthy profit for landlords who had previously seen no return from their rent-free common wastes. Many of their new tenants were immigrants from France and the Low Countries, who were both more willing and better prepared to cultivate the drained lands than the native inhabitants were. Much of the land remained untilled, however, and was still used for pasture and meadow to support vast herds of cattle, sheep, and horses.

Not all fenlanders experienced better fortunes, of course: for the many smallhold farmers and landless cottagers living in the region, the drainage was profoundly disruptive. The comparatively rapid rise in fenland population over the previous century—largely the result of in-migration, encouraged by the availability of extensive commons—had already placed a noticeable strain on the region's resources by 1630. The sudden loss of roughly one-third of their common wastes was a catastrophe for those who depended on them for their livelihood. Smallholders had no choice but to convert what little arable land they had to summer pasture, which curtailed their ability to grow winter fodder. They were forced to reduce the number of animals kept year-round, which not only left them poorer, but also diminished the supply of manure available for fertilizing the land, making it less fertile. The loss of fresh deposits of upland silt that used to arrive annually when the rivers flooded did not help matters. Many smallholders found that their tiny plots of land could no longer sustain them, and they had little choice but to sell out to their wealthier neighbors.

Cottagers and squatters with no legally recognized customary use rights in the common wastes were even worse off, and most were pushed from the land altogether. Their small herds were no longer suffered to graze on the remaining commons, nor could they rely on the traditional fenland by-employments of fishing, fowling, and gathering reeds. Instead, they became day laborers on the farms of the "better sort," tending to the more labor-intensive market crops now being sown on their estates, such as coleseed. Others found employment in the nascent manufacturing industries of the region, such as cloth- or rope-making, which fueled the demand for fenland crops such as flax and hemp. Creating productive employment for the allegedly idle fenland cottagers had been a principal justification for the drainage in the first place. Thomas Fuller wrote that such men, "whose hands are *becrampt* with laziness . . . if the *Fens* were drained, would quit their *idleness*, and betake *themselves* to more *lucrative manufactures*."[10] Such men were once seen as a threat to decent society, exploiting common lands to which they had no claim, leading sickly and squalid lives, and answering to no one; now they would at last know honest work and earn an honest wage, serving their employers and enriching the commonwealth.

Despite the social and economic upheaval caused by the drainage, the Adventurers encountered little significant resistance in the Fens after the crisis year of 1653. Minor anti-drainage rioting erupted from time to time, but it rarely rose above the level of vandalism and was suppressed through legal prosecution and the threat of military force.[11] In most cases the fenlanders expressed their lingering dissatisfaction through "complaint made in a regular way"—petitions addressed either to the company itself, or to an independent commission appointed under the 1663 General Drainage Act to hear and settle complaints.[12] Of the hundreds of individual petitions the company received through the rest of the seventeenth century, none challenged the legitimacy of the drainage. Most presented routine complaints and sought specific redress for them: adjustment of unfair allotments of land, repair of defective or decaying drainage works, the building of new bridges or highways for better access to commons, permission for private landholders to enclose some of the remaining commons, prompt payment for laborers seeking their wages, and so forth.[13] The company was not necessarily popular among the fenlanders, but what it had done could not be undone, and they had little choice but to acknowledge it as the only remaining drainage authority in the region. Indeed, the very regularity with which so many petitions were presented, considered, and decided by the appointed officials, all according to due process, only helped to reinforce the company's legitimacy and permanence.[14]

Maintaining the drainage, however, proved to be far more challenging than anyone had imagined.[15] Though the people of the Fens had been brought to heel, the land itself remained unpredictable and unruly, and the chief cause of the problem was the very drainage network the company had constructed. The topsoil throughout the Great Level was composed of a layer of peat several feet thick. The water content of peat under normal wetland conditions can be as much as eight times the amount of earthy matter it contains. Once the Fens were drained and the water table was lowered, the peat immediately began to dry out, contracting as it dried so that the surface of the drained Fens probably fell several inches just within the first few years. Moreover, peat is composed of vegetable matter naturally protected from decay by the acidic or alkaline water that permeates it. Once the water is removed, it becomes vulnerable to bacteria that devour it, so that shrinkage continues at the rate of an inch or two each year. When the shareholders and their tenants burned and plowed the dried peat to prepare it for cultivation, they accelerated the process and caused significant soil erosion.

Without sufficient water to protect and sustain it, peat shrinkage is an ongoing process; the more effective the drainage, the more rapid the decay. The Adventurers had unwittingly ignited a natural process of land subsidence that soon undermined much of what had been achieved. Fens that had once been several feet above mean sea level gradually sank to just a few feet above, and in some cases actually fell below it. The artificial drainage channels designed to carry water efficiently to the North Sea no longer functioned properly. Their diminished gradient slowed their flow and allowed silt to choke the riverbeds, while the surrounding land subsided until it lay well below the rivers intended to drain it. The problem was exacerbated by the rapid decay of the company's massive artificial riverbanks. The banks were critical for holding back surging floodwaters when they came, a key element of Vermuyden's plan for the drainage, but most had been constructed using peaty soil from the surrounding Fens and were equally subject to dehydration and bacterial decomposition. Banks that had seemed substantial and durable in 1653 began to disintegrate within just a few years, leaving the whole region vulnerable to even a minor flood surge. When floods inevitably happened, they were all the more severe because of land subsidence, which made it that much harder for excess water to drain back into the rivers and thence to the sea. The banks could only be permanently repaired using heavier soil from the surrounding uplands, but transporting sufficient quantities of such soil many miles into the Fens was hugely expensive.

Peat shrinkage placed a mounting strain on the new drainage network, so that flooding once again became a persistent problem. As early as 1661, the company's proceedings record the formation of a committee to consider solutions for flooding in the South Level, and the need for major repairs on all of the region's main riverbanks.[16] Two years later, when juries of sewers were impanelled to investigate the illegal placing of fish weirs in the company's drains, they responded with an alarming array of problems in need of urgent attention. An Ely jury alleged that the crucial banks on both sides of the Old and New Bedford Rivers were "very much defective and ought to be repaired," while a jury in the western part of the level claimed that the new north bank of Morton's Leam was "in great decay and broken in such manner that the North Level has been this last winter much overflown with water."[17] In 1665, a former rival of Vermuyden's named William Dodson presented to the company his own *Designe for the Present Draining of the Great Level of the Fens (Called Bedford Level)*—a striking proposal, given that the region was supposed to have been completely and permanently drained just twelve years earlier.[18]

Matters only grew worse over time, as the land continued to subside. The winter of 1672–73 was an especially hard one. An observer in Boston informed the Privy Council that "the fresh waters have drowned a great part of the level. . . .

Many cattle drowned, stacks of hay and oats and other grain either swimming or standing a yard deep in water . . . the poor people's houses full of water and are forced to save themselves in boats, all their coleseed lost and all they have besides."[19] The company saw a sharp increase in the number of petitions received each year, some from landowners and towns complaining about flooding on their lands, others from workmen who were hired to repair the damage but had yet to be paid for their labor.[20] Serious flooding recurred throughout the rest of the century, and by 1700, many parts of the Great Level had all but reverted to their pre-drainage state.

When Daniel Defoe toured the Fens in 1724–25, he was impressed with the "innumerable numbers of cattle, which are fed up to an extraordinary size by the richness of the soil," and he admired "the greatest improvements by planting of hemp, that, I think, is to be seen in England." But he also noted the "infinite number of wild fowl" still to be found there, as well as the abundance of fish, and at times he observed a landscape that probably looked much as it had a century earlier: "As we descended westward [toward Cambridge], we saw the fen country on our right, almost all covered with water like a sea." Defoe was conscious during his travels of the "history of the draining those Fens, by a set of gentlemen call'd the *Adventurers*," who "at a prodigious expense . . . cut new channels, and even whole rivers . . . to carry off the great flux of waters." Yet he could not help but observe that "notwithstanding all that hands could do, or art contrive, yet sometimes the waters do still prevail, the banks break, and whole levels are overflow'd together."[21]

The company struggled to cope with the worsening situation, but it did not understand the nature of the challenge. Its officials failed to recognize the fundamental problem of land subsidence, which would remain a mystery until well into the eighteenth century, and they mistook the silting up of the fenland rivers for the root cause of the trouble, rather than a symptom of it. William Dodson's proposal of 1665 had correctly observed, "it is not your dikes' bottoms which rise, but your grounds which sink," but his opinion was an outlier, and even he failed to grasp that the subsidence was an inevitable consequence of the drainage—he suggested that the problem could be solved simply by dredging and deepening the rivers once the land stopped sinking.[22] The company's preferred approach was simply to repair the riverbanks as they decayed, but because the problem was ongoing, this proved to be a large and endless expense, running to more than £10,000 annually, with much higher costs after a serious flood.[23]

The company's ability to raise funds for annual maintenance was strictly limited to its authority, granted by act of Parliament, to tax the ninety-five thousand acres of fenland allotted to the shareholders in recompense for their initial investment. The remainder of the Great Level, both severals and commons, was legally exempt from making any further contribution to the drainage—this was part of the reason for awarding the company so much land in the first place. The taxes levied to repair decayed drainage works ran between two and five shillings per acre each year, a sum the shareholders could not afford indefinitely, particularly as their allotted lands deteriorated and stopped generating revenue. Taxing the shareholders beyond their ability to pay only caused them to abandon the venture altogether, but although their forfeited shares were supposed to be sold to new investors, willing buyers were hard to find.[24] In order to provide some fiscal relief for the shareholders, in 1668 the company was permitted to introduce a graduated system of taxation, recognizing that not all of its lands were of equal quality or value. The entire allotment was surveyed and sorted into eleven classes, with class one (the lowest quality) to be taxed henceforth at eight pence per acre, and class eleven (the best quality) taxed at three shillings and eight pence per acre. It is telling that, according to the company's own survey, the vast majority of its lands were placed in the bottom four classes.[25]

The graduated taxation scheme may have been fairer, but it still yielded insufficient revenue to cover the cost of urgently needed repairs; the total value of the company's assets no longer covered expenses. The company was caught in a vicious cycle: important works were neglected for want of money to fix them, the drainage continued to decay, and land values and tax revenues fell even further. The worsening situation was not confined just to the company's allotted lands, of course, and the shareholders' fenland neighbors suffered alongside them. Something had to be done: simply allowing the floodwaters to reclaim the region would have been disastrous, since the land's subsidence would have led to more severe and longer-lasting floods than anything seen in living memory. For better or worse, the company had permanently committed the Fens to an artificial drainage network, designed and built by human beings and in constant need of their attention to maintain it. When that network began to fall apart, the only way forward was to introduce a powerful new technology for conveying the floodwaters from the lower-lying fenlands into the rivers that now flowed several feet above them. Machines would move the water where gravity could not.

Drainage mills had been in use in the Low Countries since the fifteenth century, and may have been introduced in England before the end of the sixteenth—several mills are known to have existed in the Fens by 1600, but whether they were used for drainage or another purpose is uncertain. Scattered requests for patents on new drainage engines survive among the English State Papers from the 1570s onward, as Continental engineers sought monopoly privileges for their inventions.[26] They were not widely used in England, however, until the latter part of the seventeenth century when land subsidence in the Fens made them indispensable. Though the earliest drainage mills were powered by horses, windmills soon came to dominate (fig. E.1). Mills were placed wherever a major drainage channel intersected with a secondary channel that was supposed to empty into it but had sunk too far to be able to do so. A steady wind turned the blades of the windmill and rotated a large wheel with several paddles attached to it. The paddles scooped water from the lower-lying channel, lifted it over an ele-

Figure E.1. A photograph taken by the author of the restored drainage windmill in the Wicken Fen National Nature Reserve in Cambridgeshire, March 2016. According to the Society for the Protection of Ancient Buildings, it is the last surviving wooden drainage windmill located in the Fens. It is not an early modern structure, but was probably built c. 1912 for use in nearby Adventurers' Fen, though the design is not unlike that used in much older mills. It was moved to Wicken Fen and restored in 1956 and is still used to pump water occasionally today.

vated riverbank, and emptied it into the higher channel. Scoop wheels could lift water between three and five feet, depending on their diameter; if additional elevation was required, multiple mills were placed in sequence.[27]

Drainage windmills were effective, but they had their drawbacks. Besides the challenge and expense of erecting a machine strong enough to withstand severe fenland storms, mills tended to scoop and throw water somewhat haphazardly, and they sometimes damaged fragile riverbanks, shoveled mud into rivers, or even flooded adjacent lands. The Bedford Level Corporation resisted the use of mills for these reasons, but from the 1660s onward, they proliferated nonetheless, as small groups of private landowners built them hoping to improve the drainage of their own lands when the company could not. Mills were thus a widespread local response to a centralized drainage network that had ceased to function properly but could not be undone or easily replaced. The company deemed private mills a nuisance and ordered dozens of them to be dismantled.[28] But the hard realities of land subsidence could not be ignored, even by the company, and eventually the shareholders were forced to succumb: in 1678, they ordered that "for the better and speedier cleansing and scouring of drains, the four surveyors of the level do forthwith buy each of them a mill, made for that purpose."[29]

The Bedford Level Corporation was clearly losing control of the Great Level as time went on. The company was legally responsible for maintaining the drainage network it had built, but the network no longer worked as it was designed to, and the shareholders' declining fortunes meant that they could not afford even urgent repairs. It remained the only legally recognized drainage authority in the region, but the land for which it was responsible was ever more vulnerable to flooding, very little of it could still be classified as "winter ground" as required by statute, and private landowners were forced to erect their own drainage mills as a result. The company continued to oppose the construction of private mills, but when it tried to intervene, the fenland inhabitants insisted that if the company could not keep their lands dry, they should be allowed to put up mills at their own expense free of the company's interference.

The issue came to a head in 1726–27, when the residents of the town of Haddenham in the Isle of Ely petitioned Parliament for permission to manage their own drainage.[30] They pleaded that their lands had been flooded for years because the main rivers of the area were elevated too far above the land they were supposed to drain, and mills were the only possible solution. Parliament responded with "An Act for the Effectual Draining and Preserving of Haddenham Level" (13 Geo. I, cap. 18), which created the first local drainage board in the Great Level since the Drainage Act of 1649 was passed. The act delineated the boundaries of a small district around Haddenham and authorized a local board of commissioners to build an internal drainage network for the land in question, levying taxes on area landowners to pay for it. In practical terms, this meant constructing a series of new secondary drains to carry water toward one of the company's main drainage channels, together with one or more windmills to hoist the water into it. The law explicitly permitted the construction of "engines for draining the Fens," an indication that drainage mills were now widely understood to be indispensable.[31] Within the bounds of the new Haddenham drainage district, the Bedford Level Corporation's ability to intervene or oppose the local commissioners was strictly limited.

After the success of the Haddenham petition, many more soon followed, as communities throughout the Great Level moved to establish their own local drainage boards. The company opposed such bills as a threat to its ability to maintain the central drainage network, but its failure to offer any workable alternative solution ensured that several of the bills were passed into law. Local boards proliferated and were soon responsible for building the innumerable dikes and mills needed to convey excess water from their respective districts into the company's main rivers. Over time they constructed an elaborate, decentralized network of subsidiary drains throughout the level, as well as the hundreds of drainage windmills that became a defining feature of the landscape.[32]

The bold early modern experiment of creating a single, coordinated drainage network in the Great Level, and a centralized corporate body to manage it, had largely failed. The Bedford Level Corporation continued to exist and to maintain its own drainage works well into the twentieth century, but its regional authority was undermined by a patchwork of small districts governed by independent boards, each of which insisted on the right to look after its own drainage affairs as far as possible. The return of multiple drainage authorities inevitably gave rise to internecine disputes and conflicts of interest, as the local boards disagreed with the company and with one another about the best course of action to take and who should pay for it. "In the post-drainage Fenland, as in the pre-drainage Fenland," H. C. Darby writes, "there was everywhere a chaos of authorities and an absence of authority . . . a variety of separate interests was still very apparent."[33] The drainage projects of the seventeenth century had transformed the Fens, to be sure, and made it impossible to return the land to its former state; but the ongoing problems of managing the still-troubled landscape, along with its contentious inhabitants, might have seemed depressingly familiar to Lord Burghley in the 1570s.

⌒

The draining of the Fens was an early modern quest to manage and exploit England's limited natural resources as effectively as possible. For most of the Middle Ages, the Fens were understood to be a viable, dynamic, and productive landscape that offered many opportunities for alternative agriculture to flourish, as long as the annual floods that shaped and defined the region were kept under moderate control. Robust local institutions, the commissions of sewers, had evolved over time to meet this need. But worsening conditions during the sixteenth century, together with a failure to achieve local consensus about how best to deal with the problem, convinced the Privy Council of Queen Elizabeth I that the Crown must take a more active hand in managing the region.

At first this effort was limited to brokering resolutions to intractable local disputes, but when that failed, the Crown grew ever more determined to see the whole region drained once and for all—a trend that persisted and deepened from Elizabeth's reign, through those of James I and Charles I as well. The Fens were no longer seen as viable and productive, but as a dysfunctional and disorderly landscape offering little benefit to the commonwealth and fostering instead only illness, poverty, and idleness in the few poor souls who eked out a living there. To repair the land would require human intervention, a massive and costly effort to impose order and reason on it. This was to be accomplished through projectors, men from outside the Fens who would undertake the work at their own cost and risk, with the full support of the Crown. Projectors such as Cornelius Vermuyden offered their royal patrons valuable technical expertise from the Low Countries that, combined with an entrepreneurial vision of a bountiful fenland future, would overcome and supersede the limited conservatism of the commissions of sewers.

Drainage projects were thus a manifestation of the early modern centralization of governance under a unitary English monarchy—an exercise in state building. The Crown was eager to maximize England's resources, for the good of the commonwealth as well as the royal treasury, and as it became more actively engaged in the Fens, its power and authority grew. But the state's increasing interest and involvement in the region were not necessarily welcomed there. What the king might view as effective leadership and good government, the commissioners of sewers too often saw as unwanted royal interference in local affairs. As political tensions mounted between Crown and Parliament, more radical fenland critics deemed the Crown's sponsorship of fen drainage schemes to be a tyrannous assault on the liberty and property of freeborn Englishmen. The drainage projects were always a political issue, inseparable from the wider concerns about overweening monarchical power that marked the tumultuous politics of England in the mid-seventeenth century. But though many issues divided royalist and parliamentarian leaders throughout the Civil War and Interregnum, fen drainage was not one of them. As successive rulers of an increasingly powerful and centralized English nation-state, James I, Charles I, Oliver Cromwell, and Charles II were all in agreement about the need to improve the Fens, and each was willing to invest the authority of the state in seeing it done.

The many fenlanders who did not share the interventionist view of the Fens that dominated in circles of power were themselves seen as part of the problem. Their conservatism was ascribed to ignorance, laziness, stubbornness, or private interest, the inability or unwillingness to admit just how miserable they really were, wading across the muddy "ground" of their flooded wastes. In the wider discourse of improvement and utopian reform so prevalent in the 1650s (and beyond), redeeming the Fens would require not only draining the land, but also civilizing the fenlanders in order to bring both into better alignment with the orderly government and market economy that prevailed in the rest of the realm. Fen drainage thus went beyond state building to become a quasi-imperial project, sharing much in common with improvement schemes in Ireland and the New World that were undertaken at the same time, and for the same reasons.

The land and people of the Fens were indeed transformed in the end, but the transformation had unintended ecological consequences that created at least as many problems as were solved. Ever more powerful technologies were required to keep the land dry and productive, with no apparent alternative. As the land continued to subside year by year, more drainage mills were needed (often several in sequence) to lift excess water into the artificial rivers that were supposed to carry it to the sea. Windmills were ubiquitous in the Fens for a century, but even they were not reliable enough to cope with the growing burden, and powerful steam-driven pumps eventually replaced them. These, in turn, have given way over time to diesel, and then to electric pumps. As the existing rivers and drains proved incapable of handling all the water being pumped into them, massive new drains were constructed, from the eighteenth century through the twentieth. Fenland drainage has become ever more artificial, technologically driven, and dependent on constant human intervention. The landscape historian Oliver Rackham has described the post-drainage Fens as "a somewhat precarious triumph of technology over a deteriorating situation, made possible by vigilance, expensive repairs, and an ever-increasing input of energy."[34]

Traveling though the Fens today, by railroad or automobile, one gazes at a vast, fruitful, and productive rural landscape, some of the richest farmland in all of Britain, with field after field of dark, rich soil, flourishing crops, and grazing animals. Even when other parts of England experience minor flooding, the

Figure E.2. A photograph taken by the author of the drained Fens near the intersection of Vermuyden's (Forty-Foot) Drain and Thurloe's (Sixteen-Foot) Drain in Cambridgeshire, between Chatteris and Manea, March 2016. Note the raised artificial riverbank, which prevents the water in the man-made drains from flooding into the arable land that is noticeably lower in elevation.

fenland drainage network helps to keep the land there comparatively dry and productive.[35] The region's flat landscape is carved into geometric segments by the innumerable drainage dikes that still serve to keep it dry, punctuated by the startling sight of large rivers that flow in elevated beds as much as fifteen feet above the surrounding countryside (figs. E.2, E.3). The windmills, once such a prominent feature of the region, are all but gone, with the few surviving examples now just historical artifacts. Yet without the continual work of electric pumping stations, massive sluices, and other technological wonders, the Fens would certainly be underwater much if not all of the time. The very order and fertility of the landscape communicates clearly that it is an entirely artificial construct, maintained only with great care and expense.

Yet this is not the only imagined future for the Fens. During the twentieth century, even as England's few remaining wetlands were being steadily drained and converted to arable ground, new generations of conservationists staked their

Figure E.3. A photograph taken by the author of the drained Fens, near the River Great Ouse just south of Ely, March 2016. This drainage dike helps to keep the surrounding fields dry by venting water into the Great Ouse.

own claim to expertise in the area. Believing that the land had never been well suited for arable agriculture in the first place, they viewed the drainage projects as a destructive waste of valuable natural resources and sought to undo what prior generations had worked so hard to achieve. The Fens might be best employed and enjoyed by generations to come, they suggested, by restoring and embracing them as the natural wetlands they ought to be. Far from being seen as waste, the flooded Fens would be of enormous value—not as farmland, but as an ecological preserve and natural tourist destination.

The wide strip of land between the Old and New Bedford Rivers, known today as the Ouse Washes, was intended by Vermuyden to serve as a reservoir for excess waters that would otherwise flood the entire area after strong spring tides or heavy upland rains. That land was therefore left uncultivated and was used only for grazing in the dry summer months. It remained prone to flooding each winter and so retained some of its wetland character. Today the Ouse Washes are a sanctuary of international importance for migratory birds and other wetland wildlife, with a conservancy managed by the Royal Society for the Protection of

Birds (RSPB) near Chatteris, and another by the Wildfowl & Wetlands Trust (WWT) near Welney.[36]

Wicken Fen in Cambridgeshire is another example. Like the Ouse Washes, Wicken Fen was originally intended as a receptacle for upland flood surges, and the land there was left uncultivated. Today the Wicken Fen National Nature Reserve is one of the oldest and largest wetland reserves in Britain, with more than 1,800 acres of wetland being maintained by the National Trust since 1899. It is home to some 8,500 species of plants, animals, and insects, and more than forty thousand people visit it each year. Yet in order to maintain its distinctly fenland ecology, Wicken Fen cannot be left entirely to its own devices. Because the peat there never dried out and subsided, today the reserve lies several feet above the surrounding farmland, and water actually has to be pumped into it. Moreover, even the pre-drainage Fens were never a purely natural landscape, but have always been shaped by their human inhabitants. Too many plant and animal species now considered vital to the area's wetland ecology actually depend directly or indirectly on regular human patterns of exploitation, carried on over many centuries. The reeds and sedge in the Wicken Fen reserve must therefore be cut from time to time, while grazing herds of cattle and ponies have been introduced to help promote new cycles of plant growth.[37]

More recently, nature conservancies have moved beyond the conservative goal of protecting what little remains of England's eastern wetlands to adopt a more proactive policy. They have acquired wide swaths of arable farmland in the Fens with the intent of allowing them to flood, "re-recovering" the land in order to restore a wetland habitat. The National Trust, for example, has embarked on a hundred-year mission called "The Wicken Fen Vision" to expand the reserve to an impressive twenty-two square miles, extending from the current reserve all the way to Cambridge, by purchasing and flooding adjoining lands in between.[38] Another striking example is Lakenheath Fen, a nature reserve in Suffolk owned by the RSPB right on the Norfolk border. According to the organization's website, it "has converted an area of arable farmland into a large wetland, consisting mainly of reedbeds and grazing marshes. The new reedbeds have attracted hundreds of pairs of reed warblers and sedge warblers, as well as bearded tits and march harriers," along with dozens of other varieties of birds and animals that thrive in wetland environments.[39]

Beyond promoting the conservation of a huge diversity of plant and animal species, wetland reserves can play an important role in purifying the water table and serving as reservoirs for excess water in times of flooding in an age of extreme weather and climate change. They may even help to mitigate climate change by acting as a carbon reservoir, as the peaty soil is gradually restored. Organizations such as the National Trust, the WWT, and the RSPB are also dedicated to public outreach, helping the people of Great Britain (and beyond) to understand and appreciate the ecological value of their wetlands and encouraging them to enjoy the land and wildlife for their beauty and diversity. Wicken Fen boasts a series of trails for hiking, cycling, and horseback riding, with more being built all the time. Lakenheath Fen features two nature trails and a new visitor center, with events and educational programs throughout the year, and weekend rail service deposits passengers within easy walking distance.[40]

During the seventeenth century, large landowners, drainage projectors, and enterprising investors worked in partnership with the Crown to transform England's Fens into something they were not. They imported the technical expertise of Dutch land drainers and relied on the legal and military power of the state to support their ventures even in the face of impassioned fenland opposition. Many invested great fortunes in such ventures, hoping to reap even greater rewards, though most ended up as a skeptical observer once predicted they would: slinking away "without faithful performance of what they promise . . . with more knowledge and less money."[41] In spite of the enormous difficulty and expense, they succeeded for a time in transforming the region into fertile farmland, only to see their labors and their investment dry up and blow away like the desiccated peat they had bought so dearly. With so much labor, money, and energy expended to reimagine the Fens as drained, arable farmland, it is ironic that organizations such as the RSPB, the WWT, and the National Trust are working so hard to reimagine them again as wetland reserves. Yet in working *with* the natural tendencies of the landscape, rather than against them, perhaps they (and we) will eventually enjoy an even more valuable, and more sustainable, success.

Abbreviations

AHEW	Thirsk, gen. ed., *The Agrarian History of England and Wales*
APC	Dasent et al., eds., *Acts of the Privy Council of England*
BDCE	Skempton et al., eds., *A Biographical Dictionary of Civil Engineers in Great Britain and Ireland*
BL	British Library, London
BLC	Bedford Level Corporation
C	Chancery Papers, in the National Archives
COS	Council of State
CRO	Cambridgeshire County Record Office
CUA	Cambridge University Archives
DNB	*Dictionary of National Biography*
E	Exchequer Papers, in the National Archives
EDR	Ely Diocesan Records
HMC	Historical Manuscripts Commission, United Kingdom
JHC	*Journal of the House of Commons*
NA	The National Archives, Kew
OED	*Oxford English Dictionary*
RSPB	Royal Society for the Protection of Birds
SPD	State Papers, Domestic Series, in the National Archives
VCH	*Victoria County History*
WWT	Wildfowl & Wetlands Trust

Notes

INTRODUCTION: The Unrecovered Country

1. Jonson, *The Devil Is an Ass*, 164. The dotterel is a wading bird, a species of plover, frequently found in the pre-drainage Fens during their annual migrations. It had a reputation for being curious, unsuspecting, and tame, and was thus regarded by hunters as easy prey. Especially in the early modern period, the word was also used to refer to a foolish, gullible, or stupid person, the ideal mark for a crooked projector such as Merecraft; see *OED*, s.v. "dotterel"; and "Eurasian Dotterel," *Wikipedia*, http://en.wikipedia.org/wiki/Eurasian_Dotterel, accessed 18 August 2016.

2. Evans, "Contemporary Contexts," 149–50.

3. The term *common waste* refers to marginal land that was not formally delegated as arable land, meadow, or regulated pasture, but it should not be taken in this context to imply worthlessness or barrenness.

4. "The answere to the obiections made against the drayning of the ffenns," NA, SPD 16/339/27 (n.d. [c. 1606]). All subsequent quotations from this debate are from this source.

5. Although the editors of the State Papers have suggested that the document is from the 1630s, the reference to a session of sewers at Wisbech at which more than five hundred commoners were in attendance almost certainly connects it with the proposed drainage project of Sir John Popham in 1605–6. Several similar petitions were presented to Parliament in early 1606 in connection with this project. Moreover, two copies of the list of objections exist in the British Library (one with and one without the appended responses: BL, Harleian MS 368, fols. 169–70; and Add. MS 41613, fols. 49v–51v), one of which is dated 1605–6.

6. Paul Slack has argued that these trends were especially pronounced in England; see Slack, *The Invention of Improvement*, chaps. 2–3.

7. Appleby, *Economic Thought*; Merchant, *Death of Nature*; idem, *Ecological Revolutions*; Thomas, *Man and the Natural World*; Cronon, *Changes in the Land*; Butlin, *Transformation*; idem, "Images"; Appuhn, *Forest on the Sea*; Ash, "Amending Nature"; Cooper, "Modernity"; Richards, *Unending Frontier*; Warde, *Ecology*; idem, "Environmental History"; idem, "Idea of Improvement"; McRae, "Tree-Felling"; Hoyle, "Introduction"; Cosgrove, "Elemental Division"; Schaffer, "Earth's Fertility"; Montaño, *Roots of English Colonialism*; Pluymers, "Taming the Wilderness"; Kupperman, "Controlling Nature."

This drive was certainly not unique to the early modern period. It became a central theme throughout modern Europe and its empires, and eventually in market-oriented economies the world over: Geertz, *Agricultural Involution*; Carroll, *Science*; Drayton, *Nature's Government*; Blackbourn, *Conquest of Nature*; Worster, *Rivers of Empire*; Steinberg, *Nature Incorporated*. It was not the only pattern of economic development to arise from such circumstances, however: see Richards, *Unending Frontier*, chaps. 5–6, comparing Tokugawa Japan with early modern England.

8. Cosgrove and Petts, eds., *Water, Engineering and Landscape*; van de Ven, ed., *Man-Made Lowlands*; Ciriacono, *Building on Water*; idem, ed., *Eau et développement*; Danner et al., eds., *Polder Pioneers*; Renes, "Fenlands of England"; Richards, *Unending Frontier*, 214–21.

9. On the prevalence of malaria in English wetlands, particularly salt marshes, and the dramatic impact it had on health and mortality, see Dobson, *Contours of Death*, esp. part 3.

10. Thirsk, *English Peasant Farming*; idem, "Isle of Axholme"; idem, gen. ed., *AHEW*, vol. 4, 28–48, 163–94; Darby, *Medieval Fenland*; idem, *Changing Fenland*; Holmes, *Seventeenth-Century Lincolnshire*.

11. The first formal commissions of sewers were issued by the Crown in the mid-thirteenth century, though the various local customs and arrangements upon which the early commissions were based go back much further; Darby, *Medieval Fenland*, 155.

12. My understanding of early modern politics, and of popular political culture, owes much to Tim Harris's excellent introduction in Harris, ed., *Politics of the Excluded*, and many of the other essays in that volume; and also to Andy Wood's work on local custom and popular politics in Wood, *Politics of Social Conflict* and *Riot, Rebellion and Popular Politics*. See also Hindle, *State and Social Change*; idem, "Political Culture"; Shagan, *Popular Politics*; Griffiths, Fox, and Hindle, eds., *Experience of Authority*; Braddick and Walter, eds., *Negotiating Power*; Collinson, "De Republica Anglorum" and "Monarchical Republic," reprinted in *Elizabethan Essays*; McDiarmid, ed., *Monarchical Republic*; Underdown, *Revel* and *Freeborn People*; Walter, "'The Pooremans Joy'"; Brewer and Styles, eds., *Ungovernable People*; Falvey, "Custom, Resistance, and Politics"; Scott, *Domination*.

13. In this regard, the commissions of sewers and the legal and political culture they embodied are similar to the barmoot courts of the Peak District, which developed their own legal and political culture out of the customs of free mining law, and had a long history of defending the rights of free miners against incursions from large landowners and the Crown. See Wood, *Politics of Social Conflict*.

14. On the early modern English understanding of waste land as a moral problem, and the divine and civilized imperative to improve and cultivate it, see Di Palma, *Wasteland*, introduction and chaps. 1–2.

15. Braddick, *State Formation*; Hindle, *State and Social Change*; Shagan, *Popular Politics*; Kesselring, *Mercy and Authority*.

16. Ash, *Power, Knowledge, and Expertise*; idem, *Expertise*; Smith, *Business of Alchemy*; Ashworth, *Customs and Excise*; Mukerji, *Impossible Engineering*; Wakefield, *Disordered Police State*; Appuhn, *Forest on the Sea*; Stewart, *Rise of Public Science*; Long, *Openness, Secrecy, Authorship*; idem, *Artisan/Practitioners*; Barrera-Osorio, *Experiencing Nature*; Nummedal, *Alchemy and Authority*; Cormack, "Twisting the

Lion's Tail"; Adams, "Architecture for Fish"; Sandman, "Mirroring the World";
Goodman, *Power and Penury*; Heller, *Labour, Science and Technology*; Thirsk,
Economic Policy and Projects; Cipolla, *Guns, Sails and Empires*.

17. Ash, *Power*; idem, "Expertise"; see also Rabier, ed., *Fields of Expertise*; Dear,
"Mysteries of State."

18. Thirsk, *Economic Policy and Projects*; Novak, ed., *Age of Projects*; Ratcliff,
"Art to Cheat"; Cramsie, "Commercial Projects"; Yamamoto, "Reformation"; idem,
"Distrust." Also of recent interest are the articles collected in Keller and McCormick,
eds., "Towards a History of Projects," special issue of *Early Science and Medicine*
21:5 (2016).

19. Defoe, *Essay upon Projects*, 11, 15–16. While Defoe believed that the
"original of the projecting humour that now reigns" arose only in the years after
1680, he also wrote that "by times it had indeed something of life in the time of the
late Civil War"—a period, not coincidentally, when several drainage projects were
under way (24).

20. Pérez-Ramos, *Francis Bacon's Idea*; Vickers, "Francis Bacon"; Martin, *Francis
Bacon*; Leary, Jr., *Francis Bacon*; Gaukroger, *Francis Bacon*; Harkness, *Jewel House*;
Ash, *Power*, chap. 5.

21. Swift, *Travels*, part 3, chaps. 4–6; quote on 156.

22. McRae, *God Speed the Plough*; Warde, "Idea of Improvement"; Slack, *Invention
of Improvement*; idem, *From Reformation to Improvement*; Thirsk, "Plough and Pen";
idem, *Economic Policy and Projects*; Leslie and Raylor, eds., *Culture and Cultivation*;
Hoyle, ed., *Estates*; Edelson, *Plantation Enterprise*; Kerridge, *Agricultural Revolution*;
Drayton, *Nature's Government*, chap. 3; Montaño, *Roots of English Colonialism*;
Pluymers, "Taming the Wilderness"; Kupperman, "Controlling Nature"; Yamamoto,
"Reformation"; idem, "Distrust." Some, however, have taken a more cynical view of the
commonwealth-oriented claims of those promoting improvement projects, seeing them
as little more than a fig leaf to cover naked depredation for profit; see Hoyle,
"Introduction."

CHAPTER 1: Land and Life in the Pre-drainage Fens

1. "A Relation of a Short Survey of the Westerne Counties, in Which Is Briefely
Described the Citties, Corporations, Castles, and Some Other Remarkables in Them.
Observ'd in a Seven Weekes Journey Begun at Norwich, & Thence into the West. On
Thursday August 4th 1635 and Ending att the Same Place. By the Same Lieutennant,
That with the Captaine, and Ancient of the Military Company in Norwich Made a
Journey into the North the Yeere Before." BL, Lansdowne MS 213 , fols. 347–84 (1635);
quotations on 381v–82v.

2. *The Anti-Projector*, 8. The text is anonymous but was probably written by John
Maynard; it is similar in many respects to his *Picklock of the Old Fenne Project*. Early
English Books Online tentatively dates the text to 1646, but internal evidence strongly
suggests that it must have been written after March 1653.

3. Drayton, *The Second Part*, 108.

4. Drayton, *The Second Part*, 109–10, 112.

5. Camden, *Britain*, 491, 500, 529–31, 543.

6. This phenomenon is similar to the critique of local customs for maintaining and exploiting woodlands during the same period, which were portrayed by agricultural improvers as irrational, wasteful, and destructive. See McRae, "Tree-Felling."

7. The following description of the geography and landscape of the Fens is based on Reeves and Williamson, "Marshes"; Taylor, "Fenlands"; Rackham, *History of the Countryside*, chap. 17; Darby, *Medieval Fenland*, chap. 1; idem, *Changing Fenland*, chap 1; Cook and Williamson, *Water Management*; Summers, *Great Level*, chap. 1; Renes, "Fenlands of England"; Butlin, "Images"; Miller and Skertchly, *Fenland Past and Present*; and Skertchly, *Geology of the Fenland*.

8. Reeves and Williamson, "Marshes," 152.

9. In fact, the term *fen* technically refers only to freshwater wetlands bearing neutral or alkaline peat, which tends to grow moss, reeds, and grasses. A wetland with acidic peat is called a *bog* and supports a different pattern of vegetation, mostly consisting of moss. The peat found in the Fens of eastern England is alkaline.

10. The comparative dryness and inhabitability of the silt marshes should not be overstated. Several of the smaller drainage projects of the seventeenth century, including those in Holland Fen, East and West Fens, and Wildmore Fen, took place in the northern part of the fenlands region between Boston and Lincoln in what is actually Lincolnshire silt marsh. For the history of these wetlands and the efforts to drain them, see Darby, *Changing Fenland*; Lindley, *Fenland Riots*; and Kennedy, "Charles I."

11. The modern county of Cambridgeshire still includes the Isle of Ely and has incorporated the whole of Huntingdonshire.

12. There are several rivers in Britain called "Ouse." The Great Ouse in East Anglia is the longest of them and will be referred to hereafter as the Great Ouse to distinguish it from another River Ouse in the Hatfield Level.

13. While remnants of timber causeways have been found in Cambridgeshire, the best preserved and most extensive examples are in the Somerset Levels in western England; Taylor, "Fenlands," 171.

14. Rackham, *History of the Countryside*, 387.

15. On Morton's Leam, see Darby, *Medieval Fenland*, 167–68.

16. See, for example, Warde, *Ecology*; Richards, *Unending Frontier*; and Isenberg, *Destruction of the Bison*. Warde has suggested that the very term *environment* inherently implies a division between human activity and the natural expanse in which it takes place, with humans acting only to disrupt an otherwise stable natural system. He therefore prefers the term *ecology*, which he takes to include human actors and actions as one integral part of the natural whole (*Ecology*, 9–19). While I have not adopted his strict terminological distinction here, I share his concerns and wish to stress the need to consider human beings as a constitutive part of the fenland ecology.

17. The following account of the late medieval and early modern fenland economy is based on Thirsk, *English Peasant Farming*, chaps. 1, 2, 6, 7; idem, "Isle of Axholme"; idem, "Seventeenth-Century Agriculture"; idem, *Alternative Agriculture*; idem, gen. ed., *AHEW*, vol. 4, *1500–1640*, vol. 5, pt. 1, *1640–1750: Regional Farming Systems*, and vol. 5, pt. 2, *1640–1750: Agrarian Change*; McRae, *God Speed the Plough*; Ravensdale, *Liable to Floods*; Spufford, *Contrasting Communities*; Falvey, "Custom, Resistance, and Politics"; Cunningham, "Common Rights"; Rackham, *History of the Countryside*;

Di Palma, *Wasteland*, chap. 3; Taylor, "Fenlands"; Butlin, "Images"; Darby, *Medieval Fenland*; idem, *Changing Fenland*; Summers, *Great Level*. For a slightly later period, see also Neeson, *Commoners*; Thompson, *Customs*.

18. For a very thorough account of the unhealthy nature of English wetlands and the ravages caused by endemic malaria in particular, see Dobson, *Contours of Death*. Although Dobson's study is overwhelmingly devoted to the coastal wetlands of southern England, particularly Romney Marsh in Kent and the Thames estuary, she suggests that conditions were hardly better in England's other coastal wetlands, including the Somerset Levels and the East Anglian Fens. She also suggests that malaria may have become a bigger problem in the sixteenth century, as new reclamation efforts in the salt marshes restricted the tides and created even more stagnant pools of salt water, which the malaria-carrying mosquitoes favored for breeding.

19. Thirsk, *English Peasant Farming*, 31–36.

20. Lord Willoughby to the Earl of Essex, HMC, *Report on the . . . Earl of Ancaster*, 337–38 (n.d. [1597]).

21. Thirsk, *English Peasant Farming*, chap. 1; idem, "Isle of Axholme."

22. Darby, *Medieval Fenland*, 142–46.

23. Thirsk, *English Peasant Farming*, 13.

24. The following account of the history of commissions of sewers is based on Kirkus, "Introduction"; Webb and Webb, *Statutory Authorities*, 1–106; Richardson, "Early History"; Kennedy, "So Glorious a Work"; Darby, *Medieval Fenland*, chap. 5; idem, *Changing Fenland*, 36–41; Butlin, "Images"; Renes, "Fenlands of England."

25. Kirkus, "Introduction," xv–xvi.

26. 6 Hen. VI, c. 5 (*Statutes of the Realm*, 2:236–38); Kirkus, "Introduction," xx; this act is also printed in Kirkus and Owen, *Records*, 1–3.

27. 23 Hen. VIII, c. 5 (*Statutes of the Realm*, 3:368–72); this act is also printed in Kirkus and Owen, *Records*, 3–11.

28. Webb and Webb, *Statutory Authorities*.

29. In early modern English, *quorum* referred not to a minimum number of members required to conduct official business, as it does in modern English, but rather to a select group of experienced and eminent members, at least some of whom had to be present for commission actions to be considered legitimate; OED, s.v. "quorum."

30. Kirkus and Owen, *Records*, 5.

31. On the centralizing tendencies of the early Tudor state, and especially with respect to legal institutions, see Kesselring, *Mercy and Authority*.

32. My argument concerning the local political culture of the Fens, manifested in the commissions of sewers, is indebted to Andy Wood's analysis of the political culture of the free miners of the early modern Peak District. The miners' history of autonomous self-governance through customary practices gave rise to the legal institution of the local mining court, the barmoot, which enabled the miners to develop a local political culture capable of articulating and defending their interests against outside incursions and the interference of the state. Though their laws were rooted in custom and were thus fundamentally conservative in nature, the miners also proved adept at interpreting custom dynamically, when necessary, to meet their changing needs as circumstances warranted. Though the two institutions obviously differ in many ways,

I would argue that the barmoot and the commission of sewers shared an analogous development and played similar roles within their respective communities. See Wood, *Politics of Social Conflict*; idem, *Riot, Rebellion and Popular Politics*.

33. Kennedy, "So Glorious a Work," chap. 3; Kirkus, "Introduction," xxxvi–xxxvii. See also chapter 2 in the present volume.

34. I am grateful to Keith Pluymers for emphasizing this point in his critique of an earlier draft.

35. Anonymous critique of the judgment of a commission of sewers, NA, SPD 14/18/104 (n.d. [c. 1610]).

36. Darby, *Medieval Fenland*, 93–100.

37. Unlike the much more severe Ice Age, in which glacial formation caused a drop in global sea levels, the Little Ice Age was more accurately a period of general cooling, but not extensive glaciation. Indeed, some climatologists believe that the Medieval Warm Period of the preceding centuries may have melted enough glacial ice to flood the oceans with an abundance of fresh water, raising sea levels and causing oceanic thermohaline circulation to slow down, resulting in cooler temperatures. See "Little Ice Age," *Wikipedia*, accessed 18 August 2016, http://en.wikipedia.org/wiki/Little _Ice_Age#Causes.

38. See Richards, *Unending Frontier*, chap. 2, for an analysis of Switzerland's agricultural difficulties during the same period.

39. Darby, *Changing Fenland*, 43–45.

40. Dugdale, *History of Imbanking*, 332–39; Darby, *Changing Fenland*, 43–45.

41. Unsigned letter to the Privy Council, BL, Add. MS 35171, fol. 48v (n.d.). Though undated, the document is included with a series of others dated 1570–1620.

42. Dugdale, *History of Imbanking*, 375. The argument that the dissolution of the monasteries played a key role in the deteriorating situation of the sixteenth-century Fens is long-standing, but controversial. Dugdale himself argued that the greater problem lay in the rivers' outfalls being choked up by tidal silting. Likewise, Darby also admits that "evidence does not permit any definite conclusion about the results following from the dissolution of the monasteries" and points out that the monastic houses were themselves frequently cited for not maintaining their sewers (*Changing Fenland*, 44–45). See also Hoyle, "Introduction"; and L. E. Harris, *Vermuyden*, 17–21.

43. See especially Thirsk, *English Peasant Farming*, 36–38, 112–17.

CHAPTER 2: State Building in the Fens, 1570–1607

1. APC, 27:274–76 (n.d. [4 July 1597]). Though the letter is undated, a subsequent letter (see below) refers to a previous message sent "the fourth July last."

2. APC, 27:367–68 (18 September 1597).

3. On state formation see Braddick, *State Formation*; Hindle, *State and Social Change*; Shagan, *Popular Politics*.

4. Kennedy, "So Glorious a Work," 301–3.

5. During Elizabeth's reign, members of the Privy Council (especially Burghley) frequently became investors in putatively private ventures that accorded with Crown priorities. See Thirsk, *Economic Policy and Projects*; Ash, *Power*; idem, "Queen v. Northumberland."

6. In his classic study of politics in Elizabethan Norfolk, A. Hassell Smith describes the Privy Council's use of licensed informers, church officials, and lords lieutenant to give them greater purchase and flexibility in governing the county and to weaken the power of local officials such as justices of the peace to circumvent or obstruct Crown policies. See Smith, *County and Court*, esp. chap. 6; see also Hirst, *England in Conflict*; Braddick, *State Formation*; and Ash, *Power*, chap. 2.

7. For a discussion of another early modern example of a local English political culture, see Wood, *Politics of Social Conflict*, esp. chaps. 10–12.

8. Holinshed, *Chronicles*, 3:1222. Holinshed's narrative of the flood and its aftermath is taken almost entirely from a pamphlet printed just weeks after the event: Knell, *A Declaration*. For another account, in verse, see Tarlton, *A Very Lamentable*.

9. Kennedy, "So Glorious a Work," 50–51; Darby, *Draining*, 12–22. William Dugdale's history contains detailed summaries of a large number of sewer laws passed from 1570 onward, generally in response to severe flooding; see, for example, Dugdale, *History of Imbanking*, 339ff., dealing with Cambridgeshire and the Isle of Ely.

10. See Darby, *Draining*, 13–16. Few of the patents issued resulted in any progress; the chief exception was a patent issued in 1598 to Sir Thomas Lovell, giving him exclusive right to employ various techniques for land drainage in the Fens that he had learned during his military service in the Low Countries (BL, Add. MS 33467, fol. 20r-24v [1598]). Lovell did undertake a sizable, and temporarily successful, drainage project in Deeping Fen in the late 1590s; see Kennedy, "So Glorious a Work," 88–100.

11. 13 Eliz. I, c. 9 (*Statutes of the Realm* 4:543–44); Kennedy, "So Glorious a Work," 48–49.

12. The best recent biography of Burghley is Alford, *Burghley*.

13. Kirkus, "Introduction," lvi; Kennedy, "So Glorious a Work," 47–48.

14. Francis Harrington to Burghley, NA, SPD 12/77/49 (10 April 1571); Adlard Welby to Burghley, NA SPD 12/78/3 (2 May 1571). According to the 1532 Act of Sewers, the requirement for a viable session of sewers was just six commissioners, three of whom had to be from the quorum. So long as these conditions were met, even a tiny session of sewers could act with the full authority of the entire body. This allowed larger, countywide commissions (such as the one in Lincolnshire) to subdivide by area, so that local commissioners could handle local problems without having to call the entire body into session—rather like petty sessions of the county bench. During the sixteenth and early seventeenth centuries, however, when disagreements frequently arose among commissioners in matters impacting more than one area, local factions could convene separate meetings with just a few commissioners in attendance to pass laws that were binding throughout the entire commission's jurisdiction. Of course, rival factions could then convene their own meetings to repeal the new laws. This invariably created a great deal of confusion, stagnation, and resentment, and required Crown intervention to sort out. See 23 Hen. VIII, c. 5 (*Statutes of the Realm* 3:368–72); Kirkus, "Introduction," xxiii.

15. Earl of Bedford to Burghley, BL, Lansdowne MS 25/20 (18 May 1577).

16. Holinshed, *Chronicles*, 3:1222. Dugdale, *History of Imbanking*, 339–47, contains a presentation from one jury of sewers in 1571 running to 120 items, each of which lists a specific repair to be made and often who should be responsible for making it.

17. NA, SPD 12/108/36 (3 June 1576); E 178/3047 (1577); E 134/21&22Eliz/Mich33 (1579).

18. For an excellent and thorough narrative of this dispute, see Kennedy, "Fen Drainage"; see also idem, "So Glorious a Work," 59–64; and Darby, *Draining*, 16–18.

19. NA, SPD 12/213/28 (25 July 1588); and E 134/28&29Eliz./Mich24 (27 September 1586).

20. BL, Lansdowne MS 57/11 (n.d. [c. 7 August 1588]); and 57/12 (7 August 1588).

21. APC 17:112–13 (21 March 1589).

22. Kennedy, "So Glorious a Work," 48n6; NA, SPD 10/14/52 and 52.I (27 June 1549); *Calendar of the Patent Rolls* 2:109 (26 January 1555). See also Kirkus, *Records*, passim; Kennedy, "Commissions."

23. On information gathering as a foundational component of improvement projects, see Slack, *Invention of Improvement*, chap. 2

24. APC 17:112–13 (21 March 1589). See also *BDCE*, s.v. "Ralph Agas," "John Hexham." Hexham produced a map of the northern part of the Great Level that is probably the plat in question; it can be seen in Skelton and Summerson, *Description of Maps*, 52–54 and plate 10.

25. Ash, *Power*, 77.

26. For Bradley's biography, see *DNB*, s.v. "Humphrey Bradley"; *BDCE*, s.v. "Humphrey Bradley"; Harris, *Two Netherlanders*.

27. Bradley, "Discorso sopra il stato, delle paludi, over Terre Inundate (volgarmente ffennes) nelle Prouincie di Nortfolcia, Huntingtona, Cambrigia, Northamtona, e Lincolnia composto par Humfredo Bradley: Gentilhuomo Brabantino: Aº 1589 alli 3 di Decembre," BL, Lansdowne MS 60/34 (3 December 1589). The Italian has been translated and published in Darby, *Draining*, 263–69; all quotations and references below are from the Darby translation. See also Harris, *Two Netherlanders*; Heal and Holmes, "Economic Patronage," 214–18; and Holmes, "Drainage Projects."

28. Bradley, "Discorso," 263–69.

29. Bradley, "A Proiect ffor the Drayning off the Fennes, in Her Mtes Countes, off Cambridge, Huntington, Lincolne, Norfolke and Southfolke, Contayning More Then Aight Hundrith Thousand Acers . . . ," BL, Lansdowne MS 74/65 (3 April 1593), fols. 180r–v. See also Bradley to Burghley, NA, SPD 12/244/97 (29 March 1593). As will be seen in later chapters, Bradley was completely mistaken with respect to how easily and cheaply the Great Level might be drained.

30. Bradley, "A Proiect," fol. 180v. Bradley's fellow fenland surveyor, Ralph Agas, recognized the same dilemma: "[W]ithout a general consent of the landholders there (whether voluntary or forced)," he wrote to Burghley, "the [drainage] will never be performed." Agas to Burghley, BL, Lansdowne MS 84/32 (1597).

31. BL, Harleian MS 5011, vol. 1, fols. 198–99; Kennedy, "So Glorious a Work," 68–69.

32. Bradley, "A Proiect," fols. 180v–181r and 178r. Given such a bizarrely naive assertion, one can only wonder whether Bradley had ever seen an Elizabethan act of Parliament, or what Burghley made of this suggestion.

33. Bradley, "A Proiect," fols. 180v–181v.

34. Bradley to Burghley, NA, SPD 12/244/97 (29 March 1593).

35. Both the Company of Mines Royal and the Muscovy Company, for example, were ostensibly private joint-stock ventures, but they were sponsored by the Crown and

boasted several courtiers and members of the Privy Council among their shareholders. Ash, *Power.*

36. Kennedy, "So Glorious a Work," 71–72; Heal and Holmes, "Economic Patronage," 214.

37. *APC*, 26:531–32 (2 March 1597).

38. "An Acte for the Recovering of Three Hundred Thowsand Acres the More or Lesse of Wastes Marish and Watry Groundes in the Isle of Ely and in the Counties of Cambridge, Hunt., Northon., Lincoln, Norff. and Suffolk," Parliamentary Archives, House of Lords Main Papers, HL/PO/JO/10/1/6, fols. 47–57 (13 December 1597) (my emphasis); Kennedy, "So Glorious a Work," 74–77.

39. Baron Willoughby de Eresby to the Earl of Essex, HMC, *Ancaster*, 337–39 (n.d. [late 1597]); Kennedy, "So Glorious a Work," 77–79. Burghley and Essex were bitter rivals among the queen's advisers, and it is not surprising that Essex should seek to block a bill that would generate such material and social benefits for Burghley and those in his patronage network. The reasons for the queen's sudden and negative intervention are unknown.

40. Petition to the Privy Council from the Isle of Ely, BL, Add. MS 33467, fols. 15r–v (n.d. [c. 1598]).

41. "Divers Gentlemen and Other Inhabitants of the Counties of Lincoln, Northampton, and Huntingdon to the Lords of the Council," HMC, *Salisbury*, 8:243–44 (June 1598). See also the petition from the Lincolnshire commissioners of sewers to the Privy Council, BL, Lansdowne MS 87/4 (June 1598).

42. The bill's quick passage may have been due in part to the fact that the Earl of Essex had been executed for treason earlier in the year, removing him as an obstacle and causing great harm to those in his patronage network. Burghley was also dead by 1601, but the bill was introduced and carried by those who had been his clients.

43. 43 Eliz., c. 11 (*Statutes of the Realm* 4:977–78), my emphasis; see also Kennedy, "So Glorious a Work," 65–69, 83–85.

44. On the poor reputation of projectors, see Thirsk, *Economic Policy*; Novak, *Age of Projects*; Yamamoto, "Distrust."

45. Dugdale, *History of Imbanking*, 377.

46. 23 Hen. VIII, c. 5 (*Statutes of the Realm* 3:368–72); Albright, "Entrepreneurs."

47. NA, C 147/45 (20 September 1597), C 225/2/66 (16 July 1596), C 225/2/67 (11 April 1599); EL 2531 (41 Eliz. [1598–99]), Ellesmere Manuscripts, Huntington Library, San Marino, CA; Dugdale, *History of Imbanking*, 350–54; Kennedy, "So Glorious a Work," 94; Albright "Entrepreneurs."

48. Dugdale, *History of Imbanking*, 351–52; BDCE, s.v. "John Hunt"; Kennedy, "So Glorious a Work," 100–101.

49. James I to the commission of sewers for the Great Level counties, BL, Add. MS 35171, fol. 205r (11 July 1604).

50. BDCE, s.v. "Richard Atkyns."

51. Atkyns survey, CUA, EDR A/8/1, pp. 252–95 (survey of fen lands) and 296–301 (observations of the rivers). Both documents are credited to "R.A." and are described as being compiled in January and February 1605; this is almost certainly Richard Atkyns, though Dugdale describes Atkyns's survey as beginning on 2 April 1605, a date that would make more sense, because winter floods would have made an earlier survey

exceedingly difficult (*History of Imbanking*, 378–79). The 1605 survey should not be confused with another one undertaken by Atkyns in 1619, discussed in chapter 4 in the present volume.

52. *BDCE*, s.v. "William Hayward."

53. Hayward survey, CUA, EDR A/8/1, pp. 304–7; the survey is quoted in Dugdale, *History of Imbanking*, 382–83. The original great wall map that accompanied this survey has long since been lost, but it served as the template for a number of subsequent maps of portions of the Great Level, many of which survive. Two of the most detailed and extensive of these are in the CRO ("An Exact Copy of a Plan of the Fenns as It Was Taken Anno. 1604 by William Hayward, Carefully Copy'd from ye Originall by Mr. Payler Smyth, Anno. Dom. 1727," CRO, BLC papers, R/59/31/40/1) and in the BL ("A Generall Plotte and Description of the Fenns and Other Grounds within the Isle of Ely and in the Counties of Lincolne, Northampton, Huntington, Cambridge, Suffolke, and Norffolke . . . ," Cotton MS, Augustus I.i.78). On the map's conceptual originality, see Di Palma, *Wasteland*, 86–87, 100–101.

54. "The Answer of the Jury of Burgh Soke," BL, Add. MS 35171, fol. 44v (n.d. [1604]); "Address from the Inhabitants of Ely to the Bishop," CUA, EDR A/8/1, p. 89 (n.d. [1604]); Kennedy, "So Glorious a Work," 103.

55. James I to the commission of sewers for the Great Level counties, BL, Add. MS 35171, fol. 206r (17 April 1605).

56. Privy Council to the commission of sewers for the Great Level counties, BL, Add. MS 35171, fol. 205v (21 April 1605).

57. NA, C 181/1, pp. 213–15 (20 April 1605). As required by statute, commissions of sewers always specified the geographic area to be included within their jurisdiction. This particular commission was extraordinarily verbose and precise in its rendering, suggesting that it was issued with a specific drainage project in mind and tailored to fit exactly what the projectors wanted it to cover. See also NA, C 181/1, pp. 228–29 (18 June 1605); and SPD 15/36/43 (18 June 1605).

58. Kennedy, "So Glorious a Work," 105–6.

59. CUA, EDR A/8/1, p. 90 (29 May 1605), and pp. 105–7 (31 August 1606); BL, Add. MS 35171, fol. 208r (31 Aug. 1606); NA, SPD 14/26/38–39 (2 February 1607); Dugdale, *History of Imbanking*, 379.

60. Dugdale, *History of Imbanking*, 381; a summary of the survey may be seen on pp. 380–81. In addition to being a prominent commissioner of sewers, Hunt was apparently a skilled engineer: the specific works he proposed in 1605 eventually became the basis for much of the massive drainage project undertaken by the 4th Earl of Bedford in the 1630s—see chapter 6 in the present volume.

61. CUA, EDR A/8/9 (21–27 June 1605).

62. James I to the commission of sewers for the Great Level counties, BL, Add. MS 35171, fol. 206v (24 June 1605).

63. James I to the commission of sewers for the Great Level counties, BL, Add. MS 35171, fol. 205r (11 July 1604). Mark Kennedy suggests that Popham may have been the main force behind the project from the beginning, initially staying in the background while using Hunt and Totnall as surrogates ("So Glorious a Work," 104–5). Whatever the case, Hunt was apparently the architect of the drainage scheme itself—see Dugdale, *History of Imbanking*, 380–81.

64. *DNB*, s.v. "Sir John Popham"; *History of Parliament*, s.v., "John Popham"; Rice, *Life and Achievements of Sir John Popham*.

65. CUA, EDR A/8/4 (28 June 1605); Kennedy, "So Glorious a Work," 109–11. The Isle of Ely was assessed £50,000; the rest of Cambridgeshire was to pay £12,000; Norfolk, £10,000; Suffolk, £10,000; Huntingdonshire, £15,000; Northamptonshire, £5,000; and Lincolnshire, £3,000; the source for the remaining £15,000 was not specified.

66. "The Humble Petition of the Inhabitants of the Towns on the West Side of the County of Cambridge Which Border South upon the River of [Great] Ouse," NA, SPD 14/18/102 (n.d. [1606]). References to projectors as "strangers" and "foreigners" appear frequently in the primary sources and often indicate someone not native to the Fens, rather than someone from outside England; though once Dutch land drainers got involved, after 1620, the term could obviously mean either or both.

67. "Petition from the Inhabitants of Ely to John Popham," CUA, EDR A/8/1, pp. 94–95 (July 1605).

68. BL, Harleian MS 368, fol. 170v (n.d.); NA, SPD 14/26/38 (n.d. [c. January 1607]). It should be noted that both sources are pro-drainage documents, which cite the number of protestors only to claim that, after their objections were answered, they had all been won over.

69. CUA, EDR A/8/1, pp. 96–102 (13 July 1605). Despite the prominence of Hunt and Totnall in the early phases of the scheme, this document lists only Popham, Thomas Fleming (baron of the court of exchequer), William Rumney (alderman of London), and John Eldred (citizen and clothworker of London) as the chief undertakers.

70. The stipulations were mostly taken from a list submitted to the commission by the bishop of Ely, Martin Heton: CUA, EDR A/8/1, pp. 91–93 (n.d.). The diocese was a major landowner in the Great Level, and Heton's list of concerns was probably compiled in consultation with other prominent area landowners.

71. Dugdale, *History of Imbanking*, 385. The drain ran from the old course of the River Nene near March to Well Creek near Nordelph (Darby, *Draining*, 31–32); it was the only part of the project ever built. Dugdale notes, however, that within a few months of its completion the banks of the new river were breached by severe storms, and it had to be stopped up again to avoid further flooding.

72. CUA, EDR A/8/1, pp. 96–102 (13 July 1605), and EDR A/8/5 (n.d.).

73. Kennedy, "So Glorious a Work," 117.

74. James I to the commission of sewers, BL, Add. MS 35171, fol. 207r (23 July 1605).

75. Sir John Peyton, the younger, to Salisbury, HMC, *Salisbury* 17:452 (n.d. [10 October 1605 or later]).

76. Archbishop of Canterbury to Salisbury, HMC, *Salisbury*, 17: 435–36 (27 September 1605).

77. Anonymous letter to the king, NA, SPD 14/19/47 (n.d. [14 March 1606?]); see also Kennedy, "So Glorious a Work," 130. In addition to a great deal of vitriol aimed at Popham for a variety of exploitive projects and schemes, this letter is full of criticism of James I, and particularly his lack of understanding as a foreigner of English laws and customs. His reign is explicitly and negatively compared throughout with that of

Queen Elizabeth. One can only wonder what the author hoped to achieve in sending it to the king; its reception at court is indicated by the brief endorsement on the cover sheet: "A base malicious libel."

78. Petition to the king, BL, Add. MS 35171, fol. 207v (n.d.); see also Dugdale, *History of Imbanking*, 385.

79. "The Humble Petition of the Inhabitants of the Towns on the West Side of the County of Cambridge Which Border South upon the River of [Great] Ouse," NA, SPD 14/18/102 (n.d.).

80. Privy Council to the commission of sewers, CUA, EDR A/8/1, pp. 105–7 (31 August 1606); other copies may be seen at BL, Add. MS 35171, fol. 208r; and BL, Add. MS 34218, fol. 91r. See also Dugdale, *History of Imbanking*, 385.

81. Commission of sewers to the Privy Council; BL, Add. MS 35171, fol. 208v (23 October 1606); another copy exists at BL, Add. MS 34218, fol. 92r. See also Dugdale, *History of Imbanking*, 385; Kennedy, "So Glorious a Work," 126–27.

82. "The Answere to the Obiections Made againste the Drayning of the Ffenns," NA, SPD 16/339/27 (n.d. [c. 1606]). This document is discussed in more detail in the introduction to the present volume.

83. "The Answere to the Obiections Made againste the Drayning of the Ffenns," NA, SPD 16/339/27 (n.d. [c. 1606]).

84. See chapter 3 in the present volume.

85. John Popham to Thomas Lambert, BL, Hargrave MS 33/8, fols. 220r–222v (7 September 1605); Kennedy, "So Glorious a Work," 115.

86. Callis, *Reading*; Dugdale, *History of Imbanking*, passim; Darby, *Medieval Fenland*; Kirkus, *Records*, passim; Holmes, "Statutory"; Kennedy, "So Glorious a Work," chap. 6.

87. Popham to Lambert, BL, Hargrave MS 33/8, fols. 220r–222v (7 September 1605).

88. JHC, 1:270 (18 February 1606); Dugdale, *History of Imbanking*, 386–87. For a thorough account of the parliamentary debates and negotiations, see Kennedy, "So Glorious a Work," 118–30.

89. Bowyer, *Parliamentary Diary*, 149–50.

90. JHC, 1:308 (12 May 1606).

91. JHC, 1:371–72 and 1043 (9 May 1607); Kennedy, "So Glorious a Work," 129–31.

92. Dugdale, *History of Imbanking*, 385.

CHAPTER 3: The Crisis of Local Governance, 1609–1616

1. Law of sewers from a session held at Ely, BL, Add. Charters 33088 (9 June 1609).

2. Law of sewers from a session held at Ely, BL, Add. Charters 33091 (8 August 1609). This law called for the creation of several new cuts along the River Great Ouse, largely confirming a previous law passed on 9 June 1609 (BL, Add. Charters 33089).

3. A copy of one such warrant, addressed to the constables of Haddenham in Cambridgeshire for the collection of £60, may be seen at CUA, EDR A/8/1, pp. 131–33 (17 June 1609). Similar warrants were also issued for several neighboring towns, copies of which may be found in BL, Add. MS 33467, fols. 61–95; the latter volume also contains detailed accounts of all monies collected and spent on the new works of 1609.

4. Petition from several inhabitants of Cambridgeshire and the Isle of Ely to the commission of sewers, CUA, EDR A/8/1, pp. 125–26 (10 August 1609).

5. Report from John Cutts, John Cotton, and Francis Brackin to Sir Edward Coke, CUA, EDR A/8/1, pp. 133–37 (16 August 1609). Cutts, Cotton, and Brackin were gathering information on the local attitude toward the new drains at Coke's behest, in the hope of reaching an agreeable settlement of the dispute.

6. Warrant to the bailiff of Ely, CUA, EDR A/8/1, p. 130 (29 August 1609).

7. Warrant to the sheriff of Cambridgeshire, BL, Harleian MS 5011, vol. 1, fol. 12r (12 September 1609).

8. Goldie, "Unacknowledged Republic"; Braddick, *Nerves of State*; idem, *State Formation*; Hindle, *State and Social Change*; Fletcher, *Reform*; Hirst, "Privy Council."

9. Bridges is sometimes listed in the court documents as "Briggs."

10. Cage's response to a bill of complaint, NA, E 112/71/188 (14 March 1620).

11. Committee report to the Privy Council, NA, SPD 14/130/124 (n.d. [May 1622]).

12. Privy Council to the sheriff of Cambridgeshire, CUA, EDR A/8/1, p. 124 (2 October 1609).

13. Bill of complaint against John Cage, NA, E 112/71/188 (Hillary term, 1619); see also Kennedy, "So Glorious a Work," 158–59. For a definition of *tarde venit*, see *Law Dictionary*, accessed 18 August 2016, http://www.lawyerintl.com/law-dictionary/6611 -tarde%20venit.

14. Commission of Sewers for the Isle of Ely to Lord Treasurer Salisbury, NA, SPD 14/48/107 (16 October 1609).

15. On the broader contemporary debate surrounding improvement projects, private interest, and the public good, see Slack, *Invention of Improvement*, chap. 3.

16. Royal prerogative courts were a series of courts instituted in England during the Middle Ages through which the monarch's discretionary powers and privileges were exercised; these included the Court of Chancery and the Court of Star Chamber, among others. The prerogative courts derived from the Crown's sovereign authority and operated independently of the chief common law courts (the Courts of King's Bench and Common Pleas). They were created to provide justice in the king's name in cases where the common law rendered unjust outcomes or else could not provide adequate legal remedy to a plaintiff. Because they operated independently, under the king's discretion, they were far less structured and bound by precedent than the common law courts. By the seventeenth century, they had become quite powerful and constituted a series of alternative jurisdictions in which the decisions of common law courts could be challenged and overturned. In the decades before the Civil War, some of them also came to be associated with arbitrary royal government. Sir Edward Coke strove to counter the prerogative courts' ever-increasing power and to defend the jurisdiction of the common law, and his assault on the courts of sewers was only a small part of his larger battle against the Court of Chancery, the most powerful of the royal prerogative courts.

17. "An Act for the More Speedie Recoverie of Many Hundred Thousand Acres of M[ar]sh & Other Grownds Subiect Co[m]monly to Surrounding w[ith]in the Isle of Ely & Counties of Camb. North. Lincolne Norff. Hunting. & Yorke," CUA, EDR A/8/1, pp. 50–61. The table of contents of the volume in which this copy of the bill appears dates the document to February 1605, but Mark Kennedy has shown that it

was actually presented to the House of Commons on 8 May 1604; see *JHC*, 1:202 (8 May 1604); and Kennedy, "So Glorious a Work," 132–36.

18. "Reasons of the Inhabitants of the Isle of Ely Why the Aforesayd Bill Should Not Pass," CUA, EDR A/8/1, pp. 62–65 (n.d. [mid-1604]); "An Answere to Certaine Matters in the Bill Not Spoken unto before the 2d Amendment of the Same," CUA, EDR A/8/1, pp. 66–67 (n.d. [mid-1604]); "Answers to Matters of the Bill Now Thirdly Inserted," CUA, EDR A/8/1, pp. 68–70 (n.d. [mid-1604]).

19. "The Byll Fownders Answers to the Former Arguments, & a Replication to the Said Answerers, by the Defendants, Inh[ab]itants & Co[m]moners," CUA, EDR A/8/1, pp. 71–86 (n.d. [mid-1604]); Kennedy, "So Glorious a Work," 135–36.

20. Kennedy, "So Glorious a Work," 136–46; Darby, *Draining*, 5–11.

21. "Doubts or Questions Moved at the Co[m]missio[n] of Sewers, 1609," CUA, EDR A/8/1, pp. 137–45 (1609), quotation on p. 143. For examples of Sandys's fenland lawsuits, see NA, E 134/3Jas1/Mich30 (26 September 1605); E 134/3Jas1/Mich31 (24 September 1605); E 134/3Jas1/Mich31 (27 September 1605); E 178/3609 (26 January 1606 and 26 May 1609).

22. Kennedy, "So Glorious a Work," 138–39, 142–44; Cunningham, "Common Rights"; *VCH*, Cambridge and the Isle of Ely 4:153–54; Darby, *Draining*, 59n.

23. "Doubts or Questions Moved at the Co[m]missio[n] of Sewers, 1609," CUA, EDR A/8/1, pp. 137–45 (1609); CUA, T.XII.1, section 2 (c. 17 May 1619), section 4, no. 3 (17 May 1619), section 17, no. 1 (n.d.); CUA, T.XII.3, section 1, pp. 11–19 (27–29 April 1619); CUA, T.XII.4(ii) (17 May 1619); Kennedy, "So Glorious a Work," 143–44.

24. Laws of sewers passed at a session of sewers held at Ely, BL, Add. Charters 33089 and 33088, respectively (9 June 1609); quotations taken from 33088.

25. See CUA, EDR A/8/1, pp. 131–33 (17 June 1609); and BL, Add. MS 33467, fols. 61–95 (1609).

26. Petition from several inhabitants of Cambs. and the Isle of Ely to the commission of sewers, CUA, EDR A/8/1, pp. 125–26 (10 August 1609).

27. Petition from Swaffham Prior to the Privy Council, CUA, EDR A/8/1, pp. 147–49 (n.d. [1609]).

28. "Doubts or Questions Moved at the Co[m]missio[n] of Sewers, 1609," CUA, EDR A/8/1, pp. 137–45 (1609), my emphasis.

29. John Popham to Thomas Lambert, BL, Hargrave MS 33/8, fols. 220r–222v (7 September 1605). See also Holmes, "Statutory Interpretation"; Kirkus and Owen, *Records*; Kennedy, "So Glorious a Work," 175–76.

30. Privy Council to Coke, CUA, EDR A/8/1, p. 147 (n.d. [1609]).

31. Kennedy, "So Glorious a Work," 153.

32. Report from John Cutts, John Cotton, and Francis Brackin to Sir Edward Coke, CUA, EDR A/8/1, pp. 133–37 (16 August 1609). See also the record of an argument between Sandys, Cotton, and Cutts during which Sandys lost his temper and became exceptionally heated: BL, Harleian MS 5011, vol. 2, fol. 541r (28 July 1609).

33. Summary of the hearing before Lord Chief Justice Edward Coke, CUA, EDR A/8/1, pp. 147–49 (n.d. [December 1609]), my emphasis; see also BL, Harleian MS 5011, vol. 1, fols. 13v–14r (n.d. [1609]); and Kennedy, "So Glorious a Work," 162–65.

34. Kennedy seems to argue that Coke's 1609 opinion invalidated any and all new drainage works as being against statute, but I believe this overstates things ("So

Glorious a Work," 164, 168, 181–82). The panel of justices ruled only that "the commissioners could not make *the said new river* out of the main land" (my emphasis) because they had failed to show that it would be "for the general good of the Common Weal and in that regard to be favored." Even if new works of obvious benefit were perhaps "not altogether lawful," this decision would seem to leave the door open to them if they met the justices' standard of being manifestly for the common good. Coke also maintained in his 1614 *Reports* on the case (see below) that new works were permissible if they were universally approved by the local stakeholders or were specifically sanctioned by Parliament. See the summary of the hearing before Lord Chief Justice Edward Coke, CUA, EDR A/8/1, pp. 147–49 (n.d. [December 1609]); *The Case of the Isle of Ely*, Coke, *Reports*, 10:141–43 (*English Reports*, 77:1139–43); Smith, *Sir Edward Coke*, chap. 3.

35. "Request for an Act of Parliament to Discharge the Fines Laid upon John Cage," BL, Egerton MS 2651, fol. 76r (n.d. [c. 1620]); Kennedy, "So Glorious a Work," 165.

36. A parliamentary bill introduced around the same time for the draining of some forty thousand acres of fenland in Cambridgeshire, Huntingdonshire, and Suffolk was most likely sponsored by Sandys and his allies as a means of securing legal authority for their proposed project in spite of Coke's determination, but the bill was vigorously opposed and quashed; see Kennedy, "So Glorious a Work," 165–67.

37. Much of the following discussion is based on the analyses in Holmes, "Statutory Interpretation"; and Kennedy, "So Glorious a Work," chap. 6. I include it here as part of my larger argument that during James's reign, the Crown's involvement in new, large-scale drainage projects in the Fens was a manifestation of state building—the Crown had both a political and a financial interest in promoting such schemes. As will be seen, Coke's ill-advised assault on the broad powers of commissions of sewers was related to his aversion to the abuse and unwarranted expansion of royal prerogative power generally, in the Fens and everywhere else. He was also eager to defend the superordinate jurisdiction of England's common law courts from other, prerogative-based courts, such as Chancery. The Privy Council's response to his assault, while arguably pragmatic and balanced in its intent to reaffirm the authority of the sewer commissions, only further implicated the Crown as an aggressive and partisan agent in handling fenland drainage issues. See also Smith, *Sir Edward Coke*, chap. 3; and Baker, "Common Lawyers."

38. *The Case of Chester Mill upon the River of Dee*, Coke, *Reports*, 10:137–38 (*English Reports*, 77:1134–36).

39. *Keighley's Case*, Coke, *Reports*, 10:139–40 (*English Reports*, 77:1136–39).

40. On the 1613 floods, see the letter from John Chamberlain to Dudley Carleton, NA, SPD 14/75/28 (25 November 1613); Dugdale, *History of Imbanking*, 276–77.

41. *William Hetley Plaintiff, against Sir John Boyer, Sir Anthony Mildmay, Matthews, Robinson, and All Defendants*, Bulstrode, *Reports*, 2:197–98 (*English Reports*, 80:1064–66).

42. *William Hetley Plaintiff, against Sir John Boyer, Sir Anthony Mildmay, Matthews, Robinson, and all Defendants*, Bulstrode, *Reports*, 2:197–98 (*English Reports*, 80:1064–66). The printed report contains an obvious misprint: ". . . but in commission is more abused than this is" (1065).

43. *William Hetley Plaintiff, against Sir John Boyer, Sir Anthony Mildmay, Matthews, Robinson, and all Defendants,* Bulstrode, *Reports,* 2:197–98 (*English Reports,* 80:1064–66).

44. *The King against Sir Anthony Mildmye* [sic], Bulstrode, *Reports,* 2:299 (*English Reports,* 80:1137–38).

45. David Chan Smith has argued that Coke asserted the superordinate jurisdiction of the common law not to attack or undermine royal prerogative power but to preserve it from the taint of corruption. Coke believed that the common law was the only legal means available to oversee Crown officers entrusted to act with the king's delegated powers and to punish those who acted unjustly for private gain. If left unpunished, such corruption would certainly undermine confidence in the legal system and in the king's justice, and defending the common law's superior and independent jurisdiction was therefore vital to protecting royal prerogative power from a very real threat. This created tension, however, when King James I did not share Coke's view of the beneficent role of the common law and he repeatedly sought to involve the Crown in cases to ensure that their outcomes were truly just, especially when he found the law to be inconsistent, obscure, or unfair (Smith, *Sir Edward Coke,* "Introduction").

Coke's handling of the commissions of sewers and the royal rebuke he received for it, I wish to argue, resulted from the clash between his commitment to protecting the independence of the common law and the Crown's political priorities under James I. Coke moved to restrict the commissioners' discretionary powers through the common law courts because of their perceived self-interest and corruption in wielding them. This, he believed, was the only way to protect English subjects from dishonest royal officials and to prevent royal prerogative power from seeming arbitrary. But this in turn forced the Privy Council to move aggressively in reinforcing the commissioners' broad powers as agents of Crown governance, acting with discretion in the king's name to protect the realm from harm. In so doing, they reaffirmed by implication an expansive interpretation of royal prerogative power more generally. I am very grateful to David Chan Smith for his critique and advice in revising this chapter.

46. Baker, "Common Lawyers," 255–57.

47. Equity law was an alternative legal tradition in England, operating outside of and parallel to the common law. It was developed primarily in the royal prerogative Court of Chancery during the fourteenth and fifteenth centuries in response to cases in which the common law either could not provide legal remedy to a plaintiff, or else produced a manifestly unjust outcome. In such cases the aggrieved party appealed to the king, who referred the matter to the lord chancellor (the head of Chancery) for judgment in the king's name. Equity decisions were not bound by common law precedents and were viewed by many common lawyers as being arbitrary in nature, and whereas common law judgments were the product of an independent judiciary, equity decisions derived from the king's prerogative powers to dispense justice. By the end of the sixteenth century, equity law had become much more defined and less arbitrary, and lords chancellor were expected to have legal training. As equity law grew in popularity among English litigants seeking legal remedy, the Court of Chancery started to encroach on the jurisdiction of the chief common law courts, especially King's Bench, not least in its power to compel a common law litigant to halt his suit

and have the matter decided in Chancery instead. Coke saw this as a grave threat to the integrity of the common law and to the independent judiciary who practiced it, and he waged a fierce battle to reassert the superiority of the jurisdiction of King's Bench against encroachments from Chancery.

48. Most famously in *Prohibitions del Roy* in 1607, Coke, *Reports*, 12:63–65 (*English Reports*, 77:1342–43). See also *DNB*, s.v. "Sir Edward Coke"; Smith, *Sir Edward Coke*, chaps. 7–8; Knafla, *Law and Politics*, chaps. 6–7; Holmes, "Statutory Interpretation," 113–17; Kennedy, "So Glorious a Work," chap. 6.

49. The main medieval statute governing *praemunire* is 27 Ed. III, c. 1 (*Statutes of the Realm*, 1:329–31); it had also been invoked repeatedly in the sixteenth century during the English Reformation, in order to prevent Catholic subjects, especially clergy, from appealing to the Vatican or the heads of monastic orders against the king's authority as Supreme Head of the Church of England. See "Praemunire," *Luminarium*, accessed 18 August 2016, http://www.luminarium.org/encyclopedia /praemunire.htm.

50. The fact that Coke had not promulgated the 1609 cases in the eighth volume of his *Reports*, printed in 1611, suggests that he did not believe them to be of great significance until his hearing of *Hetley v. Boyer* in 1614; see Holmes, "Statutory Interpretation," 114; Kennedy, "So Glorious a Work," 180.

51. It should be noted that only the first eleven volumes of the *Reports* were published during Coke's lifetime—the final two volumes were printed from Coke's notebooks in the 1650s, long after his death.

52. See Holmes "Statutory Interpretation," 110. Fifteenth-century sewer law statutes had declared that commissions were authorized to repair existing drainage works "and the same and other as often, and where need shall be, to make new" (*et eadem et alia quotiescunque et ubi necesse fuerit de novo facienda*). The 1532 statute, however, had for some reason omitted the short phrase "and other" (*et alia*), probably when the draftsman had translated the original language into English. The omission, Coke argued, altered the meaning of the statute so that commissions were only empowered after 1532 to repair existing works, "and the same . . . to make new."

53. *The Case of the Isle of Ely*, Coke, *Reports*, 10:141–43 (*English Reports*, 77:1139–43).

54. *The Case of the Isle of Ely*, Coke, *Reports*, 10:141–43 (*English Reports*, 77:1139–43).

55. *The Case of the Isle of Ely*, Coke, *Reports*, 10:141–43 (*English Reports*, 77:1139–43).

56. *Keighley's Case*, Coke, *Reports*, 10:139–40 (*English Reports*, 77:1136–39).

57. *Keighley's Case*, Coke, *Reports*, 10:139–40 (*English Reports*, 77:1136–39); *The Case of Chester Mill upon the River of Dee*, Coke, *Reports*, 10:137–38 (*English Reports*, 77:1134–36). See also *Rooke's Case*, Coke, *Reports*, 5:99–100 (*English Reports*, 77:209–10); Holmes "Statutory Interpretation," 110–12.

58. John Popham to Thomas Lambert, BL, Hargrave MS 33/8, fols. 220r–222v (7 September 1605). Mark Kennedy has noted that the wording of various parliamentary statutes of sewers passed during the sixteenth century shielded courts of sewers from having to submit to the oversight of any other court, as long as they did not violate the due process spelled out in the statute. During the Middle Ages, the presentments of a

jury of sewers could be removed to King's Bench for a new trial using a writ of certiorari (an order from a superior court to a lower one to send records of its proceedings for judicial review), but that was no longer the case after the 1532 statute, so that commissioners were effectively "without legal restraint in regard to the exercise of their office" ("So Glorious a Work," 177).

59. *Headley v. Sir Anth. Mildmay*, Rolle, *Reports*, 1:395 (*English Reports*, 81:560); Kennedy, "So Glorious a Work," 186.

60. *APC*, 35:44 (13 October 1616), 226–28 (12 April 1617), and 230–31 (10 April 1617); Wells, *History*, 2:43.

61. Copy of a letter from the Privy Council to the justices of the peace in Northampton, Lincoln, and Cambridge, NA, SPD 14/88/46 (6 August 1616). The text of this letter does not make explicit reference to drainage issues, and it is actually endorsed to the justices of the peace rather than the commissioners of sewers. However, the letter is also endorsed as being "about the fens," and the context certainly implies popular resistance to a drainage project. The text itself makes reference to "some of you, who heretofore have been employed in this work . . . ," implying that at least some of the justices of the peace were also involved in the project in question, most likely as commissioners of sewers (as prominent local landowners it was common for the same individuals to serve in both offices). Subsequent documents support this analysis (see below).

62. *APC*, 35:44 (13 October 1616).

63. *APC*, 35:57–58 (8 November 1616). For examples of individuals being committed to prison at the council's pleasure, see *APC*, 35:39 (12 October 1616), 44 (13 October 1616), 54–55 (3 November 1616), and 128 (29 January 1617).

64. More than a year later, in April 1618, Hetley was still in trouble with the commission of sewers. The commissioners informed the Privy Council that they had considered Hetley's most recent petition concerning his taxes and found his case to be without merit. Moreover, they had recently received a petition from John Davye, the jailer of Peterborough, claiming the former prisoner still owed him some £31 for food and board during his incarceration. The commissioners ordered Hetley to pay Davye at least £20 toward his debt, which he refused to do, and they therefore wrote to ask the council to confirm their order (NA, SPD 14/97/34, [21 April 1618]). The council did so, though they later relented and released Hetley from the debt (*APC*, 36:139 [15 May 1618], and 161–62 [7 June 1618]).

65. *APC*, 35:59 (8 November 1616), my emphasis.

66. *APC*, 35:58 (8 November 1616). Kennedy refers to this dual role of the Privy Council, both supporting and overseeing the commissions of sewers, as a "double-barreled policy" ("So Glorious a Work," 192–94).

67. *DNB*, s.v. "Sir Edward Coke."

68. *The Lord Chancellor's Speech to Sir Henry Montague, When He Was Sworn Chief Justice of the Kings-Bench*, K. B. Moore, *Reports*, 826–29 (*English Reports*, 72:931–32).

69. *DNB*, q.v. "Robert Callis"; Holmes, "Statutory Interpretation," 107; idem, *Seventeenth-Century Lincolnshire*, 48–49.

70. Callis, *Reading*, 5–7.

71. Callis, *Reading*, 67–79, 85–87, 92–98; see also Holmes "Statutory Interpretation," 107–8. Although not published until 1647, Callis's lectures circulated widely in

manuscript after their first hearing at Grey's Inn in 1622, as indicated by the large number of manuscript copies of the text that still survive. Based on the frequency with which these lectures were cited by subsequent authors, Callis's was by far the most influential reading of sewer law in early modern England.

72. Anonymous reading of the law of sewers, CUA, Dd.V.3 (n.d.), quotation from fol. 5v. The Cambridge University archival catalog speculates that the text may have been written by Sir John Popham, but internal evidence strongly suggests that it was written after Popham's death. Based on structural and syntactic similarities, this manuscript is almost certainly an independent translation of the reading given in French by John Herne at Lincoln's Inn in 1638, the first printed English translation of which appeared as Herne, *Learned Reading*, in 1659. I have quoted from the manuscript version here because it appears to offer a more expansive translation.

CHAPTER 4: The Struggle to Forge Consensus, 1617–1621

1. Clement Edmonds's report to the Privy Council, NA, SPD 14/99/52 (20 September 1618); Richard Atkyns's report to the bishop of Ely, CRO, BLC papers, R/59/31/4, vol. 1, pp. 1–22 (n.d. [1619]). For a detailed discussion of this dispute and its origins, see Kennedy, "So Glorious a Work," chap. 7.

2. Privy Council to commissioners of sewers in Norths., Cambs., and Lincs., NA, SPD 14/89/78 (13 December 1616); APC, 35:83–84 (13 December 1616).

3. Privy Council to the commissioners of sewers of the Great Level counties, NA, SPD 14/92/16 (9 May 1617); BL, Add. MS 33467, fols. 190r–191v (16 May 1617). See also APC, 35:125–26 (28 January 1617) and 246–48 (9 May 1617).

4. Privy Council to the commissioners of sewers of the Great Level counties, NA, SPD 14/92/16 (9 May 1617).

5. Draft of a law of sewers, BL, Add. MS 33467, fols. 228r–229v (n.d. [December 1617]).

6. In some cases this proved to be more difficult than expected, because of uncertainty over who owned certain lands that had once been held by local religious houses before their dissolution. For example, the Ely jury called for maintenance on certain river banks located on "lands sometimes appertaining to the monastery of Croyland." The lack of precise knowledge of present ownership and liability in the wake of post-Reformation land speculation made it difficult to enforce the call for maintenance, as evidenced by the commissioners' frustratingly vague marginal annotation: "*fiat lex*, To be done by those that ought to do it." Ely jury presentment, CRO, BLC papers, R/59/31/9/1A, fol. 129v (16 September 1617); see also Darby, *Draining*, 5–11.

7. For example, "Item we find and present that Mr. March the brewer hath encroached into and upon the river of [Great] Ouse with a stone work to enlarge his brew house which is a great hindrance to the water course in the said river." BL, Add. MS 33467, fol. 212r (16 September 1617).

8. BL, Add. MS 33467, fols. 209–25 (16 September 1617).

9. Law of sewers passed at a session at Huntingdon, NA, C 225/1/16 (24 February 1618).

10. BL, Add. MS 33467, fols. 215r and 221–22, respectively (16 September 1617).

11. Draft of a letter from the commission of sewers to the Privy Council, BL, Add. MS 33467, fols. 226r–v (10 December 1617).

336 NOTES TO PAGES 114–117

12. Law of sewers passed at a session at Huntingdon, NA, C 225/1/16 (24 February 1618).

13. Petition from the commission of sewers for Norths., Cambs., Hunts., Lincs., Norfolk, and the Isle of Ely to the Privy Council, NA, SPD 14/97/111 (19 June 1618).

14. *DNB*, s.v. "Sir Clement Edmonds."

15. Order from the Privy Council to the commissioners of sewers for Norths., Cambs., Hunts., Lincs., Norfolk, and the Isle of Ely, BL, Add. MS 33466, fol. 20r (19 June 1618). Another copy may be seen at NA, SPD 14/97/112; see also APC, 36:177 (19 June 1618).

16. Daily notes of Edmonds's survey, NA, SPD 14/98/86 (12–19 August 1618). In addition to the nominated commissioners, Edmonds noted in his final report that Francis, Lord Russell also accompanied the group throughout "and gave great assistance to the business," NA, SPD 14/99/52 (20 September 1618). Russell was the future 4th Earl of Bedford, who would undertake to drain the Great Level in the 1630s (see chapter 6 in the present volume).

17. The party traveled down the Great Ouse from Huntingdon to Ely on 13 August, and continued to the Great Ouse's outfall at Lynn on 14 August; on 15 August, they traveled from Lynn to Wisbech via Popham's Eau, and stayed in Wisbech for two days to see the Nene's outfall and to examine the many contentious local drains in the area. On 17 August, they traveled to Crowland and examined Clough's Cross, South Eau, and other sewers along the way; on 18 August, they went on to Spalding and viewed the Welland's outfall; and on 19 August, they traveled to Peterborough, where the next session of sewers was scheduled to meet the following day in order to discuss the view taken, determine what new works and repairs were required, and enact laws accordingly. Edmonds's final report, NA, SPD 14/99/52 (20 September 1618).

18. The number of commissioners endorsing any given day ranges from nine to thirteen; on no day did all eighteen appointed commissioners sign their names, and some never signed at all. The days with the most signatures tended to address the most contentious issues.

19. Daily notes of Edmonds's survey, NA, SPD 14/98/86 (13 and 17 August 1618).

20. Mark Kennedy has argued, however, that the upland faction had an inherent advantage because there were more upland counties, giving them a greater proportional representation in the survey party. The imbalance was only increased when Sir Miles Sandys, who tended to side with the uplanders against his own (hostile) neighbors, was appointed for the lowland Isle of Ely. Their numerical superiority, Kennedy believes, allowed the uplanders to win Edmonds over to their point of view, which consistently proved to be the decisive one. He also points out that the original decision to limit each county to three representatives was contentious enough to require a poll count to settle it (Kennedy, "So Glorious a Work," 224). Despite the dominance of the upland faction, however, it is notable that the minority were nonetheless given a full hearing and an opportunity to dissent in writing.

21. Edmonds's final report, NA, SPD 14/99/52 (20 September 1618). See also APC, 36:263 and 291–99 (29 September 1618).

22. Kennedy, "So Glorious a Work," 225–26.

23. Francis Bacon to the Cambs. commissioners of sewers, BL, Add. MS 35171, fol. 51v (28 November 1618); Privy Council response to a petition from the commissioners

of sewers for Cambs. and the Isle of Ely, CUA, T.XII.2, fol. 1 (29 January 1619); APC, 36:350 (29 January 1619).

24. See Kennedy, "Fen Drainage."

25. Petition to the Privy Council from several commissioners of sewers from Cambs. and the Isle of Ely, NA, SPD 14/105/57 (20 January 1619).

26. Letter from the Privy Council to the commissioners of sewers for the Great Level, NA, SPD 14/105/59 (20 January 1619). See also APC, 36:347–48 (20 January 1619); Privy Council response to a petition from the commissioners of sewers for Cambs. & the Isle of Ely; CUA, T.XII.2, fol. 1 (29 January 1619); APC, 36:350 (29 January 1619).

27. Letter from the commissioners of sewers in Cambs. and the Isle of Ely to Nicholas Massey, NA, SPD 14/105/102 (13 February 1619). Letter from the upland commissioners to the Privy Council, NA, SPD 14/107/34 (16 March 1619).

28. Order from the Privy Council to the commission of sewers for the Great Level, CUA, T.XII.2, fol. 2 (9 March 1619); APC, 36:390–91 (9 March 1619).

29. Letter from the upland commissioners to the Privy Council, NA, SPD 14/107/34 (16 March 1619).

30. Order of the Privy Council to the commissioners of sewers, CUA, T.XII.4(i) (21 April 1619); APC, 36:425 (21 April 1619).

31. Order of the Privy Council, CUA, T.XII.4(ii) (17 May 1619). See also a rough memorandum of the Privy Council hearing, BL, Lansdowne MS 162/13 (17 May 1619).

32. DNB, s.v. "Nicholas Felton."

33. A manuscript copy of Atkyns's report may be seen at the CRO, BLC papers, R/59/31/4, vol. 1, #1; it has also been transcribed and printed in Wells, History, 2:71–97. All citations below refer to the transcription in Wells. The report contains no explicit reference to having been commissioned by Felton, nor is it dated (Wells dates it, incorrectly I believe, to 1618). The 1619 date is suggested by Atkyns's reference to the Edmonds survey as having taken place "in summer last" (73). The identity of Felton as the sponsor of the survey is suggested by a mention of "your Lordship's worthy predecessor Morton" (80). John Morton had served as bishop of Ely, archbishop of Canterbury, and lord chancellor during the late fifteenth century. Atkyns may have been addressing either the sitting archbishop (George Abbot) or lord chancellor (Francis Bacon), but because the report was most likely written in 1619, and Felton had been asked by the Privy Council to undertake a survey of the Fens that summer, he is the most obvious candidate.

34. Atkyns's report, 77, 91, 89.

35. Atkyns's report, 89.

36. Edmonds's final report, NA, SPD 14/99/52 (20 September 1618).

37. Dugdale, History of Imbanking, 82 and 401; see also Badeslade, History, 31; Kennedy, "So Glorious a Work," 244–45; and History of Parliament, s.v. "Sir William Ayloffe." Sir Edward Coke, now back in the good graces of the Crown and restored to the Privy Council, was also part of Arundel's delegation.

38. Proposal of Ayloffe and Thomas, NA, SPD 14/109/153 (n.d. [22 July 1619]). See also APC, 37:15 (22 July 1619); and Dugdale, History of Imbanking, 401–8.

39. James I to the Privy Council, NA, SPD 14/110/47 (4 September 1619). See also the minutes of a session of sewers held at Ely, CUA, T.XII.3, sec. 1, pp. 40–41 (22–23

September 1619); a letter from James I to the sheriff of Norths., BL, Add. MS 33466, fol. 71r (17 August 1619); and an act of sewers from a general session held at Peterborough, BL, Add. MS 33466, fol. 76r (8 September 1619).

40. Minutes of a session of sewers held at Ely, CUA, T.XII.3, sec. 1, pp. 40–41 (22–23 September 1619).

41. James I to the Privy Council, NA, SPD 14/110/47 (4 September 1619).

42. Minutes of a session of sewers held at Ely; CUA, T.XII.3, sec. 1, pp. 40–41 (22–23 September 1619). However, the *Acts of the Privy Council of England* dates this letter somewhat earlier (*APC*, 37:27–28, 1 September 1619).

43. Act of sewers from a general session held at Peterborough, BL, Add. MS 33466, fol. 76r (8 September 1619); proceedings of a session of sewers held at St. Ives, BL, Add. MS 33466, fol. 77r (24 September 1619).

44. Minutes of a session of sewers held at Ely; CUA, T.XII.3, sec. 1, pp. 41–43 (22–23 September 1619). A copy of the printed resolution and agreed-upon concessions may be seen at NA, SPD 14/110/75.

45. Daniel Wigmore to the bishop of Winchester, CUA, EDR A/8/20 (27 September 1619).

46. It should be noted that the author of the letter believed that permanently draining the Fens was manifestly impossible and that anyone proposing such a thing must therefore be incompetent.

47. Anonymous letter to an unknown recipient, NA, SPD 14/128/105 (n.d. [c. 1619–20]). The letter is addressed to "your Wor." This document is a copy and contains marginal annotations in a later hand that tentatively date it to "abt. March 1622." By that time, however, the king had long since announced his intention to step in himself as sole undertaker (7 July 1621). The buffoonish projectors in question might be Cornelius Liens and Cornelius Vermuyden, with whom the king had subcontracted in February 1622 (see below); but the letter also contains a marginal reference to Ayloffe, and it is unlikely in any case that Liens and Vermuyden, who had experience in drainage projects, would have been fooled by an offer of coastal lands flooded at low tide. The date should therefore be rather earlier, c. 1619–20. See also Kennedy, "So Glorious a Work," 254–56.

48. Minutes of a session of sewers held at Cambridge, CUA, T.XII.3, sec. 1, pp. 45–52 (13–14 October 1619).

49. Minutes of a session of sewers held at Cambridge, CUA, T.XII.3, sec. 1, pp. 46–48 (13–14 October 1619).

50. Minutes of a session of sewers held at Cambridge, CUA, T.XII.3, sec. 1, p. 46 (13–14 October 1619).

51. Rough notes of a session of sewers held at Cambridge, CUA, CUR 3.3/94 (13–14 October 1619).

52. Minutes of a session of sewers held at Cambridge, CUA, T.XII.3, sec. 1, p. 49 (13–14 October 1619).

53. Rough notes of a session of sewers held at Cambridge, CUA, CUR 3.3/94 (13–14 October 1619). There may well be other discrepancies between this account and the official session minutes, but the writing is very difficult to make out in many places. Moreover, several towns and villages included in the official minutes are not mentioned in the alternative version.

54. Printed decree of the session of sewers held at St. Ives, NA, SPD 14/110/135 (15–16 October 1619). The £10 fine with which the sheriffs were threatened if they did not comply was spelled out in the printed notice.

55. Privy Council to Commission of Sewers, CUA, T.XII.2, fol. 25 (5 December 1619); APC, 37:84–85 (5 December 1619).

56. Draft of a letter from the commission of sewers at Cambridge to the Privy Council, CUA, CUR 3.3/85 (n.d. [c. 10 December 1619]); minutes of a session of sewers held at Cambridge, CUA, T.XII.3, sec. 1, pp. 53–54 (9–10 December 1619); commission of sewers for Huntingdon, Northampton, Lincoln, and Norfolk to the Privy Council, NA, SPD 14/111/85 (21 December 1619).

57. Anonymous letter, probably addressed to the commissioners of sewers from the Isle of Ely; BL, Harleian MS 5011, vol. 1, fol. 23 (n.d. [1620]).

58. Mr. Secretary George Calvert to the Privy Council, NA, SPD 14/112/61 (4 February 1620); see also APC, 37: 143–44 (29 February 1620).

59. Memorandum of the Privy Council's consideration of the drainage project, BL, Lansdowne MS 162/19 (11 April 1620); Edmonds's final report, NA, SPD 14/99/52 (20 September 1618).

60. "The Proposition of the Undertakers for Drayning of the Ffennes, of That W^{ch} Formerlye They Intended, & by Their First Overture to the Kings Matie Still Doe Purpose, & Resolve by Gods Helpe, & His Maties Gracious Assistaunce to Doe, & Performe," CUA, CUR 3.3/77 (n.d. [April 1620]). See also the memorandum of the Privy Council's consideration of the drainage project, BL, Lansdowne MS 162/19 (11 April 1620); order from the Privy Council to the undertakers and commissioners of sewers, CUA, CUR 3.3/86 (11 April [1620]); APC, 37:179 (11 April 1620).

61. Petition from the commission of sewers for Cambridge and Ely to the king, NA, SPD 14/112/84 (13 February 1620). A copy of the same petition may be found at CUA, T.XII.1, sec. 7, no. 3.

62. Memorandum of the Privy Council's consideration of the drainage project, BL, Lansdowne MS 162/19 (11 April 1620).

63. Commission of sewers' answer to the propositions of the undertakers, CUA, T.XII.2, fol. 21 (28 April 1620). See also NA, SPD 16/152/83, another copy of the commissioners' answers that has been misdated.

64. Warrant from the commissioners of sewers to the clerk of sewers, NA, SPD 14/113/78 (13 April 1620).

65. Privy Council to the undertakers, CUA, T.XII.2, fol. 5 (12 July 1620); Privy Council to the commission of sewers, CUA, CUR 3.3/64 (20 July 1620); APC, 37:250–52 (12 July 1620) and 257 (20 July 1620).

66. James I to the commission of sewers, CUA, CUR 3.3/64 (25 July 1620).

67. Privy Council to the commission of sewers, CUA, CUR 3.3/64 (20 July 1620); APC, 37:257 (20 July 1620).

68. Minutes of a session of sewers at Cambridge, CUA, T.XII.3, sec. 1, pp. 70–73 (17 August 1620); commission of sewers for Cambridge and Ely to various parish constables, CUA, CUR 3.3/67 (1 September 1620), my emphasis.

69. Series of land surveys in various parishes of Cambridgeshire, CUA, T.XII.2, fols. 49–76 (15–21 September 1620). A similar survey was ordered by the commissioners of sewers for the upland counties; a record of the order survives, but it does not make

explicit mention of the eight shillings per acre threshold, and there is no record of the resulting valuations (order of a session of sewers at Stilton, BL, Add. MS 33466, 75r–v, 5 September 1620).

70. Land surveys from Tydd St. Giles and Leverington, CUA, T.XII.2, fols. 74 and 64–67, respectively.

71. Land survey from Sutton, CUA, T.XII.2, fol. 72.

72. Land survey from Haddenham, CUA, T.XII.2, fol. 75.

73. Copy of the contract offered by the commission of sewers at Ely to the undertakers, CUA, T.XII.2, fols. 13–14 (28 September 1620).

74. Response of the undertakers to the Privy Council, concerning the commission of sewers' proposed contract, NA, SPD 14/117/78 (17 November 1620).

75. Petition of the undertakers to the Privy Council, NA, SPD 14/117/78.I (17 November 1620).

76. *History of Parliament*, s.v. "Sir William Ayloffe."

77. Vermuyden, *Discourse*, 1.

78. Session of sewers for the Great Level counties, BL, Add. MS 34217, fol. 8r (1 August 1621).

79. Privy Council to the commission of sewers, BL, Add. MS 33466, fol. 125 (7 July 1621); *APC*, 38:5–6 (7 July 1621).

80. Jury presentment from Norths.; BL, Add. MS 33466, fols. 140r–41r (19 September 1621).

81. Series of land valuations from throughout the Great Level, BL, Add. MS 33466, fols. 172–95 and 198–200. The certificates in this collection range in date from 5 October 1621 through 20 February 1622. They were submitted parish by parish from across the Great Level, though the quotations given here are representative of the group and could easily be multiplied.

82. James I to the commission of sewers, BL, Add. MS 33466, fol. 166r (3 October 1621).

83. James I to the commission of sewers, CUA, T.XII2, fol. 10 (24 January 1622); James I to the commission of sewers, BL, Add. MS 33466, fol. 196r (19 February 1622).

84. Copy of an agreement between the Privy Council, Cornelius Liens, and Cornelius Vermuyden, NA, SPD 14/127/146 (February 1622). Of this portion, fifty thousand acres were to go to the Dutchmen and their partners, to be held in free soccage, with the income from the remaining twenty thousand to be allotted toward the maintenance of the drainage works in perpetuity. Vermuyden also received permission to bring 204 experienced workers to England from the Low Countries, presumably to be employed in the work. *APC*, 38:393 (14 January 1623).

85. Vermuyden, *Discourse*, 1.

86. Commission of sewers order to the clerk of sewers, BL, Add. MS 33466, fols. 201r–202v (13 May 1623).

87. Privy Council to the commission of sewers, BL, Add. MS 33466, fols. 203r–204v (28 May 1623); *APC*, 38:510–11 (28 May 1623).

88. Dugdale, *History of Imbanking*, 408.

89. Edmonds's final report, NA, SPD 14/99/52 (20 September 1618).

90. Anonymous letter, probably addressed to commissioners of sewers from the Isle of Ely, BL, Harleian MS 5011, vol. 1, fol. 23 (n.d. [1620]).

91. Anonymous letter to an unknown recipient, NA, SPD 14/128/105 (n.d. [c. 1619–20]).

92. Anonymous letter to an unknown recipient, NA, SPD 14/128/105 (n.d. [c. 1619–20]).

CHAPTER 5: Draining the Hatfield Level, 1625–1636

1. Articles of agreement between Charles I and Cornelius Vermuyden, BL, Lansdowne MS 205/24 (24 May 1626).

2. Information of John Kitchen, NA, SPD 16/113/38.I (17 August 1628).

3. Information of Baldwyn Verwarmonte, William Thompson, Richard Hayford, and Aquilla Broadly, NA, SPD 16/113/38.I (17 August 1628).

4. On women's participation in grain and enclosure riots in early modern England, see Walter, "Grain Riots," 62–63; Tim Harris, "Introduction," 17–20; Capp, "Separate Domains?," 137–39; Lindley, Fenland Riots, 63; Manning, Village Revolts, 96–98; Underdown, Freeborn People, 60–62.

5. Information of William Thompson, NA, SPD 16/113/38.I (17 August 1628). On the social composition of enclosure rioters, as well as the challenges of interpreting evidence taken from depositions, see Holmes, "Drainers and Fenmen"; Wood, Riot, chaps. 3–4; Underdown, Freeborn People, chap. 3; Walter, "Grain Riots"; Sharp, In Contempt, passim; Lindley, Fenland Riots, 59–60, 253–59; Manning, Village Revolts, chap. 12.

6. Information of John Milbourne, gent., and Benjamin Brooke, NA, SPD 16/113/38.I (17 August 1628).

7. Information of Richard Hayford, John Milbourne, gent., Aquilla Broadley, and John Liversedge, NA, SPD 16/113/38.I (17 August 1628).

8. Examination of John Warren, NA, SPD 16/113/38.III (19 August 1628). "Mr. Laynes" may refer either to Vermuyden's partner Cornelius Liens, his brother Joachim Liens, or (most likely) Joachim's son Johan (John) Liens. All of them were involved in various English drainage ventures at this time, but Johan is known to have been responsible for building drainage works in the Isle of Axholme; see BDCE, s.v. "Johan or John Liens."

9. Examination of Baldwyn Vanwarmon (almost certainly the same as Baldwyn Verwarmonte, who gave information three days earlier), NA, SPD 16/113/38.III (20 August 1628).

10. Information of Richard Stockwell, NA, SPD 16/113/38.II (18 August 1628).

11. Examination of John Warren, NA, SPD 16/113/38.III (19 August 1628).

12. Sir Ralph Hansby to the Duke of Buckingham, NA, SPD 16/113/38 (21 August 1628).

13. APC, 44:114–15 (27 August 1628), my emphasis.

14. Although Vermuyden was certainly involved in the Great Level drainage during the 1650s, his connection to the first phase of that project during the 1630s, long assumed by historians, has recently been brought into doubt: see Knittl, "Design"; and chapters 6 and 8 in the present volume.

15. See Slack, Invention of Improvement, 53–62; Kennedy, "Charles I."

16. Sharp, In Contempt, 7–9; Manning, Village Revolts, 2–3.

17. Lindley, Fenland Riots, 57–65.

18. Underdown, *Freeborn People*, chaps. 1–3; Holmes, "Drainers and Fenmen"; Wrightson, "Politics of the Parish"; Hindle, *State and Social Change*; Wood, *Politics*; Falvey, "Custom, Resistance, and Politics"; Fox, "Custom."

19. Harris, "Introduction," 8.

20. Wood, *Politics*, chaps. 6–7; see also Harris, "Introduction"; Underdown *Revel* and *Freeborn People*.

21. Wood, *Riot*, chap. 3; Lindley, *Fenland Riots*, 58.

22. There are several rivers called "Ouse" in Britain. The River Ouse in Yorkshire is not the same as the much longer one in the Great Level, which is referred to in this book and elsewhere as the "Great Ouse" to distinguish it.

23. Dugdale, *History of Imbanking*, 144. See also L. E. Harris, *Vermuyden*, chap. 4; Korthals-Altes, *Sir Cornelius Vermuyden*, chap. 3; and Lindley, *Fenland Riots*, 23–24.

24. Thirsk, "Isle of Axholme"; idem, *English Peasant Farming*; Lindley, *Fenland Riots*, 17–18; Korthals-Altes, *Sir Cornelius Vermuyden*, 25–27.

25. Lindley, *Fenland Riots*, 23.

26. Thomas Suffolk and Fulke Greville to Thomas Lake, NA, SPD 14/92/58 (2 June 1617).

27. BL, Lansdowne MS 897, fol. 190v (n.d.). De la Pryme also claimed that a man called Laverock made a proposal to Queen Elizabeth to drain the lands, but that this plan came to nothing (fol. 190r). For another eighteenth-century antiquarian's description of the level and account of the drainage, see the bound manuscript of George Stovin among the University of Nottingham Manuscripts and Special Collections, Hatfield Chase Corporation Records, HCC 9111 (n.d.). Stovin was a landowner in the Hatfield Level, and his account is biased overall in favor of Vermuyden and the Participants, though a portion of it is clearly taken from anon., *State of That Part of Yorkshire*, which sympathizes with the commoners.

28. Thomas Jenkins, "The State of His Majesty's Forest or Chase of Hatfield in the County of York," NA, SPD 14/180/82 (n.d. [1625?]). Though the document is undated, internal evidence places the date at some point after December 1624. Jenkins was the man responsible for producing the survey and plat of the level.

29. Copy of an agreement between the Privy Council, Cornelius Liens, and Cornelius Vermuyden, NA, SPD 14/127/146 (February 1622).

30. There is no really adequate biography of Cornelius Vermuyden, and it may be doubted whether enough archival information still exists to write one. The best work on Vermuyden's early life is still L. E. Harris, *Vermuyden*, esp. chaps. 2–3, as well as the entry in *BDCE*, s.v. "Sir Cornelius Vermuyden." Except where otherwise indicated, this biographical section is based mostly on those two sources. Another, more detailed account of Vermuyden's Dutch origins and activities in the Hatfield Level may be found in Korthals-Altes, *Sir Cornelius Vermuyden*, though this work must be used with caution, as some of its details are unreliable. In addition, one should also consult Smiles, *Lives of the Engineers*, vol. 1, chap. 2; and *DNB*, s.v. "Sir Cornelius Vermuyden." There is also no known portrait of Vermuyden; a portrait in the Valence House Museum in Dagenham, Essex, once thought to be of Vermuyden, has been reidentified with some confidence as a portrait of Sir Philibert Vernatti, one of Vermuyden's business partners in the Hatfield Level drainage (Valence House Museum, personal communication with author, February 2016).

31. On Vermuyden's ancestors, see esp. Korthals-Altes, *Sir Cornelius Vermuyden*, 59.

32. Copy of an agreement between the Privy Council, Cornelius Liens, and Cornelius Vermuyden, NA, SPD 14/127/146 (February 1622). There is a story, almost certainly apocryphal, that places Vermuyden in England in 1609 in the exalted company of James's son Prince Henry as he enjoyed a day's hunting at Hatfield Chase. However, there is absolutely no primary source evidence to support this tale. L. E. Harris has traced it to the general history of Samuel Wells, but Wells cited no source for his claim (Wells, *History* 1:92–93). The story of the prince's hunting party may be traced to the antiquarian Abraham de la Pryme, though the latter made no mention in his manuscript history of Vermuyden's participation in the hunt, and indeed later historians have cast doubt on his claim that the prince ever visited Hatfield Chase (BL, Lansdowne MS 897, fol. 39 [n.d.]). See L. E. Harris, *Vermuyden*, 35–36; and Hunter, *South Yorkshire* 1:156n4.

33. Draft of a Privy Council resolution to the Essex commission of sewers, NA, SPD 14/134/93 (December 1622). See also *APC*, 38:377 (26 December 1622); petition to the Privy Council from sixty laborers, NA, SPD 14/135/33 (n.d. [1622]).

34. Essex commission of sewers to the Privy Council, NA, SPD 14/138/2 (1 February 1623).

35. Essex commission of sewers to the Privy Council, NA, SPD 14/147/13 (20 June 1623).

36. Thomas Jenkins, "The State of His Majesty's Forest or Chase of Hatfield in the County of York," NA, SPD 14/180/82 (n.d. [1625?]).

37. Letters patent to Cornelius Vermuyden, NA, C 225/2/41B (1 August 1625).

38. Articles of agreement between King Charles I and Cornelius Vermuyden, BL, Lansdowne MS 205/24 (24 May 1626).

39. Dugdale, *History of Imbanking*, 144.

40. BL, Lansdowne MS 897, fols. 190–92 (n.d.).

41. Dugdale, *History of Imbanking*, 144.

42. Robert Heath, "Remembrances for the King's Service at My Going to Court," NA, SPD 16/44/1 (n.d. [1626]). Heath was also a business partner of Vermuyden's in a lead mine venture in Derbyshire. See L. E. Harris, *Vermuyden*, 35; Wood, *Politics*, 225–31; *DNB*, s.v. "Sir Robert Heath."

43. Articles of agreement between King Charles I and Cornelius Vermuyden, BL, Lansdowne MS 205/24 (24 May 1626).

44. Articles of agreement between King Charles I and Cornelius Vermuyden, BL, Lansdowne MS 205/24 (24 May 1626).

45. *APC*, 40:486 (24 May 1626).

46. L. E. Harris, *Vermuyden*, 43–44; Stonehouse, *History and Topography*, 74–76.

47. Anon., *State of That Part of Yorkshire*, 4. The reference is biblical and invokes the fabled and exotic wealth of King Solomon: "For the king had at sea a navy of Tarshish with the navy of Hiram: once in three years came the navy of Tarshish, bringing gold, and silver, ivory, and apes, and peacocks" (1 Kings 10:22). The commentator is alluding not only to the large sums of money that Vermuyden was able to secure in the Low Countries, but also to the impressive array of supplies and laborers he brought back with him.

48. Dugdale, *History of Imbanking*, 145.

49. Hunter, *South Yorkshire* 1:165–66; Dugdale, *History of Imbanking*, 145.

50. L. E. Harris, *Vermuyden*, 43–44.

51. Korthals-Altes, *Sir Cornelius Vermuyden*, 30–37; Stonehouse, *History and Topography*, 78–80.

52. BL, Lansdowne MS 897, fol. 191v (n.d.); Dugdale, *History of Imbanking*, 145; L. E. Harris, *Vermuyden*, 49; Korthals-Altes, *Sir Cornelius Vermuyden*, 37; Hunter, *South Yorkshire* 1:161.

53. Draft of a petition from Crowle to the House of Commons, BL, Egerton MS 3518, fol. 87r (n.d.); Lindley, *Fenland Riots*, 24–27.

54. Hughes, "Drainage Disputes," 15–17; Stonehouse, *History and Topography*, 79; Korthals-Altes, *Sir Cornelius Vermuyden*, 36.

55. There is disagreement in the secondary sources regarding the date of the Mowbray indenture. Peck claims it is 1 May in the thirty-third year of Edward III, or 1359 (*Topographical*, vol. 1, app. 2, p. iii); Hughes says 1 May 1305 ("Drainage Disputes," 34); Lindley says 31 May 1359 (*Fenland Riots*, 26). I am using 1 May 1359, because that is the date given (1 May 33 Edw. 3) in both of Daniel Noddel's pamphlets, which include full transcriptions of the indenture, translated from Latin (see chapter 7, in the present volume).

56. "A True Copy of the Ancient Deed of John de Mowbray . . . ," in Peck, *Topographical*, vol. 1, app. 1, p. i.; see also Noddel, *Declaration*; Fox, "Custom"; idem, *Oral*, chap. 5; Lindley, *Fenland Riots*, 23–33.

57. Richard Bridges to Thomas Wentworth (6 September 1630), quoted in Hunter, *South Yorkshire*, 1:164n1.

58. APC, 46:29 (25 June 1630); see also the Privy Council decree dated the same day, NA, SPD 16/169/38.

59. Anon., *State of That Part of Yorkshire*, 4.

60. APC, 42:325–26 (7 June 1627).

61. APC, 42:326–27 (7 June 1627); see also Lindley, *Fenland Riots*, 71–72.

62. APC, 44:285–86 (29 December 1628), and 301 (19 January 1629).

63. APC, 44:393–94 (10 April 1629).

64. Charles to Richard Weston, NA, SPD 16/147/21 (21 July 1629); L. E. Harris, *Vermuyden*, 51; Korthals-Altes, *Sir Cornelius Vermuyden*, 41; Hughes, "Drainage Disputes," 17. The Crown originally mortgaged the estates to Vermuyden for £10,000 plus rent in order to raise cash quickly, but when it could not repay the principal on the loan, Vermuyden was permitted to purchase them outright for an additional £6,800, minus the interest paid. See Hoyle, "Disafforestation," 382; and Hunter, *South Yorkshire*, 1:161–62.

65. APC, 45:115–16 (31 July 1629) and 378–79 (12 May 1630).

66. Anon., *State of That Part of Yorkshire*, 7–11.

67. APC, 46:25 (23 June 1630).

68. Anon., *State of That Part of Yorkshire*, 12–13. See also Hunter, *South Yorkshire*, 1:163.

69. Anon., *State of That Part of Yorkshire*, 13–14.

70. APC, 46: 340–41 (20 May 1631) and 348–49 (25 May 1631).

71. Anon., *State of That Part of Yorkshire*, 14–15. The orders were never decreed before the Council of the North, despite the Privy Council's repeated insistence that

they should be. There is a copy of an Exchequer decree in the Hatfield Chase Corporation papers from November 1630 that addresses this dispute and apparently settled the remaining points of contention at that time. However, it does not include some of the most important rulings of Wentworth's commission, such as Vermuyden's responsibility for building a new bank and his obligation to keep the land dry in perpetuity. Moreover, the commoners' ongoing complaints through 1631 indicate that Vermuyden either failed or refused to live up to the agreement they thought they had reached in the previous year. See University of Nottingham Manuscripts and Special Collections, Hatfield Chase Corporation Records, HCC 8215 (30 November 1630).

72. Petition of Sir Philibert Vernatti, Matthew Valkenburgh, Luke Valkenburgh, and John Corselis, to the king, NA, SPD 16/279/93 (n.d.). See also an order of the Privy Council regarding Low Countries Participants who were also in arrears, NA, SPD 16/243/27 (24 July 1633).

73. Petition from the inhabitants of Fishlake and Sykehouse to the Privy Council, NA, SPD 16/270/59 (n.d. [1634]); petition from the inhabitants of Drax, Newland, Great Husholme, Little Husholme, West Armyn, Lanehouses, Lanevack, Woodhouses, Sharphill, Brockholes, and Camblesford to the justices of the peace at Doncaster, NA, SPD 16/299/58 (13 October 1635); petition of Ruben Eastropp to Edward Osborne, NA, SPD 16/300/40 (26 October 1635).

74. Anon., *State of That Part of Yorkshire*, 16–20; L. E. Harris, *Vermuyden*, 54–58.

75. Anon., *State of That Part of Yorkshire*, 22; L. E. Harris, *Vermuyden*, 52–53; Hughes, "Drainage Disputes," 15. Korthals-Altes writes that the additional work cost £33,000 (*Sir Cornelius Vermuyden*, 109).

76. APC, 46:29 (25 June 1630); see also the Privy Council decree dated the same day, NA, SPD 16/169/38.

77. The Earl of Clare and Sir Gervase Clifton to the Privy Council, NA, SPD 16/246/65 (24 September 1633).

78. Petition of Gringley, Everton, Scaftworth, Scrooby, Mattersey, Wiston, Sutton, and Lownd to the king, NA, SPD 16/250/55 (13 November 1633); and the same to the Privy Council, NA, SPD 16/257/26 (n.d. [1633]).

79. Petition from Misterton to the Privy Council, NA, SPD 16/279/102 (n.d. [late 1633]).

80. William Noye to the Privy Council, NA, SPD 16/252/17 (4 December 1633).

81. Commissioners to the king, NA, SPD 16/268/95 (May 1634).

82. "Reasons Why No Commission of Sewers Should Get Issue for Ordering of the Late Surrounded Grounds in the Counties of Lincoln and Nott.," NA, SPD 16/279/93.1 (n.d. [1634]).

83. Commissioners to the king, NA, SPD 16/268/95 (May 1634); petition of Philibert Vernatti, Matthew Valkenburgh, and John Corselis to the king, NA, SPD 16/279/94 (n.d. [1634]); notes from a session of sewers, NA, SPD 16/279/95 (n.d. [1634]); Edward Osborne et al. to the king, NA, SPD 16/296/21 (24 August 1635).

84. For three such decrees of sewers, see NA, C 225/1/24 (8–15 July 1635), C 225/1/23 (October 1635), and C 225/1/19 (10 November 1635).

85. Noddel, *Declaration*, 14.

86. Lilburne, *Case of the Tenants*, 2.

87. Lindley, *Fenland Riots*, 23–27.

88. Articles of agreement between King Charles I and Cornelius Vermuyden, BL, Lansdowne MS 205/24 (24 May 1626); Dugdale, *History of Imbanking*, 144.

89. Information of Richard Hayford, John Milbourne, gent., and John Liversedge, NA, SPD 16/113/38.I (17 August 1628).

90. APC, 44:114–15 (27 August 1628).

91. Francis Thornhill to Cornelius Vermuyden, NA, SPD 16/117/67 (25 September 1628).

92. APC, 44:160–61 (26 September 1628), 164 (26 September 1628), 166 (28 September 1628), and 171–72 (29 September 1628).

93. Philibert Vernatti to M. St. Gillis, NA, SPD 16/119/73 (n.d. [1628]); Lindley, *Fenland Riots*, 75–76.

94. APC, 45:73 (7 July 1629), 81–82 (12 July 1629), and 129 (27 August 1629); Lindley, *Fenland Riots*, 77.

95. APC, 44:192–93 (10 October 1628).

96. Charles I to the barons of the court of Exchequer, BL, Egerton MS 2553, fol. 39 (n.d. [1629?]); 1636 Exchequer decree, printed in Hughes, "Drainage Disputes," 36–37. The chief judges in the Court of Exchequer were known as "barons," whereas those in the other central common law courts were called "justices."

97. Special Exchequer commission of inquiry, *Heath v. Popplewell and Torksey*, 15 April 1629, NA, E 178/5412 (27 April 1630).

98. 1636 Exchequer decree, printed in Hughes, "Drainage Disputes," 37–38.

99. Special Exchequer commission of inquiry, *Heath v. Popplewell and Torksey*, NA, E 178/5412 (27 April 1630).

100. 1636 Exchequer decree, printed in Hughes, "Drainage Disputes," 38–39; Charles I to the Exchequer special commission, BL, Egerton MS 2553, fols. 43v–44r (5 August 1631); Special Exchequer commission of inquiry, NA, E 178/5430 (3 September 1631).

101. Bill of complaint and answer, NA, E 112/198/104 (February 1632). See also the 1636 Exchequer decree, printed in Hughes, "Drainage Disputes," 39–41.

102. Anon., *Mannor of Epworth*; Lindley, *Fenland Riots*, 28–29. The English mark was equivalent to thirteen shillings and four pence, or two-thirds of a pound sterling; it did not exist as a coin but was a unit of account.

103. 1636 Exchequer decree, printed in Hughes, "Drainage Disputes," 42–44; Anon., *Case of the Mannor of Epworth*; Noddel, *Great Complaint*, 3; Lilburne, *Case of the Tenants*, 3.

104. 1636 Exchequer decree, printed in Hughes, "Drainage Disputes," 42–43; Lindley, *Fenland Riots*, 30; Hughes, "Drainage Disputes," 19–20.

105. Anon., *Case of the Mannor of Epworth*; 1636 Exchequer decree, printed in Hughes, "Drainage Disputes," 42–43; Lindley, *Fenland Riots*, 28–29; Hughes, "Drainage Disputes," 18–20.

106. Lilburne, *Case of the Tenants*, 2; Noddel, *Declaration*, 2; Lindley, *Fenland Riots*, 31–32; "Replevin," *Law Dictionary*, accessed 18 August 2016, http://www .lawyerintl.com/law-dictionary/9033-replevin.

107. Lilburne, *Case of the Tenants*, 2.

108. Notes by Francis Windebank, NA, SPD 16/242/62 (12 July 1633); Noddel, *Declaration*, 2.

109. Noddel, *Declaration*, 2; Lindley, *Fenland Riots*, 78–79.

110. Lilburne, *Case of the Tenants*, 3; 1636 Exchequer decree, printed in Hughes, "Drainage Disputes," 44–45.

111. Anon., *Case of the Mannor of Epworth*; Lilburne, *Case of the Tenants*, 3–4; Lindley *Fenland Riots*, 30–31.

112. Dugdale, *History of Imbanking*, 145; Lindley, *Fenland Riots*, 31–32; Hughes, "Drainage Disputes," 21–22.

113. Dugdale, *History of Imbanking*, 144–45.

114. Noddel, *Declaration*, 13.

115. Lilburne, *Case of the Tenants*, 4–5.

116. For a similar case highlighting Charles's style of governance, in the area of prerogative taxation, see Cust, *Forced Loan*.

117. APC, 44:192–93 (10 October 1628). There are extensive surviving proceedings of a commission of sewers for at least the West Riding (Hatfield Chase) portion of the Hatfield Level from the middle of 1635 through October 1640, though by that time most of the construction work on the drainage project was already completed. That commission included several shareholders (Dutch and English) and tenants of the Hatfield Chase Corporation. The record of their proceedings is composed of mostly routine business, including the maintenance and repair of drainage works, the collection of taxes, and hearing dozens of complaints from inhabitants in the West Riding that their grounds had been cut through, to their great loss. The commission of sewers thus resumed its customary role in managing drainage affairs in the level only when the drainage project was all but finished and succeeded in incorporating the projectors in the process of local governance. Moreover, insofar as the commission challenged any aspect of the drainage, it only served to reinforce the Crown's authority by requiring the projectors to secure the king's subjects from unprecedented flooding caused by their new drainage works. See the University of Nottingham Manuscripts and Special Collections, Hatfield Chase Corporation Records, HCC 6001, vol. 1 (1635–52).

118. See Kishlansky, *Monarchy Transformed*, chaps. 4–5.

CHAPTER 6: The First Great Level Drainage, 1630–1642

1. Anonymous letter to an unknown recipient, NA, SPD 14/128/105 (n.d. [c. 1619–20]); see also chapter 4 in the present volume.

2. H. C., *Discovrse*, fol. A2r.

3. H. C., *Discovrse*, fols. B3r–[B4r].

4. H. C., *Discovrse*, fols. A3v, C3r.

5. H. C., *Discovrse*, fols. A3v, [A4v], Br–B2r.

6. Charles I to the commissions of sewers in the Great Level counties, NA, SPD 16/144/84 (16 June 1629).

7. Commission of sewers for Lincolnshire to the king, NA, SPD 16/148/96 (29 August 1629).

8. Charles I to the commission of sewers for Lincolnshire, NA, SPD 16/153/30 (8 December 1629).

9. Results of a Lincolnshire session of sewers held at Boston, NA, SPD 16/158/56 (11–15 January 1630).

10. Charles I to the Lincolnshire commission of sewers, NA, SPD 16/161/34 (20 February 1630).

11. Lincolnshire decree of sewers, NA, C 225/2/44 (15 May 1630); see also Kennedy, "Charles I," and Lindley, *Fenland Riots*, 48.

12. Commission of sewers for the Great Level to the king, NA, SPD 16/150/2 (1 October 1629).

13. Charles I to the Lincolnshire commission of sewers, NA, SPD 16/161/34 (20 February 1630).

14. *DNB*, s.v. "Francis Russell, fourth earl of Bedford."

15. *DNB*, s.v. "William Russell, first Baron Russell of Thornhaugh." Russell would later serve as lord deputy of Ireland in the 1590s.

16. William Russell to the lord chancellor and lord treasurer, BL, Lansdowne MS 63/15 (21 May 1590); petition from William Russell to the Privy Council, BL, Lansdowne MS 110/4 (n.d. [early 1590s]); William Russell to the Privy Council, BL, Lansdowne MS 110/6 (n.d. [early 1590s]); L. E. Harris, *Vermuyden*, 19–21.

17. Summers, *Great Level*, 57–58.

18. *DNB*, s.v. "Francis Russell, fourth earl of Bedford."

19. Petition from the commissioners of sewers for the Great Level to the Privy Council, NA, SPD 14/97/111 (19 June 1618); Clement Edmonds's report to the Privy Council, NA, SPD 14/99/52 (20 September 1618).

20. Francis Bedford to Henry Vane, NA, SPD 16/171/30 (25 July 1630).

21. Robert Heath to Secretary of State Dorchester, NA, SPD 16/171/41 (27 July 1630); Robert Heath to Secretary of State Dorchester, NA, SPD 16/172/13 (3 August 1630).

22. Minutes of a Great Level session of sewers held at King's Lynn, Norfolk Record Office, Hare MS 5136, 219x3 (2 September 1630); John Carleton to Secretary of State Dorchester, NA, SPD 16/173/29 (14 September 1630). See also Knittl, "Design."

23. Wells, *History*, 2:101; see also the original manuscript copy of the law, NA, C 225/1/40 (13 January 1631); and another copy at NA, SPD 231/27, which has been misdated.

24. Dugdale, *History of Imbanking*, 408, original emphasis.

25. Thomas Blechynden to Henry Vane, NA, SPD 16/175/8 (2–3 November 1630). Knittl suggests that "College of Drainers" is an English rendering of a Dutch term for the principal investors in a drainage project ("Design," 29n22).

26. Knittl, "Design," 28–30.

27. Thomas Blechynden to Henry Vane, NA, SPD 16/175/8 (2–3 November 1630).

28. The drainage project authorized by the Lynn Law encompassed only the Great Level; Deeping Fen was not included in it.

29. Wells, *History*, 2:98–110.

30. Wells, *History*, 2:106–7.

31. Privy Council order, NA, SPD 16/197/1 (20 July 1631).

32. Wells, *History*, 2:111–20.

33. The original text of the indenture lists the owners of only nineteen shares. Knittl speculates that the remaining share may have been reserved for Vermuyden, on the assumption that he would soon be named director of the works; in any case, the final share eventually came into Vernatti's possession ("Design," 32).

34. Although he was the only foreigner listed among the original fourteen shareholders, Vernatti subdivided his two shares among several other Dutch investors; he later petitioned the king that his fellow Dutchmen should be made denizens of England in order that they might legally purchase and possess land in the realm. Vernatti's petition to the king, NA, SPD 16/257/23 (n.d.).

35. Contract between the Great Level undertakers and the king, NA, SPD 16/204/39 (13 December 1631).

36. "A Relation of a Short Survey of 26 Counties, Briefly Describing the Citties and Their Scytuations, and the Corporate Townes, and Castles Therein. Observ'd in a Seven Weekes Journey Begun at the City of Norwich and from Thence into the North. On Monday August 11th 1634 and Ending att the same Place. By a Captaine, a Lieutenn᷄ᵗ, and an Ancient. All Three of the Military Company in Norwich," BL, Lansdowne MS 213, fols. 315–16 (1634).

37. "A Relation of a Short Survey of the Westerne Counties, in Which Is Briefely Described the Citties, Corporations, Castles, and Some Other Remarkables in Them. Observ'd in a Seven Weekes Journey Begun at Norwich, & Thence into the West. On Thursday August 4th 1635 and ending att the Same Place. By the Same Lieutennant, That with the Captaine, and Ancient of the Military Company in Norwich Made a Journey into the North the Yeere Before," BL, Lansdowne MS 213, fol. 382v (1635).

38. Dugdale, *History of Imbanking*, 410; Darby, *Changing Fenland*, 65–67. See also *BDCE*, s.v. "Andrewes Burrell."

39. The earliest account that does assert Vermuyden's involvement with the 1630s project is Wells, *History*, 1:120, published in 1830.

40. Knittl, "Design," 33–34.

41. Knittl, "Design," 38–46; *BDCE*, s.v. "Andrewes Burrell," "John Hunt."

42. Knittl, "Design," 37, 41.

43. Wells, *History*, 2:102.

44. H. C., *Discovrse*, fols. C1r–C2r.

45. Anonymous letter to an unknown recipient; NA, SPD 14/128/105 (n.d. [c. 1619–20]).

46. "Charter of Incorporation," printed in Wells, *History*, 2:120–40.

47. Decree of sewers from a session at Peterborough, CRO, R 59/31/9/1A, fol. 179r (13 June 1636).

48. Two MS copies of Hayward's survey can be seen at CRO, R/59/31/4, vols. 2–3 (14 June 1636); the document has also been printed in Wells, *History*, 2:141–235.

49. Charles I to the commissioners for the Great Level, Bodleian Library, Oxford University, Bankes MS 46/1 (24 June 1636); petition from the inhabitants of Wretton to the Earl of Bedford, NA, SPD 16/362/23 (21 June 1637); "Directions to Be Observed by yᵉ Survayers with in the Great Levill," CUA, Add. MS 40/18 (12 March 1637).

50. "A Relation of a Short Survey of the Westerne Counties . . . ," BL, Lansdowne MS 213, fols. 381–82 (1635).

51. Memorandum from Henrick van Cranhals to the Privy Council, BL, Harleian MS 6838, fol. 181r (10 September 1636).

52. Warrant to arrest several rioters in Upwell and Lynn, NA, SPD 16/351/80 (n.d. [March 1637]). See also William Sames to Edward Nicholas, clerk of the Privy Council, NA, SPD 16/357/74 (25 May 1637); Privy Council to the magistrates of

Norfolk, NA, SPD 16/357/152 (31 May 1637); bond of good behavior for nine Norfolk rioters, NA, SPD 16/362/10 (20 June 1637).

53. Order of the Privy Council and draft letter to the lord lieutenant, NA, SPD 16/355/173–74 (10 May 1637).

54. Statements to the Privy Council, NA, SPD 16/230/50–51 (n.d. [June 1637]). Keith Lindley correctly points out that these documents are misdated in the Calendar of State Papers (*Fenland Riots*, 94–95).

55. Warrant to arrest several rioters in Upwell and Lynn, NA, SPD 16/351/80 (n.d. [March 1637]).

56. Petition from the inhabitants of Wretton to the Earl of Bedford, NA, SPD 16/362/23 (21 June 1637).

57. "The Greivances of yᵉ Towne of Chatterie in yᵉ Ysle of Ely," CUA, CUR 3.3/88 (n.d. [1637?]); draft of a petition from Thomas Wendy to the commission of sewers at Huntingdon, CUA, CUR 3.3/81 (n.d. [1637?]).

58. Petition from the bishop & deans of Ely and the inhabitants and commoners of parts of Suffolk, Norfolk, Huntingdon, and the Isle of Ely to the king, CUA, Add. MS 22/87 (25 June 1637); record of a Privy Council session, CUA, Add. MS 22/88 (9 July 1637).

59. Record of a Privy Council session, CUA, Add. MS 22/88 (9 July 1637).

60. Petition of 250 laborers to the Privy Council, NA, SPD 16/533/122 (n.d. [1637]).

61. Special commission of inquiry, NA, E 163/24/33 (19 February 1639). It should be noted that this inquiry was entirely one-sided and was intended to prove that Bedford was negligent in not completing the works he had agreed to build.

62. St. Ives Law, NA C 225/2/54 (12 October 1637). This is the original copy of the law sent to London, and it is highly unusual in that it does not bear the king's manuscript endorsement. The law is also printed in Wells, *History*, 2:236–339.

63. Petition from the Earl of Bedford to the king, NA, SPD 16/323 (30 November 1637); petition from the Earl of Bedford and the adventurers to the king, NA, SPD 16/323 (13 February 1638).

64. Secretary of State Francis Windebank's response to the Earl of Bedford and the adventurers, NA, SPD 16/323 (13 February 1638).

65. Certificate from the lord treasurer and other referees concerning the progress of the drainage, Bodleian Library, Oxford University, Bankes MS 46/5 (14 March 1638). The Crown's willingness in this case to ignore the Privy Council ruling of 1616, which declared that commissions of sewers did have the power to tax whole communities and levels generally without specifying each plot of land to be taxed, shows an impressive degree of royal hypocrisy. See also Knittl, "Design," 47–50; and L. E. Harris, *Vermuyden*, 71.

66. Petition from the Earl of Bedford and his partners to the king, Bodleian Library, Oxford University, Bankes MS 46/10 (n.d. [1638]).

67. Order from the commission of sewers for the Great Level, CUA, CUR 3.3/130 (31 March 1638); Knittl, "Design," 49; Barclay, *Electing Cromwell*, 82–83; Kennedy, "Commissions," 86.

68. Act of sewers from a session held at Huntingdon, CRO, R 59/31/9/1A, fol. 181r (17 April 1638).

69. Charles to the commission of sewers, CUA, T.XII.1/26 (13 April 1638).

70. Act of sewers from a session held at Wisbech, CRO, R 59/31/9/1A, fols. 185–98 (26 May 1638). The additional lands included Deeping Level, among others.

71. Royal writ allocating revenues for the king's drainage project, BL, Harleian MS 5011, vol. 2, fol. 622v (27 September 1638).

72. Act of sewers from a session held at Huntingdon, CRO, R 59/31/9/1A, fols. 198–214 ([28] July 1638).

73. John Bankes to Windebank, NA, SPD 16/395/77 (21 July 1638); see also commission of sewers to the lord treasurer, BL, Harleian MS 5011, vol. 2, fols. 423–24 (21 July 1638).

74. Warrant from Windebank to Hugh Pechy, NA, SPD 16/390/89 (16 May 1638).

75. John Bankes to Windebank, NA, SPD 16/395/77 (21 July 1638).

76. Miles Sandys to his son, Miles Sandys, NA, SPD 16/392/28 (6 June 1638).

77. Report of two messengers to the Privy Council, NA, SPD 16/375/46 (n.d. [June 1638]).

78. Cambridgeshire magistrates to the Privy Council, NA, SPD 16/392/54 (11 June 1638).

79. Earl of Exeter to the king, NA, SPD 16/392/42 (9 June [1638]).

80. Daniel Wigmore, William Marche, and John Goodricke to the Privy Council, NA, SPD 16/392/45 (9 June 1638); Bishop of Ely to the Privy Council, NA, SPD 16/409/50 (9 January 1639).

81. John Bankes to Windebank, NA, SPD 16/395/77 (21 July 1638).

82. Royal writ allocating revenues for the king's drainage project, BL, Harleian MS 5011, vol. 2, fols. 622–25 (27 September 1638); petition of Sir Oliver Nicholas to the king, NA, SPD 16/397/22 (7 August 1638); petition of James, Lord Kintyre, to the king, NA, SPD 16/404/92 (18 December 1638).

83. Dugdale, *History of Imbanking*, 414–15; BL, Harleian MS 5011, vol. 2, fol. 536r (n.d.). Manea is located roughly seven miles northwest of Ely; whether or not Charles really thought it might be possible to erect an artificial "mont" in the lowland Fens is an interesting question.

84. Warrant allocating revenues for the king's drainage project, Bodleian Library, Oxford University, Bankes MS 60/21 (15 September 1639); see also NA, SPD 38/18 (19 September 1639).

85. Record of a session of sewers held at Peterborough, CUA, CUR 3.1/63 (5 August 1640).

86. See Kishlansky, *Monarchy Transformed*, chaps. 5–6.

87. Record of the proceedings of a commission of sewers at Peterborough, NA, SPD 16/463/46 (5 August 1640).

88. Charles's decision to rule without Parliament left him in need of alternative sources of revenue that did not require Parliamentary sanction. Ship money was a traditional levy that English kings could use to respond to naval emergencies: port towns were required to furnish for the Crown's use either a ship or, much more commonly, the money needed to hire one. Because it was meant to ensure the safety of the realm, the levy did not require Parliamentary approval. In 1634, Charles declared that piracy in the English Channel had become an emergency situation, and he began collecting ship money to address it. In the following year, he determined that because the entire realm benefited from overseas trade and safe seas, ship money

should be assessed in every English county, even though it had never before been paid by inland communities. Charles continued to collect ship money from all of England throughout the 1630s, and the policy was quite successful from a fiscal standpoint, raising nearly £800,000 over six years. It was extremely controversial, however, and did much to inflame resentments among Englishmen who already opposed the king's comparatively heavy-handed style of governance and especially his collection of taxes not sanctioned by Parliament. Ship money, along with other royal efforts to raise money without Parliament during the Personal Rule, was thus a key cause of political strife leading up to the Civil War. See Kishlansky, *Monarchy Transformed*, 121–22.

89. Lindley, *Fenland Riots*, 112–19. However, Andrew Barclay (*Electing Cromwell*, chap. 6) has argued persuasively that Cromwell's views regarding the Great Level drainage were complex and that whatever opposition he expressed during the 1630s and 1640s had more to do with a concern that the projectors were treating the fenland commoners unfairly, than from any principled opposition to draining the Great Level per se. Certainly his actions during the 1650s were all firmly in support of fen drainage projects—see chapters seven and eight in this volume.

90. "The Grand Remonstrance, with the Petition Accompanying It," para. 32, Constitution Society, accessed 18 August 2016, http://www.constitution.org/eng /conpur043.htm; see also Kishlansky, *Monarchy Transformed*, chap. 6.

91. Lindley, *Fenland Riots*, 129–30; Holmes, "Drainers and Fenmen," 172–79.

92. *DNB*, s.v. "Francis Russell, fourth earl of Bedford"; "Francis Russell, 4th Earl of Bedford," *Wikipedia*, accessed 18 August 2016, https://en.wikipedia.org/wiki/Francis _Russell,_4th_Earl_of_Bedford.

93. "The Worx of the Fennes for this Yeare," Bodleian Library, Oxford University, Bankes MS 46/15 (n.d. [c. 1641]); Vermuyden to John Bankes, Bodleian Library, Oxford University, Bankes MS 65/75 (5 October 1641); petition from Great Level workmasters to the king, NA, SPD 16/491/94 (28 July 1642); Dugdale, *History of Imbanking*, 415; *BDCE*, s.v. "Sir Cornelius Vermuyden"; Knittl, "Design," 49–50.

94. Lindley, *Fenland Riots*, 130.

95. See Kishlansky, *Monarchy Transformed*, chap. 7.

96. Wells, *History*, 2:367.

97. Vermuyden, *Discourse*, 32.

98. Vermuyden, *Discourse*, 1–3.

99. Vermuyden, *Discourse*, 13–32.

100. Vermuyden, *Discourse*, 12–13.

101. Scotten, *Desperate*, fols. A2r, 16–17, 24. The Irish rebellion had erupted in October 1641 with the slaughter of thousands of Irish Protestants by Catholic rebels, igniting a firestorm of anti-Catholic hysteria in London and much of the rest of England. To liken Vermuyden to the "popish clergy" in 1642 was therefore provocative, to say the least.

102. *BDCE*, s.v. "Andrewes Burrell."

103. Burrell, *Explanation*; idem, *Briefe Relation*; idem, *Exceptions*.

104. Burrell, *Exceptions*, fol. A2.

105. Burrell, *Explanation*, 2.

106. Burrell, *Briefe Relation*, fol. A2v, my emphasis.

107. Burrell, *Explanation*, 2.

108. Testimony of Mr. Glapthorne, Committee for the Fens, rough minutes, CRO, R/59/31/9/3, fols. 5v–6r (25 June 1646).

109. Testimony of Major Underwood, Captain Roger Pratt, and Robert Barton, Committee for the Fens, rough minutes, CRO, R/59/31/9/3, fols. 7r–10r (25 June 1646).

110. Testimony of Andrewes Burrell, Committee for the Fens, rough minutes, CRO, R/59/31/9/3, fols. 20r–v (16 February 1647 [misdated 26 February]).

111. "Exceptions Offered by Matthew Bpp. of Ely . . . to the Bill Intituled an Act for the Dreyneing of the Great Levell, Extending It Selfe into the Countyes of North-hampton, Norffolke, Suffolke, Lincolne, Cambridgeshire, Huntington, and the Isle of Ely," CUA, EDR A/8/7 (n.d. [1641–42]).

112. Committee for the Fens, rough minutes, CRO, R/59/31/9/3, fols. 3r–5v (10 June 1646).

113. Order of the Committee for the Fens, official proceedings, CRO, R/59/31/9/1, fol. v (30 June 1646).

114. Committee for the Fens, official proceedings, CRO, R/59/31/9/1, fol. x (11 November 1647); and Committee for the Fens, rough minutes, CRO, R/59/31/9/3, fols. 21v–22r (11 November 1647); anon., *To the Honourable.*

115. Committee for the Fens, official proceedings, CRO, R/59/31/9/1, fol. xi (22 November 1647).

116. Wells, *History,* 2:367. Since the Restoration of the English monarchy in 1660, the law has been more commonly known as the "Pretended Act," because it was passed after Charles's execution and never received royal assent.

117. Wells, *History,* 2:367–70, 380.

118. Wells, *History,* 2:371, my emphasis.

119. Wells, *History,* 2:373–75. The shareholders' fiscal powers were limited as a commission of sewers, however, in that they could levy taxes only on the adventurers' allotted ninety-five thousand acres to pay for all future maintenance of the works.

120. "The Grand Remonstrance, with the Petition Accompanying It," para. 32, Constitution Society, accessed 18 August 2016, http://www.constitution.org/eng /conpur043.htm, my emphasis.

CHAPTER 7: Riot, Civil War, and Popular Politics in the Hatfield Level, 1640–1656

1. Deposition of Susan Nante, NA, SPD 18/37/11.III, fol. 25 (29 February 1652), my emphasis.

2. Parliamentary committee report, NA, SPD 18/37/11 (2 June 1653).

3. *OED,* s.v. "clout," def. 4c: "*fig. man of clouts, king of clouts,* etc.: a mere 'doll' in the garb of a man, a king, etc.; a 'lay-figure.'" I am grateful to Steve Hindle for encouraging me to interrogate this term more closely.

4. Lindley, *Fenland Riots,* 253, 258; Holmes, "Drainers and Fenmen," 168, 171.

5. Petition from John Stripe, merchant, to the Privy Council, NA, SPD 16/430/65 (11 October 1639); see also the petition of the inhabitants of Hatfield to the justices of the peace in the West Riding of Yorkshire, NA, SPD 16/451/108 (n.d. [April 1640?]).

6. *Robert Earl of Kingston-upon-Hull v. Hugh Drinkall, William Drinkall, Henry Fidling,* NA, E 134/14Chas1/Mich11 (11 October 1638).

7. Archbishop of York to the archbishop of Canterbury, NA, SPD 16/327/47 (23 June 1636).

8. Petition of Robert Long and John Gibbon to the king, NA, SPD 16/323 (6 June 1637).

9. Petition of Vermuyden and his partners to the king, NA, SPD 16/323 (8 December 1637).

10. John Brooke to Robert Long, NA, SPD 16/451/65 (28 April 1640); Edward Walpole to Robert Long, NA, SPD 16/450/90 (14 April 1640).

11. Edward Walpole to Robert Long, NA, SPD 16/450/90 (14 April 1640).

12. Keeler, *Long Parliament*, 54–55; see also Holmes, *Seventeenth-Century Lincolnshire*, 138; Lindley, *Fenland Riots*, 108–9.

13. Cope and Coates, *Proceedings*, 227–28, 305–6; Lindley, *Fenland Riots*, 108–9.

14. Complaint against rioters in Lincolnshire, NA, SPD 16/452/19 (3 May 1640).

15. Draft of a petition from the Earl of Lindsey to the House of Lords, NA, SPD 16/452/32 (5 May 1640).

16. Order from the Privy Council to the Lincolnshire justices of the peace, NA, SPD 16/453/32 (13 May 1640).

17. Petition from the mayor and burgesses of Boston to the Privy Council, NA, SPD 16/461/71 (n.d. [July 1640]). The trained bands were county militias made up of nonprofessional soldiers, the only standing military force England had before the start of the Civil War.

18. D'Ewes, *Journal of Sir Simonds D'Ewes*, 19.

19. Drainage-related complaints in the Lincolnshire petitions, NA, SPD 16/480/87 (n.d. [May 1641]).

20. "The Grand Remonstrance, with the Petition Accompanying It," para. 32, Constitution Society, accessed 18 August 2016, http://www.constitution.org/eng /conpuro43.htm; see also Kishlansky, *Monarchy Transformed*, chap. 6.

21. Parliamentary committee report, NA, SPD 18/37/11 (2 June 1653); depositions of William Wroote, David Tratard, Robert Bernard, and Francis Barber, NA, SPD 18/37/11.III, fols. 12–13, 17, 24–25 (29 February 1652); *Thomas Abdy et al. v. Gregory Torr et al.*, NA, E 134/24Chas1/East4 (7 April 1648).

22. Parliamentary committee report, NA, SPD 18/37/11 (2 June 1653); deposition of Thomas Pettinger, NA, SPD 18/37/11.III, fol. 34 (6 May 1652); Lindley, *Fenland Riots*, 146–47.

23. Depositions of Timothy Ellis, John Thompson, and John Phillingham, NA, SPD 18/37/11.III, fols. 33–34 (6 and 14 May 1652); *Thomas Abdy et al. v. Gregory Torr et al.*, NA, E 134/1649/Mich11 (24 September 1649); Lindley, *Fenland Riots*, 147.

24. Depositions of William Wroote, Robert Bernard, and Francis Barber, NA, SPD 18/37/11.III, fols. 12, 24–25 (29 February 1652).

25. Petition from the Isle of Axholme commoners to the Council of State, NA, SPD 46/96/705–6 (9 July 1651).

26. The violence directed toward the Participants during the Civil War period was not limited to English commoners seeking to reoccupy and defend their commons. Many of the Dutch tenants at this time also refused to pay the Participants the taxes (scots) they owed for maintaining the drainage works and allegedly hired soldiers to defend them. Because most of the original Dutch Participants had sold their shares to

English investors, the Dutch tenants apparently felt under no particular obligation to them in such troubled times. See George Stovin's manuscript volume, University of Nottingham Manuscripts and Special Collections, Hatfield Chase Corporation Records, HCC 9111, 143–59 (n.d.).

27. Parliamentary committee report, NA, SPD 18/37/11 (2 June 1653); examination of Daniel Noddel, NA, SPD 18/37/11.I (20 July 1647); deposition of Mr. Geery, NA, SPD 18/37/11.III, fol. 4 (29 February 1652). The Participants alleged that Noddel was accompanied by at least four hundred armed men; Noddel countered that it was only two hundred.

28. Depositions of Captain Robert Dyneley and John Spittlehouse, NA, SPD 18/37/11.III, fol. 27 (29 February 1652); Lindley, *Fenland Riots*, 140–41.

29. Examination of Daniel Noddel, NA, SPD 18/37/11.I (20 July 1647).

30. The Court of Exchequer had a complicated jurisdiction during the early modern period, and it could try cases using both common law and equity law. A bill of equity was the means by which a party initiated a suit in equity law—in this case, the Participants' motion to have the commoners' common law suit dismissed and to have their legal claim to the disputed 7,400 acres affirmed.

31. Anon., *Case and Proceedings*, 4; Lindley, *Fenland Riots*, 150–51. The anonymous *Case and Proceedings* pamphlet, printed in 1656, is easily confused with another anonymous pamphlet with a very similar title, printed in 1654. The two documents contain some overlapping text but are far from identical. In order to distinguish between them, I will refer to the 1654 pamphlet as *To the Parliament*, and the 1656 pamphlet as *Case and Proceedings*.

32. *Thomas Abdy et al. v. Gregory Torr et al.*, NA, E 134/24Chas1/East4 (7 April 1648).

33. *Thomas Abdy et al. v. Gregory Torr et al.*, NA, E 134/1649/Mich11 (24 September 1649); Noddel, *Great Complaint*, 9 (mislabeled)–12.

34. Noddel, *Great Complaint*, 9 [mislabeled]–12; Lindley, *Fenland Riots*, 151–52.

35. It should be noted that the unrest in the Isle of Axholme at this time did not seriously impact the remainder of Hatfield Level. Except for a gap in 1640–46 (possibly caused by the political and social unrest in the period just before and during the Civil War) and another in 1652–55 (probably resulting from the loss of one volume of their records), the proceedings of the commissioners of sewers for Hatfield Chase show that they continued to manage the drainage in the areas outside Axholme, in Yorkshire and Nottinghamshire. On the whole, they conducted business there as usual, maintaining and repairing drainage works, assessing taxes to pay for it, and hearing various complaints and petitions. See University of Nottingham Manuscripts and Special Collections, Hatfield Chase Corporation Records, HCC 6001 (1635–52) and 6002 (1655–69).

36. The New Model Army was the name for the reorganized military force fighting for the Parliamentary cause, 1645–60. The reorganization combined various local and regional units into a single, national army that would fight throughout England (as well as in Scotland and Ireland) as needed, rather than having each unit tied to a particular area. It also streamlined and centralized the army's command and funding structures, and promoted the professionalization of both the officer corps and the rank-and-file soldiers. Under the command of Cromwell and Sir Thomas Fairfax,

it was one of the most effective armies of seventeenth-century Europe. See Kishlansky, *Monarchy Transformed*, 162–64; Braddick, *God's Fury*, 350–55, 372–74; Roberts, *Cromwell's War Machine*.

37. Anon., *Case and Proceedings*, 5.

38. Witness depositions, NA, SPD 18/37/11.III, fols. 1–46 (29 February 1652–14 May 1652); Lindley, *Fenland Riots*, chap. 6.

39. Gregg, *Free-Born John*; *DNB*, s.v. "John Lilburne."

40. *DNB*, s.v. "Sir John Wildman."

41. Kishlansky, *Monarchy Transformed*, chaps. 7–8; Hill, *World Turned Upside Down*.

42. Hughes, "Drainage Disputes," 31–34; Lindley, *Fenland Riots*, 196; Holmes, "Drainers and Fenmen," 166–67, 194–95.

43. Depositions of George Wood and John Thorpe, NA, SPD 18/37/11.III, fols. 21–22, 42 (29 February 1652 and 14 May 1652); Lindley, *Fenland Riots*, 141.

44. Parliamentary committee report, NA, SPD 18/37/11 (2 June 1653). Wildman later denied having signed the deed, but when confronted with the document bearing his signature and seal, he was forced to admit his involvement. Daniel Noddel was also to receive 200 acres for his services as solicitor, which would have left the commoners in possession of 11,200 acres out of their original 13,400.

45. Deposition of John Thorpe, NA, SPD 18/37/11.III, fols. 8–10 (29 February 1652).

46. Depositions of John Amory, Edmund Griffith, and John Thorpe, NA, SPD 18/37/11.III, fols. 5, 24 (29 February 1652).

47. Depositions of Jacob Descamps, John Clesby, and Richard Glewe, NA, SPD 18/37/11.III, fols. 5–6, 17–18, 44–45 (29 February 1652 and 14 May 1652).

48. Lilburne, *Case of the Tenants*, 4.

49. Anon. [Lilburne], *To the Right Honorable*, 5; Lindley, *Fenland Riots*, 195.

50. Depositions of Edmund Griffith, John Amory, Thomas Fisher, Anthony Massengarb, James Davies, William Wroote, Francis Briggs, John Mylner, John Clesby, and Timothy Steward, NA, SPD 18/37/11.III, fols. 3–4, 7–8, 11–16, 19–20 (29 February 1652).

51. "Club law" refers to the actions of the so-called clubmen during the English Civil War, when large bodies of commoners armed mostly with clubs, scythes, and other rudimentary weapons refused to join either the royalist or Parliamentary forces, seeking instead to protect their local communities from the depredations of both armies. Historians originally viewed their actions as a sort of conservative, provincial armed neutrality, but more recent work has shown that clubmen were seldom, if ever, truly neutral and that many such groups tended to be anti-royalist in their sympathies (See Underdown, *Revel*, chap. 6; Morrill, *Revolt*, 132–51, 197–204; Stoyle, *Loyalty*, chap. 6). At the same time, however, even the most anti-royalist clubmen refused to join with the Parliamentary armies, insisting that they fought only to defend their homes from looting by royalist troops and they would not fight outside their own communities. The clubmen of the clothing districts of the West Riding, several miles west of the Hatfield Level, were widely known as fervent anti-royalists from 1642 onward (see Hopper, "Clubmen"). However, the term's use in this instance suggests the Axholme commoners' aggressive assertion of independence from any central

authority, whether monarchy or Commonwealth—more vigilante justice than armed neutrality, and explicitly aimed at Commonwealth forces.

52. Depositions of Jacob Descamps, Edmund Griffith, Francis Briggs, John Bracer, Thomas Johnson, Susan Nante, and Richard Glewe, NA, SPD 18/37/11.III, fols. 5–6, 12–13, 22–26 (29 February 1652).

53. The Axholme commoners' tactical alliance with, and political affinity for, the Leveller leaders may be compared to the very similar reaction of free miners in the Peak District, who sought to preserve and defend their customary common use rights during the Civil War period. See Wood, *Politics of Social Conflict*, chap. 12.

54. Anon., *To the Parliament*, 1–2.

55. "An Act of General Pardon and Oblivion," in Firth and Rait, *Acts and Ordinances*, 2:567–77, reference to Axholme rioters on 575. The act shielded from prosecution and punishment all those who had committed illegal acts before 3 September 1651, with a number of explicit exceptions, usually involving serious crimes such as murder, rape, and treason. The point of the act was to settle the ongoing uncertainty over the possibility of prosecution for wartime acts committed when the normal legal institutions of the state were unable to function. The Axholme rioters' exclusion from the general pardon for "Offenses committed at Sandtoft" is probably attributable to the Participants' political influence and their success in linking the commoners to the Levellers' radical activities.

56. Lindley, *Fenland Riots*, 191–92.

57. Noddel, *Great Complaint*, 19.

58. Depositions of Robert Dynely, John Spittlehouse, Roger White, John Phillingham, William Clarke, William Mawe, Isaac Chapman, Thomas Farre, and Thomas Todd, NA, SPD 18/37/11.III , fols. 26–27, 30, 33–38 (29 February 1652; 6 and 14 May 1652).

59. Anon., *To the Parliament*, 3–6; parliamentary committee report, NA, SPD 18/37/11 (2 June 1653).

60. Anon. [Daniel Noddel?], *Brief Remembrance*; Noddel, *Declaration*; idem, *Great Complaint*. The anonymous pamphlet is shorter than the other two but matches them closely in language and argument; internal evidence dates it to just after the compilation of Say's final report.

61. Noddel, *Great Complaint*, 3.

62. Noddel, *Declaration* 1–3; idem, *Great Complaint*, 13, 24.

63. Noddel, *Declaration*, 20–21, 28. Throughout his defense, Noddel never denied that the commoners spoke such "unbeseeming" words, a possible indication of their genuineness.

64. Noddel, *Great Complaint*, 23–24.

65. "The Order of the Councel of State" (31 August 1653), printed in anon., *To the Parliament*, 6–8.

66. Noddel, *Great Complaint*, 11; anon., *To the Parliament*, 10.

67. University of Nottingham Manuscripts and Special Collections, Hatfield Chase Corporation Records, HCC 6002, passim (1655–69); and George Stovin's manuscript volume, HCC 9111 (n.d.). See also Hunter, *South Yorkshire*, 1:167–73.

68. University of Nottingham Manuscripts and Special Collections, Hatfield Chase Corporation Records, HCC 6002, fol. 26 (24 January 1656). See also the information of John Nodes, NA, SPD 18/129/144.VI (4 May 1654); remonstrance of the

commission of sewers, NA, SPD 18/129/144.IV (25 January 1656); petition of the French and Dutch settlers to the lord protector, NA, SPD 18/126/57 (n.d. [March 1656]).

69. Petition of John Dillingham to the Council of State, NA, SPD 18/128/80 (n.d. [19 June 1656]).

70. Information of John Starkey, NA, SPD 18/129/144.V (22 November 1655).

71. Council of State to Edward Whalley (21 August 1656), printed in anon., *Case and Proceedings*, 14.

72. Lindley, *Fenland Riots*, 219–22; Burton, *Diary*, 1:199–200.

73. Inhabitants of the Isle of Axholme to Parliament, BL, Egerton MS 3518, fol. 86r (n.d. [1646–49]).

74. Lilburne, *Case of the Tenants*, 5.

75. Noddel, *Great Complaint*, 24–26.

76. Noddel, *Great Complaint*, 27.

77. Noddel, *Declaration*, 27.

CHAPTER 8: The Second Great Level Drainage, 1649–1656

1. Moore, *Mapp of ye Great Levell of ye Fenns*.

2. Proceedings of the Adventurers, CRO, BLC papers, R/59/31/9/4, fols. 52v–53r (23 and 26 August 1650). The Bedford Level Corporation was not formally incorporated until the passage of the General Drainage Act of 1663 (Summers, *Great Level*, 79–85). During the 1650s, the group usually called themselves the "Adventurers," or "Participants." I refer to them in the body of this chapter as "the Adventurers" or "the company," to distinguish them from the Participants in the Hatfield Level drainage; their manuscript records are cited, in accordance with the title engraved on the cover of each volume, as the "Proceedings of the Adventurers."

3. *BDCE*, s.v. "Sir Jonas Moore."

4. Willmoth, *Sir Jonas Moore*, 117. See also Willmoth's excellent introduction to the facsimile edition of Moore's 1658 *Mapp*, soon to be published by the Cambridgeshire Records Society. I am grateful to her for sharing with me a typescript of her text.

5. BL, Maps 184.L.1 (1706). More than once, as I examined this map spread out over two large tables in the map room of the BL, other readers paused in their own work to come and have a look at it.

6. Willmoth, *Sir Jonas Moore*, 113. In her forthcoming introduction to the facsimile of Moore's *Mapp*, Willmoth notes that contemporary Dutch cartographers also visually emphasized the geometrical rationality of their own land drainage projects. On the central importance of maps in reimagining and transforming the English Fens, see also Di Palma, *Wasteland*, 110–11.

7. Darby, *Changing Fenland*, 85.

8. NA, Maps and Plans, MPC 1/88 (1658). The BL copy is a third edition of the map, printed in 1706. While the third edition is relatively common, the first (1658) and second (1684) editions exist in only one surviving copy apiece—the former in the National Archives, the latter at the Bodleian Library, Oxford University (Gough Maps Cambridgeshire 2). The latter two editions of the map were printed from the same engraving as the first, with the cartographic portions unaltered; but the 1658 edition is unique in bearing the officers' and shareholders' coats of arms.

9. Proceedings of the Adventurers, CRO, BLC papers, R/59/31/9/1, fol. 1r (1 June 1649).

10. Vermuyden, *Discourse.*

11. Proceedings of the Adventurers, CRO, BLC papers, R/59/31/9/1, fol. 1v (1 June 1649).

12. Proceedings of the Adventurers, CRO, BLC papers, R/59/31/9/1 and 2, passim (1 June 1649–25 January 1650). The negotiations have been exhaustively recounted in Wells, *History*, 1:chap. 8; and in L. E. Harris, *Vermuyden and the Fens*, chaps. 10–13. Wells's analysis is strongly biased in the company's favor, while Harris is firmly in Vermuyden's camp; both accounts should be used with caution, but they balance one another nicely.

13. Proceedings of the Adventurers, CRO, BLC papers, R/59/31/9/1, fols. 9v–11v (19 June 1649).

14. Proceedings of the Adventurers, CRO, BLC papers, R/59/31/9/1, fol. 9v (19 June 1649); and R/59/31/9/2, fol. [6v] (8 December 1649).

15. Proceedings of the Adventurers, CRO, BLC papers, R/59/31/9/2, fol. [10v] (1 January 1650). The original vellum manuscript of the company's agreement with Vermuyden, complete with the latter's signature and the former's seal, survives in the CRO collection of uncataloged papers from the Bedford Level Corporation; a more legible manuscript copy may be found in the first volume of the company's Register of Conveyances, CRO, BLC papers, R/59/31/1A, fols. 110v–112r (25 January 1650).

16. Register of Conveyances, CRO, BLC papers, R/59/31/1A, fols. 110v–112r (25 January 1650); Proceedings of the Adventurers, CRO, BLC papers, R/59/31/9/2, [fol. 12r] (25 January 1650).

17. Having seen the Ouse Washes (the area between the two Bedford Rivers) under several feet of water in March 2016, I can attest that it is an impressive sight. When I had last seen the washes, in the summer of 2004, cattle were grazing on the dry land there.

18. Dugdale, *History of Imbanking*, 415–16; Darby, *Changing Fenland*, 75–77.

19. Proceedings of the Adventurers, CRO, BLC papers, R/59/31/9/3, fols. 35v–36v (14 February 1650).

20. Proceedings of the Adventurers, CRO, BLC papers, R/59/31/9/5, fols. 105r–106v (29 August 1651).

21. Proceedings of the Adventurers, CRO, BLC papers, R/59/31/9/5, fols. 11r–12r (4 March 1651).

22. Proceedings of the Adventurers, CRO, BLC papers, R/59/31/9/5, fols. 249r–251r (2 August 1652).

23. Proceedings of the Adventurers, CRO, BLC papers, R/59/31/9/5, fols. 256v–259r (14 August 1652).

24. Proceedings of the Adventurers, CRO, BLC papers, R/59/31/9/6, fols. 20r–v (29 October 1652).

25. Proceedings of the Adventurers, CRO, BLC papers, R/59/31/9/5, fols. 259v–264r (16 August 1652).

26. Proceedings of the Adventurers, CRO, BLC papers, R/59/31/9/6, fols. 31r–32r (16 December 1652).

27. Proceedings of the Adventurers, CRO, BLC papers, R/59/31/9/6, fols. 6v–7v (16 September 1652), my emphasis.

28. The most prominent critics of Vermuyden's drainage scheme include Badeslade, *History*; Elstobb, *Observations*; and Wells, *History*, 1:290–97. Others have offered a more positive interpretation, rebutting the dominant negative view of the eighteenth and nineteenth centuries; see especially L. E. Harris, *Vermuyden and the Fens*, chaps. 15–16. For the most balanced assessment, see Darby, *Changing Fenland*, chap. 4.

29. Ash, *Power*, chap. 1.

30. *BDCE*, s.v. "Sir Cornelius Vermuyden." William Dugdale reported that there were at least eleven thousand men employed in digging the New Bedford River alone ("Things Observable in Our Itinerarie Begun from London 19 Maij 1657," BL, Lansdowne MS 722, fol. 29).

31. Proceedings of the Commissioners, CRO, BLC papers, R/59/31/9/9.I, fols. 5r–13v (21–26 March 1651).

32. Proceedings of the Commissioners, CRO, BLC papers, R/59/31/9/9.I, fols. 26v–47v (24–26 March 1653).

33. Proceedings of the Adventurers, CRO, BLC papers, R/59/31/9/3, fol. 21 (12 December 1649). The manuscript actually records a total of £109,700, but this is clearly an arithmetical error; the correct total should be £119,700.

34. Anon., *An Answer*, 12. Other sources place the total expenditure at roughly £250,000; see *BDCE*, s.v. "Sir Jonas Moore."

35. Proceedings of the Adventurers, CRO, BLC papers, R/59/31/9/6, fols. 181–83 (21 February 1655).

36. Wells, *History*, 2:367–68.

37. Wells, *History*, 2:373.

38. Wells, *History*, 2:376–77, 381.

39. Proceedings of the Commissioners under the Drainage Act of 1649, CRO, BLC papers, R/59/31/9/9, vols. 1–2 (1650–56).

40. Grainger, *Cromwell*, 55–58, 144–46.

41. Proceedings of the Adventurers, CRO, BLC papers, R/59/31/9/5, fols. 114v–116r (11 and 14 October 1651).

42. Earl of Bedford & co. to Thurloe, Bodleian Library, Oxford University, Rawlinson MS A/4, fols. 232–36 (18 July 1653).

43. Proceedings of the Adventurers, CRO, BLC papers, R/59/31/9/5, fols. 139v–142r (19 November 1651), and fol. 166 (14 January 1652).

44. Proceedings of the Adventurers, CRO, BLC papers, R/59/31/9/5, fol. 224v (29 May 1652); Wells, *History*, 1:244.

45. Proceedings of the Adventurers, CRO, BLC papers, R/59/31/9/5, fols. 104v–105r (15 August 1651).

46. COS to Col. Walton, NA, SPD 25/96, p. 483 (2 September 1651).

47. Proceedings of the Adventurers, CRO, BLC papers, R/59/31/9/5, fol. 165v (12 January 1652), and fols. 172v–173r (4 February 1652). Both of these entries concern the removal of troops from one location to another, as quartering them had become an unacceptable burden to the inhabitants.

48. Proceedings of the Adventurers, CRO, BLC papers, R/59/31/9/5, fols. 127r–128r (3 November 1651); and R/59/31/9/6, fols. 67v–68v (19 March 1653).

49. Proceedings of the Adventurers, CRO, BLC papers, R/59/31/9/6, fols. 42v–46v (31 January 1653).

50. Proceedings of the Commissioners, CRO, BLC papers, R/59/31/9/9.I, fols. 36, 38, 44 (24 March 1653).

51. Lindley, *Fenland Riots*, 176–87, contains a thorough account of the various Great Level riots of 1653.

52. Proceedings of the Adventurers, CRO, BLC papers, R/59/31/9/6, fol. 76 (23 April 1653); BLC to Robert Henley, NA, SPD 18/39/92 (30 August 1653).

53. Earl of Bedford & co. to Thurloe, Bodleian Library, Oxford University, Rawlinson MS A/4, fols. 232–36 (18 July 1653).

54. Proceedings of the Adventurers, CRO, BLC papers, R/59/31/9/6, fols. 97v–99r (22 June 1653).

55. BLC to Robert Henley, NA, SPD 18/39/92 (30 August 1653).

56. Hamond to Thurloe, NA, SPD 18/39/96 (31 August 1653). Lilburne was prosecuted for treason in October 1649, but after he ably defended himself by challenging every detail of the prosecution's proceedings and evidence against him, the jury acquitted him of the charge. His acquittal and release were widely celebrated by the people of London, an embarrassment for the ruling Commonwealth regime as it struggled to consolidate its newly won power. *DNB*, s.v. "John Lilburne."

57. Hamond to Thurloe, NA, SPD 18/39/96 (31 August 1653).

58. COS, *By His Excellency.*

59. Proceedings of the Adventurers, CRO, BLC papers, R/59/31/9/6, fol. 76 (23 April 1653).

60. COS to Edward Whalley, NA, SPD 25/69, p. 10 (2 May 1653); COS to the independent commissioners for the fens, NA, SPD 25/69, p. 170 (31 May 1653).

61. Proceedings of the Adventurers, CRO, BLC papers, R/59/31/9/6, fols. 93v–94r (8 June 1653); COS to Whalley and to Col. Humphreys, NA, SPD 25/70, pp. 302–3 (30 August 1653); Maj. Robert Swallow to Whalley, NA, SPD 18/40/19 (5 September 1653); BLC to the COS, NA, SPD 18/40/58 (10 September 1653). See also Lindley, *Fenland Riots*, 184.

62. COS, *Commission Impowering.*

63. COS, *By His Excellency.*

64. For example, a "Dr. Rich. Stanes" was named to the committee, probably the same "Mr. Doctor Stanes" in Ely, of whom the company wrote to Anthony Hamond that "they have heretofore been obliged for his former civilities and do gratefully acknowledge the same" (Proceedings of the Adventurers, CRO, BLC papers, R/59/31/9/6, fol. 42v [31 January 1653]). On the other hand, a "John Cleypool Esq." was also appointed, who is almost certainly the justice of the peace in the Soke of Peterborough who sought to raise contributions in that area to prevent the allotment of common lands there to the company (Anon. [Maynard?], *Humble Petition*, 7).

65. Proceedings of the Commissioners, CRO, BLC papers, R/59/31/9/9.I, fols. 53–56 (5 December 1653); Proceedings of the Adventurers, CRO, BLC papers, R/59/31/9/6, fols. 118–20 (3 December 1653); Hamond to Thurloe, Bodleian Library, Oxford University, Rawlinson MS A/9, fol. 112 (21 December 1653).

66. COS, *By His Excellency.*

67. Proceedings of the Adventurers, CRO, BLC papers, R/59/31/9/6, fol. 102v (30 June 1653).

68. COS, *Ordinance*; Proceedings of the Adventurers, CRO, BLC papers, R/59/31/9/6, fol. 161 (12 June 1654).

69. *DNB*, s.v. "Sir John Maynard"; Lindley, *Fenland Riots*, 166–68; Proceedings of the Adventurers, CRO, BLC papers, R/59/31/9/3, fol. 5r (10 June 1646).

70. Prynne, *Declaration*, 5.

71. Maynard, *Picklock*; Anon. [Maynard?], *Humble Petition*; Anon. [Maynard?], *Reply*; Anon. [Maynard?], *To the Supreme Authority*; Anon. [Maynard?], *Anti-Projector.* Most of these pamphlets were published anonymously, but the arguments and language they contain are all quite similar, repeated almost verbatim in many cases. They were either written by the same individual, or else the anonymous author(s) and Maynard borrowed very heavily from one another. Early English Books Online tentatively lists *The Anti-Projector* as being published in 1646, but internal reference to the two adjudications of the Great Level (p. 5) indicates that it must have been written after 26 March 1653.

72. Anon. [Maynard?], *Anti-Projector*, 8.

73. Anon. [Maynard?], *Anti-Projector*, 1–4 (original emphasis).

74. Anon. [Maynard?], *Humble Petition*, 6.

75. Anon. [Maynard?], *Anti-Projector*, 1, 5.

76. Maynard, *Picklock*, 5, 10.

77. Anon. [Maynard?], *Anti-Projector*, 3.

78. Anon. [Maynard?], *Humble Petition*, 2.

79. Maynard, *Picklock*, 15 (original emphasis).

80. Anon. [Maynard?], *Anti-Projector*, 6–7.

81. Shareholders to Robert Henley, NA, SPD 18/39/92 (30 August 1653).

82. Hamond to Thurloe, NA, SPD 18/39/96 (31 August 1653).

83. Anon., *State of the Adventurers' Case*; Anon., *An Answer.* The latter document is quite similar to a manuscript version, copies of which may be seen at NA, SPD 18/39/97, and at BL, Add. MS 25302, fols. 129–40. The manuscript version covers exactly the same topics as the printed pamphlet, though the language differs significantly.

84. Anon., *An Answer*, 2, 5, 12–14.

85. Anon., *An Answer*, 6.

86. Anon., *An Answer*, 2, 12 (original emphasis).

87. The literature on Samuel Hartlib and the Hartlib Circle is voluminous, especially with respect to the group's aspirations toward education reform, pansophism, and experimental natural philosophy. Sources addressing Hartlib's correspondence and agricultural improvement in particular include McRae, *God Speed the Plough*; Thirsk, "Plough and Pen"; idem, gen. ed., *AHEW*, vol. 5, pt. 2:542–59; Greengrass et al., *Samuel Hartlib*; Di Palma, *Wasteland*, chap. 2; Yamamoto, "Reformation"; idem, "Distrust," chap. 2; Leng, "'Potent Plantation'"; Raylor, "Samuel Hartlib"; Slack, *From Reformation*, chap. 4; see also *DNB*, s.v. "Samuel Hartlib."

88. Hartlib, *Legacie*, 52–53 (original emphasis). Child's letter, printed here anonymously, makes up the vast majority of the volume. For the ascription of the letter to Child, see Thirsk, gen. ed., *AHEW*, vol. 5, pt. 2:324.

89. Blith, *English Improver*; idem, *English Improver Improved*. See also Thirsk, "Plough and Pen," 306–13; Yamamoto, "Distrust," 150–54; *DNB*, s.v. "Walter Blith."

90. Blith, "To Those of the High and Honourable Houses of Parliament . . ." and "The Epistle to the Ingenuous Reader," in *English Improver*.

91. Blith, *English Improver Improved*, 45. For Blith's probable experience in the Great Level, see Thirsk, "Plough and Pen," 309–10.

92. Blith, *English Improver Improved*, fols. A2v, 48, 51.

93. Blith, *English Improver Improved*, 47, 52 (original emphasis), 16, 45.

94. Yamamoto, "Distrust," 117–28; *DNB*, s.v. "Cressy Dymock"; Hunt, "Hortulan Affairs."

95. Dymock, *Discoverie*, 1–2 and title page.

96. Dymock, *Discoverie*, 17–22.

97. Dymock, *Discoverie*, 11, 19, 22.

98. Boate, *Irelands Natvrall History*. See also Bernard, "Hartlib Circle"; and Coughlan, "Natural History."

99. Leng, "'Potent Plantation.'"

100. "Richard Gorges, 2nd Baron Gorges of Dundalk," *History of Parliament*, accessed 18 August 2016, http://www.historyofparliamentonline.org/volume/1660-1690/member/gorges-richard-1619-1712.

101. NA, SPD 25/35, p. 123 (29 November 1652).

102. William Goodsonn to Thurloe, Bodleian Library, Oxford University, Rawlinson MS A/34, fol. 733 (24 January 1655). In this letter, written nearly two years after the Great Level was declared to be drained, Goodsonn laments to Thurloe that his lands there had been sown with coleseed, "but I have heard, that by a flood God was pleased to take that away."

103. Fortrey, *Englands Interest*; see also Appleby, *Economic Thought*; *BDCE*, s.v. "Samuel Fortrey." Fortrey was not a wholehearted supporter of English colonization projects, however, believing them to be a waste of resources unless they provided a means to obtain commodities that England would otherwise be forced to buy from other nations; see Fortrey, *Englands Interest*, 39.

104. Kupperman, "Controlling Nature." See also Montaño, *Roots of English Colonialism*; and Bernard, "Hartlib Circle."

105. Montaño, *Roots of English Colonialism*, chap. 4; quotations from 156–57.

106. NA, Maps and Plans, MPF 1/305. I am grateful to Keith Pluymers for this reference and for his generously sharing with me a digital image of this map.

107. Child to Hartlib (23 June 1652), *The Hartlib Papers*, accessed 2 January 2017, https://www.hrionline.ac.uk/hartlib/view?docset=main&docname=15B_05_12&termo.

108. Dugdale, "To the Reader," in *History of Imbanking*, fol. A2r.

109. Dugdale, "A Mapp of the Great Levell, Representing It as It Lay Drowned," in *History of Imbanking*, bound with p. 375. North is at the bottom of this map.

110. Dugdale, "The Map of the Great Levell Drayned," in *History of Imbanking*, bound with p. 416. North is at the top of this map. The effect of the company's

364 NOTES TO PAGES 299–304

ownership is heightened in Dugdale's manuscript original of this map in the British Library, on which the company's allotted lands are traced in red ink, very similar to the copy of Moore's wall map in the same library. BL, Harleian MS 5011, vol. 2, fols. 628–29.

EPILOGUE: The Once and Future Fens

1. "Things Observable in Our Itinerarie Begun from London 19 Maij 1657," BL, Lansdowne MS 722, fols. 29–39. See also Willmoth, "Dugdale's *History*," 296–97.

2. Fuller, *History of the University of Cambridge*, 72 (mislabeled 82; original emphasis).

3. Petition from the justices of the peace of the Isle of Ely, CRO, BLC petitions, S/B/SP/42 (9 October 1669).

4. Anon., "True and Natural Description," transcribed in Anon. [Fortrey or Moore?], *History or Narrative*, 72.

5. Anon., *An Answer*, 5.

6. Proceedings of the Adventurers, CRO, BLC papers, R/59/31/9/6, fols. 188–89 (4 and 18 April 1655); and R/59/31/9/8, fols. 28–30 (20 February 1661).

7. COS, *Ordinance*; idem, *Exemplification*.

8. Wells, *History*, 2:383–425.

9. On the economy of the post-drainage Fens, see Thirsk, *English Peasant Farming*, chap. 5; idem, "Seventeenth-Century Agriculture"; idem, *AHEW*, vol. 5, pt. 2, chap. 16; Spufford, *Contrasting Communities*. Economic and social trends in the seventeenth-century Fens reflect the broader trends prevalent throughout much of rural England at that time, as larger landowners were better able to profit from agricultural innovations and market conditions while smallholders were gradually squeezed out; see Wrightson, *English Society*, esp. chap. 5.

10. Fuller, *History of the University of Cambridge*, 71 (original emphasis).

11. Lindley, *Fenland Riots*, 184–87.

12. Council of State to Edward Whalley, NA, SPD 25/69, p. 10 (2 May 1653).

13. For a wide variety of petitions both to and about the Bedford Level Corporation, 1660–1700, see the Proceedings of the Conservators, CRO, BLC papers, R/59/31/9/9.I and II, passim; Petitions Catalog, CRO, BLC petitions, S/B/SP, bundles 1–2; and Decrees of the Commission for Complaints against the Bedford Level Corporation, NA, C 299.

14. Bowring, "Between the Corporation."

15. The following account of the decay of the Great Level drainage is based on Darby, *Changing Fenland*, chap. 4; Summers, *Great Level*, chap. 5; Taylor, "Fenlands"; Hills, *Machines*; Williamson, "Dutch Engineers"; and Renes, "Fenlands of England."

16. Proceedings of the Adventurers, CRO, BLC papers, R/59/31/9/8, fols. 34v–35r (23 July 1661).

17. Proceedings of the Adventurers, CRO, BLC papers, R/59/31/9/8, fols. 53–58 (22–23 April 1663).

18. Dodson, *Designe*.

19. John Butler to Joseph Williamson, NA, SPD 29/332/3 (1 January 1673).

20. Petitions Catalog, CRO, BLC petitions, S/B/SP, bundle 1.

21. Defoe, *Tour*, 1:119 (letter 1); 2:145–46, 151 (letter 4), original emphasis.

22. Dodson, *Designe*, 4, 17.

23. Anon., *An Answer*, 12; Proceedings of the Adventurers, CRO, BLC papers, R/59/31/9/6–8, passim.

24. On the company's chronic financial difficulties in the late seventeenth century, see Summers, *Great Level*, 85–91.

25. Wells, *History*, 1:400–403; for the text of the act of Parliament changing the company's tax structure and the full survey and classification of its allotted lands, see Wells, *History*, 2:474–518.

26. One such request was presented to Francis Walsingham by Peter Morrice, NA, SPD 12/106/62 [c. 1575]. See also Hills, *Machines*, chap. 2; Williamson, "Dutch Engineers," 113.

27. Darby, *Changing Fenland*, 107–10.

28. Darby, *Changing Fenland*, 107–13; Summers, *Great Level*, 100–103.

29. Quoted in Wells, *History*, 1:426.

30. On the Haddenham drainage act and the many similar acts that followed it, see Darby, *Changing Fenland*, 112–14; Summers, *Great Level*, 119–23.

31. Quoted from Darby, *Changing Fenland*, 113.

32. Nicholas James has argued that the Bedford Level Corporation's devolution of responsibility in managing the regional drainage was gradual, with a significant degree of cooperation between local and corporate authorities in the early years, though the local commissions took on more responsibility as conditions worsened and the corporation's fortunes continued to decline. See James, "The 'Age of the Windmill.'"

33. Darby, *Changing Fenland*, 118.

34. Rackham, *History of the Countryside*, 390.

35. As I traveled from one archive to another in March 2016 to complete the research for this book, I could not help but notice that the fields I saw from the train in Oxfordshire had pools of standing water on them, while those near Peterborough and in the Isle of Ely were dry, with crops already planted in many places. The Ouse Washes between the two Bedford Rivers, however, were under several feet of water.

36. Darby, *Changing Fenland*, 238–40; "Ouse Washes," Royal Society for the Protection of Birds, accessed 18 August 2016, http://www.rspb.org.uk /discoverandenjoynature/seenature/reserves/guide/o/ousewashes/about.aspx; "Welney Wetland Centre," Wildfowl & Wetlands Trust, accessed 18 August 2016, http://www .wwt.org.uk/wetland-centres/welney.

37. "Wicken Fen Nature Reserve," National Trust, accessed 18 August 2016, https://www.nationaltrust.org.uk/wicken-fen-nature-reserve; "Wicken Fen," *Wikipedia*, accessed 18 August 2016, http://en.wikipedia.org/wiki/Wicken_Fen; "Wicken Fen National Nature Reserve," Wildlife Extra, accessed 18 August 2016, http://www .wildlifeextra.com/go/uk/wicken-fen.html#cr; "Wicken Fen Windpump," Society for the Protection of Ancient Buildings, Mills Section, accessed 18 August 2016, http:// www.nationalmillsweekend.co.uk/pages_wind/wicken%20Fen.htm. Walking the nature trails along the periphery of the reserve in 2004, I was struck by the surreal sensation of standing in a wetland and gazing several feet down upon dry, tilled fields.

38. "Wicken Fen National Nature Reserve," Wildlife Extra, accessed 18 August 2016, http://www.wildlifeextra.com/go/uk/wicken-fen.html#cr; "Wicken Fen," *Wikipedia*, accessed 18 August 2016, http://en.wikipedia.org/wiki/Wicken_Fen.

39. "Lakenheath Fen," Royal Society for the Protection of Birds, accessed 18 August 2016, http://www.rspb.org.uk/discoverandenjoynature/seenature/reserves/guide /l/lakenheathfen/about.aspx.

40. "Wicken Fen National Nature Reserve," Wildlife Extra, accessed 18 August 2016, http://www.wildlifeextra.com/go/uk/wicken-fen.html#cr; "Lakenheath Fen," Royal Society for the Protection of Birds, accessed 18 August 2016, http://www.rspb.org .uk/discoverandenjoynature/seenature/reserves/guide/l/lakenheathfen/directions.aspx.

41. Anonymous letter to an unknown recipient, NA, SPD 14/128/105 (n.d. [c. 1619–20]).

Glossary

The following list of definitions is intended to clarify some of the more obscure terms used in this book. Many refer to specific types of wetland or tools used for draining them; some are specific to the English Fens. Others deal with more technical aspects of English agriculture, land tenure, and customary and/or common law practice. Most of the definitions are adapted from the *Oxford English Dictionary*.

brovage: The right of a landlord to take in cattle from outside his estate to graze on the estate's common waste in return for a rent or fee, commonly exploited by landlords throughout the Fens.

carr: Wet, boggy ground; a fen. More specifically, a fen with shallower water that has become overgrown with sedge, bushes, and small trees.

clow: A dam for water, sometimes containing a sluice gate.

common, or common waste: An area of unenclosed land, usually less suitable for arable agriculture, customarily available for use by all members of a community, most often for grazing livestock.

copyhold: A type of land tenure, generally of limited duration, at the will of the lord of the manor and according to the custom of the manor; renewal of a copyhold lease might involve a substantial fee.

dike: A long, narrow excavation, usually intended to contain or conduct water; a ditch or trench.

disafforestation: The act of freeing land from the operation of the strict medieval forest laws that governed how it could be used and exploited, changing its formal legal state from forest to ordinary land. Disafforesting land was a legal action and had no necessary connection with the cutting or removal of trees—during the early modern period, much land that was legally considered "forest" was no longer wooded but was nevertheless still bound by the restrictive forest laws.

distrain: To confine, constrain, or hold captive; in law, to seize the property or livestock of an individual as a means of holding him or her responsible for the performance of some obligation. *Distraint* is the act of distraining.

fen: A type of wetland frequently covered with fresh water from nearby rivers and covered with layers of peat.

free and common soccage: In medieval feudalism, a form of free and certain land tenure requiring some form of honorable service to the lord of the manor; by the early modern period, roughly equivalent to a freehold lease.

freehold: Permanent, heritable tenure of land, with the freedom to dispose of it at will.

intercommoning: The practice, common in the pre-drainage Fens, of multiple communities sharing a large area of common waste located between them; usually practiced according to a series of customary rules that determined how far each community's rights extended.

leam: A drain or artificial watercourse. Commonly used only in the Fens.

level: A broad, very flat expanse of land. Commonly used in the Fens, especially as part of a proper name (e.g., the Great Level).

lockspit: To mark out the boundaries of an area of land by digging a trench around the periphery.

lode: A drain or artificial watercourse. Commonly used only in the Fens.

marsh: A type of wetland composed of marine silt and frequently flooded with salt water; more colloquially, any permanently waterlogged land.

meadow: An area of grassy land intended to be mown for hay, and thus for feeding livestock at some future date, especially when grazing land is unavailable or less productive.

mere: A lake, pond, or pool; a body of standing water or an area covered with standing water most or all of the time.

oyer and terminer: A writ or commission authorizing a judge traveling on circuit to hold courts. In the early modern period, usually issued in regard to certain serious crimes such as rioting.

pasture: An area of grassy land used for grazing livestock, especially during the summer months.

***praemunire*:** The crime of introducing a foreign authority into the jurisdiction claimed by the king or queen of England, originally intended to prevent English clergy from appealing to the Vatican against civil authorities.

replevin: An action brought by the owner of distrained property or livestock to regain possession when the property in question was wrongfully seized. The action required that the original issue that provoked the distraint be heard and decided in court.

sasse: A navigable sluice set in a river or canal, with gates at either end for raising or lowering the water level in between, to allow boats to pass from one level to the other; a lock.

severals: Land that is privately owned, occupied, and/or exploited, usually enclosed; as opposed to common waste. Such lands are held "in severalty."

sewer: An artificial watercourse for draining flooded or marshy land; a drainage channel or ditch.

sluice: A structure made of wood or stone designed to hold back the water of a river or canal and to control the release of that water by means of an adjustable gate or gates.

surrounded: Another word for "flooded" or "overflown," common in the early modern period but now obscure.

***tarde venit*:** A return made by a sheriff to a writ received too late to be executed before the required return date.

Bibliography

Manuscript Collections
Bodleian Library, Oxford University
Bankes Manuscripts
Gough Maps
Rawlinson Manuscripts

British Library

Additional Charters	Hargrave Manuscripts
Additional Manuscripts	Harleian Manuscripts
Cotton Manuscripts	Lansdowne Manuscripts
Egerton Manuscripts	Maps Collection

Cambridgeshire County Record Office
Bedford Level Corporation papers
Bedford Level Corporation petitions

Cambridge University Library
Ely Diocesan Records
Manuscripts and University Archives
Records Relating to Sewers, Drainage, and Waterways (T.XII)
University Registry Guard Books: Commons, Drains, Sewers (CUR 3)

Huntington Library
Ellesmere Manuscripts

National Archives
Chancery Papers
Exchequer Papers
Maps and Plans
State Papers, Domestic Series, for the reigns of Elizabeth I, James I, Charles I, and the
    Council of State

Norfolk Record Office
Hare Manuscripts

Parliamentary Archives
House of Lords Main Papers

University of Nottingham Manuscripts and Special Collections
Hatfield Chase Corporation Records

University of Sheffield, Humanities Research Institute
Hartlib Papers, accessed through HRI Online, http://www.hrionline.ac.uk/hartlib
/context

Printed Primary Sources

Anon. *An Answer to a Printed Paper Dispersed by Sir John Maynard.* London, 1653.
Anon. [John Maynard?]. *The Anti-Projector: Or The History of the Fen Project.* [London?], [1653?].
Anon. [Daniel Noddel?]. *A Brief Remembrance When the Report concerning the Pretended Ryot in the Isle of Axholm Shall Be Read . . . .* [London?], [1653?].
Anon. *The Case and Proceedings of at Least Sixty Gentlemen, Participants and Purchasers, for Valuable Consideration, of Lands, in the Levell of Hatfield Chase, the Counties of York, Lincolne, and Nottingham . . . .* London, 1656.
Anon. *The Case of the Mannor of Epworth in the Isle of Axholm . . . .* [London], [1695?].
Anon. [Samuel Fortrey or Jonas Moore?]. *The History or Narrative of the Great Level of the Fenns, Called Bedford Level, with a Large Map of the Said Level, as Drained, Surveyed, & Described by Sir Jonas Moore, Knight, His Late Majesties Surveyor-General of His Ordnance.* London, 1685.
Anon. [John Maynard?]. *The Humble Petition of the Inhabitants of the Soake of Peterborow.* [London?], 1650.
Anon. [John Maynard?]. *A Reply to a Printed Paper Intituled the State of the Adventurers Case.* [London?], 1650.
Anon. *The State of That Part of Yorkshire, Adjacent to the Level of Hatfield Chase: Truly and Impartially Represented: By a Lover of His Country.* York, 1701.
Anon. *The State of the Adventurers' Case, in Answer to a Petition Exhibited against them by the Inhabitants of the Soke of Peterborough.* [London?], [1650].
Anon. *To the Honourable the Knights, Citizens, and Burgesses, Assembled in the Parliament of England: The Humble Petition of Diverse Inhabitants in Norfolke, Suffolk, Huntinton[,] Northampton, Cambridge, and the Isle of Ely on the Behalfe of Themselves, and Many Thousand Inhabitants of Those Counties There Neighbours.* London, n.d. [1647].
Anon. *To the Parliament of the Common-wealth of England, Scotland, & Ireland: The Case and Proceedings of at Least 60 Gentlemen, Participants and Purchasers for Valuable Consideration of Lands, in the Levell of Hatfield Chase, the Counties of York, Lincoln, and Nottingham . . . .* London, 1654.
Anon. [John Lilburne]. *To the Right Honorable the Commons of England in Parliament Assembled.* London, 1648.
Anon. [John Maynard?]. *To the Supreme Authority, the Parliament of the Commonwealth of England: The Humble Petition of the Owners and Commoners of the Town of Islelham [sic] in the County of Cambridg.* [London?], [1652?].

Anon. "A True and Natural Description of the Great Level of the Fenns." Transcribed in Anon. [Samuel Fortrey or Jonas Moore?], *The History or Narrative of the Great Level of the Fenns, Called Bedford Level, with a Large Map of the Said Level, as Drained, Surveyed, & Described by Sir Jonas Moore, Knight, His Late Majesties Surveyor-General of His Ordnance*, 71–81. London, 1685.

Anon. *A True Report of Certaine Wonderfull Ouerflowings of Waters, Now Lately in Summerset-shire, Norfolke, and Other Places of England: Destroying Many Thousands of Men, Women, and Children, Ouerthrowing and Bearing Downe Whole Townes and Villages, and Drowning Infinite Numbers of Sheepe and Other Cattle.* London, 1607.

Badeslade, Thomas. *The History of the Ancient and Present State of the Navigation of the Port of King's-Lyn, and of Cambridge, and the Rest of the Trading-Towns in Those Parts: And of the Navigable Rivers, That Have Their Course through the Great-Level of the Fens, Called Bedford Level.* London, 1725.

Blith, Walter. *The English Improver, or a New Survey of Husbandry.* London, 1649.

———. *The English Improver Improved or the Svrvey of Hvsbandry Svrveyed.* London, 1652.

Boate, Gerard. *Irelands Natvrall History: Being a True and Ample Description of Its Situation, Greatness, Shape, and Nature. . . .* London, 1652.

Bowyer, Robert. *The Parliamentary Diary of Robert Bowyer, 1606–1607.* Edited by David Harris Willson. Minneapolis: University of Minnesota Press, 1931.

Bulstrode, Edward. *The King against Sir Anthony Mildmye [sic].* In *The English Reports*, vol. 80, edited by Max. A. Robertson and Geoffrey Ellis, 1137–38. Edinburgh: William Green & Sons, 1907.

———. *William Hetley Plaintiff, against Sir John Boyer, Sir Anthony Mildmay, Matthews, Robinson, and All Defendants.* In *The English Reports*, vol. 80, edited by Max. A. Robertson and Geoffrey Ellis, 1064–66. Edinburgh: William Green & Sons, 1907.

Burrell, Andrewes. *A Briefe Relation Discovering Plainely the True Causes Why the Great Levell of Fenns in the Severall Counties of Norfolk, Suffolk, Cambridge, Huntington, Northampton, and Lincolne Shires; Being Three Hundred and Seven Thousand Acres of Low-Lands, Have Been Drowned, and Made Unfruitfull for Many Yeares Past: And as briefly How They May Be Drained, and Preserved from Inundation in the Times to Come.* London, 1642.

———. *Exceptions against Sir Cornelius Virmudens Discourse for the Draining of the Great Fennes, &c. Which in Ianuary 1638. He Presented to the King for His Designe: Wherein His Majesty Was Mis-informed and Abused, in Regard It Wanteth All the Essentiall Parts of a Designe; And the Great and Advantagious Workes Made by the Late Earle of Bedford, Slighted; And the Whole Adventure Disparaged.* London, 1642.

———. *An Explanation of the Drayning Workes Which Have Beene Lately Made for the Kings Maiestie in Cambridge Shire, by the Direction of Sir Cornelius Virmuden: Written by Andrewes Burrell, Gent. Wherein Is Discovered How the Said Sir Cornelius Hath Abused the Kings Maiestie, and Many of His Loving Subjects.* London, 1641.

Burton, Thomas. *Diary of Thomas Burton, Esq., Member in the Parliaments of Oliver and Richard Cromwell, from 1656–59. . . .* Edited by John Towill Rutt. 4 vols. London: Henry Colburn, 1828.

C., H. A Discovrse concerning the Drayning of the Fennes and Svrrovnded Grovnds in the Sixe Counteys of Norfolke, Suffolke, Cambridge with the Isle of Ely, Huntington, Northampton, and Lincolne. London, 1629.

Callis, Robert. The Reading of That Famous and Learned Gentleman Robert Callis, Esq., Sergeant at Law, upon the Statute of 23 H.8 Cap. 5 of Sewers: As It Was Delivered by Him at Grays Inn in August, 1622. London, 1647. 2nd ed., 1685.

Camden, William. Britain, or a Chorographicall Description of the Most Flourishing Kingdomes, England, Scotland, and Ireland, and the Ilands Adioyning, out of the Depth of Antiquitie: Beavtified with Mappes of the Severall Shires of England. Translated by Philémon Holland. London, 1610.

Coke, Edward. The Case of Chester Mill upon the River of Dee. In The English Reports, vol. 77, edited by Max. A. Robertson and Geoffrey Ellis, 1134–36. Edinburgh: William Green & Sons, 1907.

———. The Case of the Isle of Ely. In The English Reports, vol. 77, edited by Max. A. Robertson and Geoffrey Ellis, 1139–43. Edinburgh: William Green & Sons, 1907.

———. Keighley's Case. In The English Reports, vol. 77, edited by Max. A. Robertson and Geoffrey Ellis, 1136–39. Edinburgh: William Green & Sons, 1907.

———. Prohibitions del Roy. In The English Reports, vol. 77, edited by Max. A. Robertson and Geoffrey Ellis, 1342–43. Edinburgh: William Green & Sons, 1907.

———. Rooke's Case. In The English Reports, vol. 77, edited by Max. A. Robertson and Geoffrey Ellis, 209–10. Edinburgh: William Green & Sons, 1907.

Council of State. By His Excellency the Lord General and the Council of State. London, 1653.

———. A Commission Impowering the Persons Therin Named, to Hear and Determine Severall Matters and Things concerning the Work of Dreyning the Great Levell of the Fenns. London, 1653.

———. An Exemplification under the Great Seal of England, of an Ordinance of His Highness the Lord Protector. . . . London, 1657.

———. An Ordinance for the Preservation of the Works of the Great Level of the Fenns. London, 1654.

Dasent, John Roche, et al., eds. Acts of the Privy Council of England. New series. 46 vols. London: H. M. Stationery Office, 1890–1964.

Defoe, Daniel. An Essay upon Projects. London, 1697.

———. A Tour thro' the Whole Island of Great Britain, Divided into Circuits or Journeys. 3 vols. London, 1724–27.

D'Ewes, Sir Simonds. The Journal of Sir Simonds D'Ewes, from the Beginning of the Long Parliament to the Opening of the Trial of the Earl of Stratford. Edited by Wallace Notestein. New Haven, CT: Yale University Press, 1923.

Dodson, William, The Designe for the Present Draining of the Great Level of the Fens, (Called Bedford Level) Lying in Norfolk, Suffolk, Cambridgeshire, Huntingtonshire, Northamptonshire, Lincolnshire, and the Isle of Ely. London, 1665.

Drayton, Michael. The Second Part, or A Continvance of Poly-olbion from the Eighteenth Song. London, 1622.

Dugdale, William. The History of Imbanking and Drayning of Divers Fenns and Marshes, Both in Forein Parts, and in This Kingdom; and of the Improvements

Thereby: Extracted from Records, Manuscripts, and Other Authentick Testimonies. London: Printed by Alice Warren, 1662.

Dymock, Cressy. *A Discoverie for Division or Setting Out of Land, as to the Best Form.* Edited by Samuel Hartlib. London, 1653.

Elstobb, William. *Observations on an Address to the Public, Dated April, 20, 1775, Superscribed Bedford Level, and Sign'd Charles Nalson Cole, Register. . . .* Lynn, 1776.

Firth, C. H., and R. S. Rait, eds. *Acts and Ordinances of the Interregnum, 1642–1660.* 3 vols. London: H. M. Stationery Office, 1911.

Fortrey, Samuel. *Englands Interest and Improvement: Consisting in the Increase of the Store, and Trade of This Kingdom.* Cambridge, 1663.

Fuller, Thomas. *The History of the University of Cambridge, since the Conquest.* Printed with *The Church-History of Britain, from the Birth of Jesus Christ untill the Year MDCXLVIII.* London, 1655.

Hartlib, Samuel, ed. *Samuel Hartlib His Legacie: Or An Enlargement of the Discourse of Husbandry Used in Brabant and Flaunders.* London, 1651.

Herne, John. *The Learned Reading of John Herne Esq; Late of the Honourable Society of Lincolns-Inne upon the State of 23 H.8 Cap. 3. [sic] concerning Commissions of Sewers, Translated out of the French Manuscript.* London, 1659.

Historical Manuscripts Commission. *Calendar of the Manuscripts of the Most Honourable, the Marquess of Salisbury, Preserved at Hatfield House, Hertfordshire.* 24 vols. London: H. M. Stationery Office, 1883–1976.

———. *Report on the Manuscripts of the Earl of Ancaster Preserved at Grimsthorpe.* Dublin: H. M. Stationery Office, 1907.

Holinshed, Raphael, et al. *The First and Second Volumes of Chronicles Comprising 1 The Description and Historie of England, 2 The Description and Historie of Ireland, 3 The Description and Historie of Scotland: First Collected and Published by Raphaell Holinshed, William Harrison, and Others; Now Newlie Augmented and Continued (with Manifold Matters of Singular Note and Worthie Memorie) to the Yeare 1586. By Iohn Hooker aliàs Vowell Gent and Others.* 3 vols. London, 1587.

Jonson, Ben. *The Devil Is an Ass* (London, 1616). In *The Complete Plays of Ben Jonson*, vol. 4, edited by G. A. Wilkes. Oxford: Oxford University Press, 1982.

Kirkus, Mary A., and A.E.B. Owen, eds. *The Records of the Commissioners of Sewers in the Parts of Holland, 1547–1603.* In *Publications of the Lincoln Record Society* 54 (1959), 63 (1968), and 71 (1977).

Knell, Thomas. *A Declaration of Such Tempestious, and Outragious Fluddes, as Hath Been in Divers Places of England.* London, 1570.

Lilburne, John. *The Case of the Tenants of the Mannor of Epworth. . . .* London, 1651.

Maynard, John. *The Picklock of the Old Fenne Project: Or, Heads of Sir John Maynard His Severall Speeches, Taken in Short-Hand, at the Committee for Lincolnshire Fens, in Exchequer Chamber.* London, 1650.

Moore, Francis. *The Lord Chancellor's Speech to Sir Henry Montague, When He Was Sworn Chief Justice of the Kings-Bench.* In *The English Reports*, vol. 72, edited by Max. A. Robertson and Geoffrey Ellis, 931–32. Edinburgh: William Green & Sons, 1907.

Moore, Jonas. *A Mapp of ye Great Levell of ye Fenns Extending into ye Covntyes of Northampton, Norfolk, Suffolke, Lyncolne, Cambridg & Huntington & the Isle of Ely as It Is Now Drained, Described by Sr Jonas Moore Surveyr Genll.* London, 1658. 3rd ed., 1706.

Noddel, Daniel. *To the Parliament of the Commonwealth of England, and Every Individual Member Thereof: The Declaration of Daniel Noddel. . . .* London, 1653.

———. *To the Parliament of the Commonwealth of England, and Every Individual Member Thereof: The Great Complaint and Declaration of about 1200. Free-Holders and Commoners, within the Mannor of Epworth. . . .* London, 1654.

Prynne, William. *A Declaration of the Officers and Armies, Illegal, Injurious, Proceedings and Practices against the XI. Impeached Members.* London, 1647.

Rolle, Henry. *Headley v. Sir Anth. Mildmay.* In *The English Reports,* vol. 81, edited by Max. A. Robertson and Geoffrey Ellis, 560. Edinburgh: William Green & Sons, 1907.

Scotten, Edmund. *A Desperate and Dangerovs Designe Discovered concerning the Fen-Countries, by a Faithfull Friend, Who as Soone as It Came to His Knowledge, Hath Taken Some Pains, Not Only to Discover, but to Prevent the Same: By Order of the Committee for the Fens. Published for the Common-Good, and in All Humility Presented to the High Court of Parliament. And in particular to Some Noble Personages Especially Interested and Concerned Therein.* London, 1642.

Swift, Jonathan. *Travels into Several Remote Nations of the World. In Four Parts. By Lemuel Gulliver, First a Surgeon, and Then a Captain of Several Ships.* Dublin, 1726.

Tarlton, Richard. *A Very Lamentable and Woful Discours of the Fierce Fluds. . . .* London, 1570.

Vermuyden, Cornelius. *A Discourse Touching the Drayning the Great Fennes, Lying within the Several Covnties of Lincolne, Northampton, Huntington, Norfolke, Suffolke, Cambridge, and the Isle of Ely, as It Was Presented to His Majestie.* London, 1642.

Wells, Samuel, ed. *The History of the Drainage of the Great Level of the Fens, Called Bedford Level: With the Constitution and Laws of the Bedford Level Corporation,* vol. 2. London: R. Pheney, 1830.

Secondary Sources

Adams, Nicholas. "Architecture for Fish: The Sienese Dam on the Bruna River—Structures and Designs, 1468–ca. 1530." *Technology and Culture* 25 (1984): 768–97.

Albright, Margaret. "The Entrepreneurs of Fen Draining in England under James I and Charles I: An Illustration of the Uses of Influence." *Explorations in Entrepreneurial History* 8:2 (1955): 51–65.

Alford, Stephen. *Burghley: William Cecil at the Court of Elizabeth I.* New Haven, CT: Yale University Press, 2008.

Appleby, Joyce Oldham. *Economic Thought and Ideology in Seventeenth-Century England.* Princeton, NJ: Princeton University Press, 1978.

Appuhn, Karl. *A Forest on the Sea: Environmental Expertise in Renaissance Venice.* Baltimore, MD: Johns Hopkins University Press, 2009.

Ash, Eric H. "Amending Nature: Draining the English Fens." In *The Mindful Hand: Inquiry and Invention from the Late Renaissance to Early Industrialisation,* edited by Lissa Roberts, Simon Schaffer, and Peter Dear, 117–43. Amsterdam: Edita; Chicago: University of Chicago Press, 2007.

———. "Expertise and the Early Modern State." *Osiris* 25 (2010): 1–24.

———, ed. *Expertise: Practical Knowledge and the Early Modern State, Osiris* 25 (2010).

——. *Power, Knowledge, and Expertise in Elizabethan England*. Baltimore, MD: Johns Hopkins University Press, 2004.

——. "Queen v. Northumberland, and the Control of Technical Expertise." *History of Science* 39 (2001): 215–40.

Ashworth, William J. *Customs and Excise: Trade, Production and Consumption in England 1640–1845*. Oxford: Oxford University Press, 2003.

Baker, Sir John. "The Common Lawyers and the Chancery: 1616." In *Law, Liberty, and Parliament: Selected Essays on the Writings of Sir Edward Coke*, edited by Allen D. Boyer, 254–81. Indianapolis, IN: Liberty Fund, 2004.

Barclay, Andrew. *Electing Cromwell: The Making of a Politician*. London: Pickering and Chatto, 2011.

Barrera-Osorio, Antonio. *Experiencing Nature: The Spanish American Empire and the Early Scientific Revolution*. Austin: University of Texas Press, 2006.

Bernard, T. C. "The Hartlib Circle and the Cult and Culture of Improvement in Ireland." In *Samuel Hartlib and Universal Reformation*, edited by Mark Greengrass, Michael Leslie, and Timothy Raylor, 281–97. Cambridge: Cambridge University Press, 1994.

Blackbourn, David. *The Conquest of Nature: Water, Landscape, and the Making of Modern Germany*. New York: W. W. Norton, 2006.

Bowring, Julie. "Between the Corporation and Captain Flood: The Fens and Drainage after 1663." In *Custom, Improvement and the Landscape in Early Modern Britain*, edited by Richard W. Hoyle, 235–61. Farnham: Ashgate, 2011.

Braddick, Michael J. *God's Fury, England's Fire: A New History of the English Civil Wars*. London: Penguin Books, 2009.

——. *The Nerves of State: Taxation and the Financing of the English State, 1558–1714*. Manchester and New York: Manchester University Press, 1996.

——. *State Formation in Early Modern England, c. 1550–1700*. Cambridge: Cambridge University Press, 2000.

Braddick, Michael J., and John Walter, eds. *Negotiating Power in Early Modern Society: Order, Hierarchy and Subordination in Britain and Ireland*. Cambridge: Cambridge University Press, 2001.

Brewer, John, and John Styles, eds. *An Ungovernable People: The English and Their Law in the Seventeenth and Eighteenth Centuries*. New Brunswick, NJ: Rutgers University Press, 1980.

Butlin, Robin A. "Images of the Fenland Region." In *Issues of Regional Identity: In Honour of John Marshall*, edited by Edward Royle, 25–43. Manchester: Manchester University Press, 1998.

——. *The Transformation of Rural England, c. 1500–1800: A Study in Historical Geography*. Oxford: Oxford University Press, 1982.

Capp, Bernard. "Separate Domains? Women and Authority in Early Modern England." In *The Experience of Authority in Early Modern England*, edited by Paul Griffiths, Adam Fox, and Steve Hindle, 117–45. New York: St. Martin's Press, 1996.

Carroll, Patrick. *Science, Culture and Modern State Formation*. Berkeley: University of California Press, 2006.

Cipolla, Carlo M. *Guns, Sails and Empires: Technological Innovation and the Early Phases of European Expansion 1400–1700*. New York: Pantheon Books, 1965.

Ciriacono, Salvatore. *Building on Water: Venice, Holland and the Construction of the European Landscape in Early Modern Times.* Translated by Jeremy Scott. New York: Berghahn Books, 2006.

——, ed. *Eau et développement dans l'Europe modern.* Paris: Éditions de la Maison des Sciences de l'Homme, 2004.

Collinson, Patrick. "*De Republica Anglorum*: Or, History with the Politics Put Back" and "The Monarchical Republic of Queen Elizabeth I." In *Elizabethan Essays.* London: Hambledon Press, 1994.

Cope, Esther S., and Willson H. Coates, eds. *Proceedings of the Short Parliament of 1640. Camden Fourth Series* 19 (1977).

Cook, Hadrian, and Tom Williamson, eds. *Water Management in the English Landscape: Field, Marsh, Meadow.* Edinburgh: Edinburgh University Press, 1999.

Cooper, Tim. "Modernity and the Politics of Waste in Britain." In *Nature's End: History and the Environment*, edited by Sverker Sörlin and Paul Warde, 247–72. Houndmills: Palgrave Macmillan, 2009.

Cormack, Lesley B. "Twisting the Lion's Tail: Practice and Theory at the Court of Henry Prince of Wales." In *Patronage and Institutions: Science, Technology, and Medicine at the European Court, 1500–1700*, edited by Brice T. Moran, 67–83. Rochester, NY: Boydell Press, 1991.

Cosgrove, Denis. "An Elemental Division: Water Control and Engineered Landscape." In *Water, Engineering and Landscape: Water Control and Landscape Transformation in the Modern Period*, edited by Denis Cosgrove and Geoff Petts, 1–11. London: Belhaven Press, 1990.

Cosgrove, Denis, and Geoff Petts, eds. *Water, Engineering and Landscape: Water Control and Landscape Transformation in the Modern Period.* London: Belhaven Press, 1990.

Coughlan, Patricia. "Natural History and Historical Nature: The Project for a Natural History of Ireland." In *Samuel Hartlib and Universal Reformation*, edited by Mark Greengrass, Michael Leslie, and Timothy Raylor, 298–318. Cambridge: Cambridge University Press, 1994.

Cramsie, John. "Commercial Projects and the Fiscal Policy of James VI and I." *Historical Journal* 43 (2000): 345–64.

Cronon, William. *Changes in the Land: Indians, Colonists, and the Ecology of New England.* New York: Hill and Wang, 1983.

Cunningham, W. "Common Rights at Cottenham and Stretham in Cambridgeshire." *Camden Miscellany* 12 (1910): 173–287.

Cust, Richard. *The Forced Loan and English Politics, 1626–1628.* Oxford: Clarendon Press, 1987.

Danner, Helga S., et al., eds. *Polder Pioneers: The Influence of Dutch Engineers on Water Management in Europe, 1600–2000.* Utrecht: Koninklijk Nederlands Aardrijkskundig Genootschap Faculteit Geowetenschappen Universiteit Utrecht, 2005.

Darby, H. C. *The Changing Fenland.* Cambridge: Cambridge University Press, 1983.

——. *The Draining of the Fens.* Cambridge: Cambridge University Press, 1940. 2nd ed., 1956.

——. *The Medieval Fenland.* Cambridge: Cambridge University Press, 1940. 2nd ed., Newton Abbot: David and Charles, 1974.

Dear, Peter. "Mysteries of State, Mysteries of Nature: Authority, Knowledge and Expertise in the Seventeenth Century." In *States of Knowledge: The Co-production of Science and Social Order*, edited by Sheila Jasanoff, 206–24. London: Routledge, 2004.

Di Palma, Vittoria. *Wasteland: A History*. New Haven, CT: Yale University Press, 2014.

Dobson, Mary J. *Contours of Death and Disease in Early Modern England*. Cambridge: Cambridge University Press, 1997.

Drayton, Richard. *Nature's Government: Science, Imperial Britain, and the "Improvement" of the World*. New Haven, CT: Yale University Press, 2000.

Edelson, S. Max. *Plantation Enterprise in Colonial South Carolina*. Cambridge, MA: Harvard University Press, 2006.

Evans, Robert C. "Contemporary Contexts of Jonson's *The Devil Is an Ass*." *Comparative Drama* 26 (1992): 140–76.

Falvey, Heather. "Custom, Resistance, and Politics: Local Experiences of Improvement in Early Modern England." PhD diss., University of Warwick, 2007.

Fletcher, Anthony. *Reform in the Provinces: The Government of Stuart England*. New Haven, CT: Yale University Press, 1986.

Fox, Adam. "Custom, Memory and the Authority of Writing." In *The Experience of Authority in Early Modern England*, edited by Paul Griffiths, Adam Fox, and Steve Hindle, 89–116. New York: St. Martin's Press, 1996.

———. *Oral and Literate Culture in England, 1500–1700*. Oxford: Clarendon Press, 2000.

Gaukroger, Stephen. *Francis Bacon and the Transformation of Early-Modern Philosophy*. Cambridge: Cambridge University Press, 2001.

Geertz, Clifford. *Agricultural Involution: The Process of Ecological Change in Indonesia*. Berkeley: University of California Press, 1968.

Goldie, Mark. "The Unacknowledged Republic: Officeholding in Early Modern England." In *The Politics of the Excluded, c. 1500–1850*, edited by Tim Harris, 153–94. Houndmills: Palgrave, 2001.

Goodman, David C. *Power and Penury: Government, Technology, and Science in Philip II's Spain*. Cambridge: Cambridge University Press, 1988.

Grainger, John D. *Cromwell against the Scots: The Last Anglo-Scottish War, 1650–1652*. East Lothian, Scotland: Tuckwell Press, 1997.

Greengrass, Mark, Michael Leslie, and Timothy Raylor, eds. *Samuel Hartlib and Universal Reformation*. Cambridge: Cambridge University Press, 1994.

Gregg, Pauline. *Free-Born John: A Biography of John Lilburne*. London: J. M. Dent, 1961. Reprint, London: Phoenix, 2001.

Griffiths, Paul, Adam Fox, and Steve Hindle, eds. *The Experience of Authority in Early Modern England*. New York: St. Martin's Press, 1996.

Harkness, Deborah E. *The Jewel House: Elizabethan London and the Scientific Revolution*. New Haven, CT: Yale University Press, 2007.

Harris, L. E. *The Two Netherlanders: Humphrey Bradley and Cornelis Drebbel*. Leiden: E. J. Brill, 1961.

———. *Vermuyden and the Fens: A Study of Sir Cornelius Vermuyden and the Great Level*. London: Cleaver-Hume Press, 1953.

Harris, Tim. "Introduction." In *The Politics of the Excluded, c. 1500–1850*, edited by Tim Harris, 1–25. Basingstoke: Palgrave, 2001.

Heal, Felicity, and Clive Holmes. "The Economic Patronage of William Cecil." In *Patronage, Culture and Power: The Early Cecils*, edited by Pauline Croft. New Haven, CT: Yale University Press, 2002.

Heller, Henry. *Labour, Science and Technology in France, 1500–1620.* Cambridge: Cambridge University Press, 1996.

Hill, Christopher. *The World Turned Upside Down: Radical Ideas during the English Revolution.* London: Maurice Temple Smith, 1972. Reprint, New York: Penguin Books, 1991.

Hills, Richard L. *Machines, Mills, and Uncountable Costly Necessities: A Short History of the Drainage of the Fens.* Norwich: Goose and Son, 1967.

Hindle, Steve. "The Political Culture of the Middling Sort in English Rural Communities, c. 1550–1700." In *The Politics of the Excluded, c. 1500–1850*, edited by Tim Harris, 125–52. Basingstoke: Palgrave, 2001.

——. *The State and Social Change in Early Modern England, 1550–1640.* Basingstoke: Palgrave, 2000.

Hirst, Derek. *England in Conflict, 1603–1660: Kingdom, Community, Commonwealth.* London: Arnold, 1999.

——. "The Privy Council and Problems of Enforcement in the 1620s." *Journal of British Studies* 18:1 (1978): 46–66.

Holmes, Clive. "Drainage Projects in Elizabethan England: The European Dimension." In *Eau et développement dans l'Europe moderne*, edited by Salvatore Ciriacono. Paris: Éditions de la Maison des Sciences de l'Homme, 2004.

——. "Drainers and Fenmen: The Problem of Popular Political Consciousness in the Seventeenth Century." In *Order and Disorder in Early Modern England*, edited by Anthony Fletcher and John Stevenson, 166–95. Cambridge: Cambridge University Press, 1985.

——. *Seventeenth-Century Lincolnshire.* Lincoln: History of Lincolnshire Committee, 1980.

——. "Statutory Interpretation in the Early Seventeenth Century: The Courts, the Council, and the Commissioners of Sewers." In *Law and Social Change in British History: Papers Presented to the Bristol Legal History Conference, 14–17 July 1981*, edited by J. A. Guy and H. G. Beale, 107–17. London: Royal Historical Society, 1984.

Hopper, A. J. "The Clubmen of the West Riding of Yorkshire during the First Civil War: 'Bradford Club-Law.'" *Northern History* 36 (2000): 59–72.

Hoyle, R[ichard] W. "Disafforestation and Drainage: The Crown as Entrepreneur?" In *The Estates of the English Crown, 1558–1640*, edited by R. W. Hoyle, 353–88. Cambridge: Cambridge University Press, 1992.

——, ed. *The Estates of the English Crown, 1558–1640.* Cambridge: Cambridge University Press, 1992.

——. "Introduction: Custom, Improvement, and Anti-Improvement." In *Custom, Improvement and the Landscape in Early Modern Britain*, edited by Richard W. Hoyle, 1–38. Farnham: Ashgate, 2011.

Hughes, J. D. "The Drainage Disputes in the Isle of Axholme and the Connexion with the Leveller Movement: A Re-examination." *Lincolnshire Historian* 2 (1954): 13–45.

Hunt, John Dixon. "Hortulan Affairs." In *Samuel Hartlib and Universal Reformation*, edited by Mark Greengrass, Michael Leslie, and Timothy Raylor, 321–42. Cambridge: Cambridge University Press, 1994.

Hunter, Joseph. *South Yorkshire: The History and Topography of the Deanery of Doncaster, in the Diocese and County of York*. 2 vols. London, 1828–31.

Isenberg, Andrew C. *The Destruction of the Bison: An Environmental History, 1750–1920*. Cambridge: Cambridge University Press, 2000.

James, N. "The 'Age of the Windmill' in the Haddenham Level." *Proceedings of the Cambridge Antiquarian Society* 98 (2009): 113–20.

Keeler, Mary Frear. *The Long Parliament, 1640–1641: A Biographical Study of Its Members*. Philadelphia: American Philosophical Society, 1954.

Keller, Vera, and Ted McCormick, eds. "Towards a History of Projects." Special issue of *Early Science and Medicine* 21:5 (2016).

Kennedy, Mark Edmund. "Charles I and Local Government: The Drainage of the East and West Fens." *Albion* 15 (1983): 19–31.

———. "Commissions of Sewers for Lincolnshire, 1509–1649: An Annotated List." *Lincolnshire History and Archaeology* 19 (1984): 83–88.

———. "Fen Drainage, the Central Government, and Local Interest: Carleton and the Gentlemen of South Holland." *Historical Journal* 26 (1983): 15–37.

———. "So Glorious a Work as This Draining of the Fens: The Impact of Royal Government on Local Political Culture in Elizabethan and Jacobean England." PhD diss., Cornell University, 1985.

Kerridge, Eric. *The Agricultural Revolution*. London: George Allen and Unwin, 1967.

Kesselring, K. J. *Mercy and Authority in the Tudor State*. Cambridge: Cambridge University Press, 2003.

Kirkus, A. Mary. "Introduction." In *The Records of the Commissioners of Sewers in the Parts of Holland, 1547–1603*, edited by Mary A. Kirkus. *Publications of the Lincoln Record Society* 54 (1959): vii–xxxvii.

Kishlansky, Mark. *A Monarchy Transformed: Britain, 1603–1714*. New York: Penguin Books, 1997.

Knafla, Louis A. *Law and Politics in Jacobean England: The Tracts of Lord Chancellor Ellesmere*. Cambridge: Cambridge University Press, 1977.

Knittl, Margaret Albright. "The Design for the Initial Drainage of the Great Level of the Fens: An Historical Whodunit in Three Parts." *Agricultural History Review* 55 (2007): 23–50.

Korthals-Altes, J. *Sir Cornelius Vermuyden: The Lifework of a Great Anglo-Dutchman in Land-Reclamation and Drainage*. London: Williams and Norgate; The Hague: W. P. Stockum and Son, 1925.

Kupperman, Karen. "Controlling Nature and Colonial Projects in Early America." In *Colonial Encounters: Essays in Early American History and Culture*, edited by Hans-Jürgen Grabbe, 69–88. Heidelberg: Universitätsverlag Winter, 2003.

Leary, John E., Jr. *Francis Bacon and the Politics of Science*. Ames: Iowa State University Press, 1994.

Leng, Thomas. "'A Potent Plantation Well Armed and Policed': Huguenots, the Hartlib Circle, and British Colonization in the 1640s." *William and Mary Quarterly* 66 (2009): 173–94.

Leslie, Michael, and Timothy Raylor, eds. *Culture and Cultivation in Early Modern England: Writing and the Land.* Leicester: Leicester University Press, 1992.

Lindley, Keith. *Fenland Riots and the English Revolution.* London: Heinemann Educational Books, 1982.

Long, Pamela O. *Artisan/Practitioners and the Rise of the New Sciences, 1400–1600.* Corvallis: Oregon State University Press, 2011.

———. *Openness, Secrecy, Authorship: Technical Arts and the Culture of Knowledge from Antiquity to the Renaissance.* Baltimore, MD: Johns Hopkins University Press, 2001.

Manning, Roger B. *Village Revolts: Social Protest and Popular Disturbances in England, 1509–1640.* Oxford: Clarendon Press, 1988.

Martin, Julian. *Francis Bacon, the State, and the Reform of Natural Philosophy.* Cambridge: Cambridge University Press, 1992.

McDiarmid, John F., ed. *The Monarchical Republic of Early Modern England: Essays in Response to Patrick Collinson.* Aldershot: Ashgate, 2007.

McRae, Andrew. *God Speed the Plough: The Representation of Agrarian England, 1550–1660.* Cambridge: Cambridge University Press, 1996.

———. "Tree-Felling in Early Modern England: Michael Drayton's Environmentalism." *Review of English Studies* 63 (2011): 410–30.

Merchant, Carolyn. *The Death of Nature: Women, Ecology, and the Scientific Revolution.* San Francisco: Harper and Row, 1980.

———. *Ecological Revolutions: Nature, Gender, and Science in New England.* Chapel Hill: University of North Carolina Press, 1989.

Miller, Samuel H., and Sydney B. J. Skertchly. *The Fenland Past and Present.* Wisbech: Leach and Son; London: Longmans, Green, 1878.

Montaño, John Patrick. *The Roots of English Colonialism in Ireland.* Cambridge: Cambridge University Press, 2011.

Morrill, John. *Revolt in the Provinces: The People of England and the Tragedies of War, 1630–1648.* 2nd ed. London: Longman, 1999.

Mukerji, Chandra. *Impossible Engineering: Technology and Territoriality on the Canal du Midi.* Princeton, NJ: Princeton University Press, 2009.

Neeson, J. M. *Commoners: Common Rights, Enclosure, and Social Change in England, 1700–1820.* Cambridge: Cambridge University Press, 1993.

Novak, Maximillian E., ed. *The Age of Projects.* Toronto: University of Toronto Press, 2008.

Nummedal, Tara. *Alchemy and Authority in the Holy Roman Empire.* Chicago: University of Chicago Press, 2007.

Peck, W. *A Topographical Account of the Isle of Axholme, Being the West Division of the Wapentake of Manley, in the County of Lincoln.* 2 vols. Doncaster, 1815.

Pérez-Ramos, Antonio. *Francis Bacon's Idea of Science and the Maker's Knowledge Tradition.* Oxford: Clarendon Press, 1988.

Pluymers, Keith. "Taming the Wilderness in Sixteenth- and Seventeenth-Century Ireland and Virginia." *Environmental History* 16 (2011): 610–32.

Rabier, Christelle, ed. *Fields of Expertise: A Comparative History of Expert Procedures in Paris and London, 1600 to Present*. Newcastle: Cambridge Scholars Publishing, 2007.

Rackham, Oliver. *The History of the Countryside: The Classic History of Britain's Landscape, Flora, and Fauna*. London: J. M. Dent, 1986.

Ratcliff, Jessica. "Art to Cheat the Common-Weale: Inventors, Projectors, and Patentees in English Satire, ca. 1630–70." *Technology and Culture* 53 (2012): 337–65.

Ravensdale, J. R. *Liable to Floods: Village Landscape on the Edge of the Fens, AD 450–1850*. Cambridge: Cambridge University Press, 1974.

Raylor, Timothy. "Samuel Hartlib and the Commonwealth of Bees." In *Culture and Cultivation in Early Modern England: Writing and the Land*, edited by Michael Leslie and Timothy Raylor, 91–129. Leicester: Leicester University Press, 1992.

Reeves, Anne, and Tom Williamson. "Marshes." In *The English Rural Landscape*, edited by Joan Thirsk, 150–66. Oxford: Oxford University Press, 2000.

Renes, Johannes. "The Fenlands of England and the Netherlands: Some Thoughts on Their Different Histories." In *European Landscapes: From Mountain to Sea*, edited by Tim Unwin and Theo Speck, 101–7. Tallinn, Estonia: Human Publishers, 2003.

Rice, Douglas Walthew. *The Life and Achievements of Sir John Popham, 1531–1607: Leading to the Establishment of the First English Colony in New England*. Madison, NJ: Fairleigh Dickinson University Press, 2005.

Richards, John F. *The Unending Frontier: An Environmental History of the Early Modern World*. Berkeley: University of California Press, 2003.

Richardson, H. G. "The Early History of Commissions of Sewers." *English Historical Review* 34 (1919): 385–93.

Roberts, Keith. *Cromwell's War Machine: The New Model Army, 1645–1660*. Barnsley, UK: Pen and Sword Books, 2005.

Sandman, Alison. "Mirroring the World: Sea Charts, Navigation, and Territorial Claims in Sixteenth-Century Spain." In *Merchants and Marvels: Commerce, Science, and Art in Early Modern Europe*, edited by Pamela H. Smith and Paula Findlen, 83–108. New York: Routledge, 2001.

Schaffer, Simon. "The Earth's Fertility as a Social Fact in Early Modern Britain." In *Nature and Society in Historical Context*, edited by Mikulas Teich, Roy Porter, and Bo Gustafsson, 124–47. Cambridge: Cambridge University Press, 1997.

Scott, James C. *Domination and the Arts of Resistance: Hidden Transcripts*. New Haven, CT: Yale University Press, 1992.

Shagan, Ethan. *Popular Politics and the English Reformation*. Cambridge: Cambridge University Press, 2003.

Sharp, Buchanan. *In Contempt of All Authority: Rural Artisans and Riot in the West of England, 1586–1660*. Berkeley: University of California Press, 1980.

Skelton, R. A., and John Summerson, eds. *A Description of Maps and Architectural Drawings in the Collection Made by William Cecil, First Baron Burghley, Now at Hatfield House*. Oxford: Oxford University Press, 1971.

Skertchly, Sydney B. J. *The Geology of the Fenland*. London: H. M. Stationery Office, 1877.

Slack, Paul. *From Reformation to Improvement: Public Welfare in Early Modern England*. Oxford: Clarendon Press, 1999.

———. *The Invention of Improvement: Information and Material Progress in Seventeenth-Century England*. Oxford: Oxford University Press, 2015.

Smiles, Samuel. *Lives of the Engineers*. New and rev. ed. 5 vols. Volume 1, *Early Engineering*. London: John Murray, 1874.

Smith, A. Hassell. *County and Court: Government and Politics in Norfolk, 1558–1603*. Oxford: Clarendon Press, 1974.

Smith, David Chan. *Sir Edward Coke and the Reformation of the Laws: Religion, Politics and Jurisprudence, 1578–1616*. Cambridge: Cambridge University Press, 2014.

Smith, Pamela H. *The Business of Alchemy: Science and Culture in the Holy Roman Empire*. Princeton, NJ: Princeton University Press, 1994.

Spufford, Margaret. *Contrasting Communities: English Villagers in the Sixteenth and Seventeenth Centuries*. Cambridge: Cambridge University Press, 1974.

Steinberg, Theodore. *Nature Incorporated: Industrialization and the Waters of New England*. Cambridge: Cambridge University Press, 1991.

Stewart, Larry. *The Rise of Public Science: Rhetoric, Technology, and Natural Philosophy*. Cambridge: Cambridge University Press, 1992.

Stonehouse, W. B. *The History and Topography of the Isle of Axholme: Being That Part of Lincolnshire Which Is West of Trent*. London, 1839.

Stoyle, Mark. *Loyalty and Locality: Popular Allegiance in Devon during the English Civil War*. Exeter: University of Exeter Press, 1994.

Summers, Dorothy. *The Great Level: A History of Drainage and Land Reclamation in the Fens*. Newton Abbot: David and Charles, 1976.

Taylor, Christopher. "Fenlands." In *The English Rural Landscape*, edited by Joan Thirsk, 167–87. Oxford: Oxford University Press, 2000.

Thirsk, Joan, gen. ed. *The Agrarian History of England and Wales*. Vol. 4: 1500–1640. Cambridge: Cambridge University Press, 1967.

———. *The Agrarian History of England and Wales*. Vol. 5, pt. 1: *Regional Farming Systems, 1640–1750*. Cambridge: Cambridge University Press, 1984.

———. *The Agrarian History of England and Wales*. Vol. 5, pt. 2: *Agrarian Change, 1640–1750*. Cambridge: Cambridge University Press, 1985.

———. *Alternative Agriculture: A History from the Black Death to the Present Day*. Oxford: Oxford University Press, 1997.

———. *Economic Policy and Projects: The Development of a Consumer Society in Early Modern England*. Oxford: Clarendon Press, 1978.

———. *English Peasant Farming: The Agrarian History of Lincolnshire from Tudor to Recent Times*. London: Routledge and Kegan Paul, 1957.

———. "The Isle of Axholme before Vermuyden." *Agricultural History Review* 1 (1953): 16–28.

———. "Plough and Pen: Agricultural Writers in the Seventeenth Century." In *Social Relations and Ideas: Essays in Honour of R. H. Hinton*, edited by T. H. Aston et al., 295–318. Cambridge: Cambridge University Press, 1983.

———. *The Rural Economy of England: Collected Essays*. London: Hambledon Press, 1984.

———. "Seventeenth-Century Agriculture and Social Change." In *Land, Church, and People: Essays Presented to Professor H.P.R. Finberg*, edited by Joan Thirsk, 148–77. Reading: British Agricultural History Society, 1970.

Thomas, Keith. *Man and the Natural World: Changing Attitudes in England, 1500–1800.* London: Allen Lane, 1983.

Thompson, E. P. *Customs in Common: Studies in Traditional Popular Culture.* New York: New Press, 1993.

Underdown, David. *A Freeborn People: Politics and the Nation in Seventeenth-Century England.* Oxford: Clarendon Press, 1996.

———. *Revel, Riot, and Rebellion: Popular Politics and Culture in England, 1603–1660.* Oxford: Clarendon Press, 1985.

van de Ven, G. P., ed. *Man-Made Lowlands: History of Water Management and Land Reclamation in the Netherlands.* Utrecht: Matrijs, 1993. 3rd rev. ed., 1996.

Vickers, Brian. "Francis Bacon and the Progress of Knowledge." *Journal of the History of Ideas* 53 (1992): 495–518.

Wakefield, Andre. *The Disordered Police State: German Cameralism as Science and Practice.* Chicago: University of Chicago Press, 2009.

Walter, John. "Grain Riots and Popular Attitudes to the Law: Maldon and the Crisis of 1629." In *An Ungovernable People: The English and Their Law in the Seventeenth and Eighteenth Centuries*, edited by John Brewer and John Styles, 47–84. New Brunswick, NJ: Rutgers University Press, 1980.

———. "'The Pooremans Joy and the Gentlemans Plague': A Lincolnshire Libel and the Politics of Sedition in Early Modern England." *Past & Present* 203 (2009): 29–67.

Warde, Paul. *Ecology, Economy and State Formation in Early Modern Germany.* Cambridge: Cambridge University Press, 2006.

———. "The Environmental History of Pre-industrial Agriculture in Europe." In *Nature's End: History and the Environment*, edited by Sverker Sörlin and Paul Warde, 70–92. Houndmills: Palgrave Macmillan, 2009.

———. "The Idea of Improvement, c. 1520–1700." In *Custom, Improvement and the Landscape in Early Modern Britain*, edited by Richard W. Hoyle, 127–48. Farnham: Ashgate, 2011.

Webb, Sidney, and Beatrice Webb. *English Local Government.* Vol. 4: *Statutory Authorities for Special Purposes.* London: Longmans, Green, 1922.

Wells, Samuel. *The History of the Drainage of the Great Level of the Fens, Called Bedford Level: With the Constitution and Laws of the Bedford Level Corporation.* 2 vols. London: R. Pheney, 1830.

Williamson, Tom. "Dutch Engineers and the Drainage of the Fens in Eastern England." In *Polder Pioneers: The Influence of Dutch Engineers on Water Management in Europe, 1600–2000*, edited by Helga S. Danner, Johannes Renes, Bert Toussaint, Gerard P. van de Ven, and Frits David Zeiler, 103–19. Utrecht: Koninklijk Nederlands Aardrijkskundig Genootschap Faculteit Geowetenschappen Universiteit Utrecht, 2005.

Willmoth, Frances. "Dugdale's *History of Imbanking and Drayning*: A 'Royalist' Antiquarian in the Sixteen-Fifties." *Historical Research* 71 (1998): 281–302.

———. "Jonas Moore's Mapp of the Great Level" (tentative title). Introduction to facsimile ed. of Moore's *A Mapp of ye Great Levell of ye Fenns*. . . . Cambridge: Cambridgeshire Records Society, forthcoming.

———. *Sir Jonas Moore: Practical Mathematics and Restoration Science.* Woodbridge, UK: Boydell, 1993.

Wood, Andy. *The Politics of Social Conflict: The Peak Country, 1520–1770.* Cambridge: Cambridge University Press, 1999.

———. *Riot, Rebellion and Popular Politics in Early Modern England.* Houndmills: Palgrave, 2002.

Worster, Donald. *Rivers of Empire: Water, Aridity, and the Growth of the American West.* Oxford: Oxford University Press, 1985.

Wrightson, Keith. *English Society, 1580–1680.* New Brunswick, NJ: Rutgers University Press, 1982.

———. "The Politics of the Parish in Early Modern England." In *The Experience of Authority in Early Modern England,* edited by Paul Griffiths, Adam Fox, and Steve Hindle, 10–46. New York: St. Martin's Press, 1996.

Yamamoto, Koji. "Distrust, Innovations, and Public Service: 'Projecting' in Seventeenth- and Early Eighteenth-Century England." PhD diss., University of York, 2009.

———. "Reformation and the Distrust of the Projector in the Hartlib Circle." *Historical Journal* 55 (2012): 375–97.

Online Sources

"Eurasian Dotterel." *Wikipedia.* Accessed 18 August 2016. http://en.wikipedia.org/wiki/Eurasian_Dotterel.

"Francis Russell, 4th Earl of Bedford." *Wikipedia.* Accessed 18 August 2016. https://en.wikipedia.org/wiki/Francis_Russell,_4th_Earl_of_Bedford.

"The Grand Remonstrance, with the Petition Accompanying It." Constitution Society. Accessed 18 August 2016. http://www.constitution.org/eng/conpur043.htm.

The Hartlib Papers. HRI Online Publications, Sheffield. Accessed 2 January 2017. https://www.hrionline.ac.uk/hartlib/context.

"Lakenheath Fen." Royal Society for the Protection of Birds. Accessed 18 August 2016. http://www.rspb.org.uk/discoverandenjoynature/seenature/reserves/guide/l/lakenheathfen/.

"Little Ice Age." *Wikipedia.* Accessed 18 August 2016. http://en.wikipedia.org/wiki/Little_Ice_Age.

"Ouse Washes." Royal Society for the Protection of Birds. Accessed 18 August 2016. http://www.rspb.org.uk/discoverandenjoynature/seenature/reserves/guide/o/ousewashes/about.aspx.

"Praemunire." *Luminarium.* Accessed 18 August 2016. http://www.luminarium.org/encyclopedia/praemunire.htm.

"Replevin." *Law Dictionary.* Accessed 18 August 2016. http://www.lawyerintl.com/law-dictionary/9033-replevin.

"Tarde Venit." *Law Dictionary.* Accessed 18 August 2016. http://www.lawyerintl.com/law-dictionary/6611-tarde%20venit.

"Welney Wetland Centre." Wildfowl & Wetlands Trust. Accessed 18 August 2016. http://www.wwt.org.uk/wetland-centres/welney.

"Wicken Fen." *Wikipedia.* Accessed 18 August 2016. http://en.wikipedia.org/wiki/Wicken_Fen.

"Wicken Fen Nature Reserve." National Trust. Accessed 18 August 2016. https://www.nationaltrust.org.uk/wicken-fen-nature-reserve.

"Wicken Fen National Nature Reserve." Wildlife Extra. Accessed 18 August 2016. http://www.wildlifeextra.com/go/uk/wicken-fen.html#cr.

"Wicken Fen Windpump." Society for the Protection of Ancient Buildings, Mills Section. Accessed 18 August 2016. http://www.nationalmillsweekend.co.uk/pages _wind/wicken%20Fen.htm.

Reference Works

A *Biographical Dictionary of Civil Engineers in Great Britain and Ireland*. Vol. 1: 1500–1830. Edited by A. W. Skempton et al. London: Thomas Telford, 2002.

Calendar of the Patent Rolls Preserved in the Public Record Office: Philip and Mary, 1553–1558. 4 vols. London: H. M. Stationery Office, 1936–39.

Dictionary of National Biography. Online version, http://www.oxforddnb.com.

The English Reports. 176 vols. HeinOnline, http://www.heinonline.org.

The History of Parliament. Online version, http://www.historyofparliamentonline.org.

Journal of the House of Commons. Vol. 1: 1547–1629. London: H. M. Stationery Office, 1802.

The King James Bible. Online version, http://www.kingjamesbibleonline.org.

Oxford English Dictionary. Online version, http://www.oed.com.

Statutes of the Realm. 11 vols. HeinOnline, http://www.heinonline.org.

The Victoria History of the County of Cambridge and the Isle of Ely. Edited by L. F. Salzman et al. 10 vols. Oxford: Oxford University Press, 1938–2002.

Index

Page numbers in *italics* refer to figures.

natural philosophy, Baconian, 11
New Bedford River, 250, 257
Newland, John, 143, 170, 171
New Model Army, 230, 231, 245, 268, 355–56n36
new year, date of, xiii
Noddel, Daniel: defense of commoners by, 225–26, 240, 241–42, 248; Gibbon and, 226–27; on Hatfield project, 166, 174; Lilburne, Wildman, and, 229, 233–35; pamphlets of, 247; riots and, 224, 236; on Say, 240
Northamptonshire commission of sewers, 95–97, 102
Nottinghamshire: complaints about drainage project from, 177–78; flooding in, 157–58, 163–66
Noye, William, 164, 165

officials, local: Charles I and, 168; commissions of sewers and, 82–83; Crown reliance on, 138, 144; loyalty to Crown and, 175; riots and, 196, 201–2, 236. See also commissions of sewers
Old Bedford River, 191, 196, 257
Ouse Washes, 312–13
oyer and terminer, definition of, 368

Parliament: Charles I and, 177, 204, 205, 222; committee investigating riots, 217–19, 237, 240–43; commoner support for cause of, 225, 237; Cromwell and, 243; drainage bills presented to, 86–88; Independent faction of, 231; law regarding Great Level sent to, 72–73, 77–78; Noddel support of cause of, 226; Presbyterian faction of, 275. See also Long Parliament; Short Parliament; specific bills
Partherich, Edward, 252, 256
partible inheritance, 48
Participants: Epworth rioters and, 218; as investors, 153–54, 156, 161–65, 166, 174–75; land awarded to, 170–71, 172, 220, 225, 227–29, 243–44; leasing of land to foreigners by, 220–21; Noddel and, 225–26; petition for relief from damages of, 238–40; statements on riots

of, 237; state support for, 173; tenants of, 229–30, 236, 354–55n26. See also Hatfield Level drainage
pasture, fenland, 28, 48, 368
patents, royal, 11
Peakirk Drain, 191
peat, 2, 3, 28, 33
peat fens, 21, 22, 23–24, 25. See also Great/Bedford Level; Hatfield Level
peat shrinkage, 302–3
Pelham, Henry, 207
Personal Rule period, 204
Petition of Right of 1628, 177, 204, 205
political culture of Fens, 40–41, 130, 134, 145–46, 219–20, 233, 247–48
Popham, John, 70; career of, 69; Coke on, 99; complaints against, 73–74; death of, 78; on legal objections to project, 76, 77; as Lord Chief Justice, 91, 101; as projector, 53, 65, 69–71, 72, 78–79
Popham's Eau, 72, 78
popular politics, price of, 238–45
Portington, Robert, 141, 143, 159, 160
Powell, Edward, 203
praemunire, statute of, 97–98, 101, 368
Pratt, Roger, 211–12
Pretended Act (Drainage Act of 1649), 213–15, 253, 263, 264–67, 275, 276–77, 279–80
prisoners of war, as conscripted labor, 268–69, 272
Privy Council: approach to Fens of, 136; as arbiters of drainage disputes, 63, 83, 107–8, 112, 114; on Ayloffe-Thomas project, 130–31; under Charles I, 151–52; Coke and, 92; commissions of sewers and, 68, 72–73, 102–5, 106, 107, 109–10; Elizabethan, 41, 49, 50–52, 55–57, 60; on Hatfield Level project, 143–44; under James I, 151; on lack of progress, 126–27; need for information of, 120, 136; proactive policy of, 80; regional drainage project and, 64; riots and, 167–68, 222; sewer laws and, 117. See also Popham, John
projectors: as agents of state, 52; Ayloffe and Thomas, 121–29, 133–34; Blith on,